POLYMER SCIENCE AND TECHNOLOGY

RADICAL POLYMERIZATION

NEW DEVELOPMENTS

POLYMER SCIENCE AND TECHNOLOGY

Additional books in this series can be found on Nova's website under the Series tab.

Additional E-books in this series can be found on Nova's website under the E-book tab.

POLYMER SCIENCE AND TECHNOLOGY

RADICAL POLYMERIZATION

NEW DEVELOPMENTS

IRENA O. PAULAUSKAS
AND
LUKAS A. URBONAS
EDITORS

Nova Science Publishers, Inc.
New York

For permission to use material from this book please contact us:
Telephone 631-231-7269; Fax 631-231-8175
Web Site: http://www.novapublishers.com

LIBRARY OF CONGRESS CATALOGING-IN-PUBLICATION DATA

Radical polymerization : new developments / [edited by] Irena O. Paulauskas and Lukas A. Urbonas.
p. cm.
Includes bibliographical references and index.
ISBN 978-1-62100-406-6 (hardcover)
1. Polymerization. 2. Free radicals (Chemistry) 3. Free radical reactions. I. Paulauskas, Irena O. II. Urbonas, Lukas A.
QD281.P6R245 2011
547'.28--dc23

2011034220

Published by Nova Science Publishers, Inc. † New York

Contents

PREFACE

In this book, the authors present topical research on new developments in radical polymerization. Topics discussed include interface radical reactions of functional polyperoxides for fabrication of three-dimensional polymeric structures; biomacromolecules in radical processes; nitroxide-mediated photo controlled/living radical polymerization; alkaline anion-exchange membranes prepared by plasma polymerization; advancements in controlled radical polymerization for functional polymers and thermal redox and photoinduced ring opening polymerization reactions.

Chapter 1 – Original way of solid surface activation by covalent grafting of functional polyperoxides (FPPs) and further formation of three-dimensional polymeric structures is described. The FPPs contain simultaneously fragments with a high affinity for polymer and mineral surfaces, and reactive fragments with peroxide groups – di-tertriary, primary-tertriary, peresteric. The FPPs can be deposited on a polymer or mineral surface and further covalently attached to the substrate. Localization of peroxide groups at the interface offers interesting aspects in constructing grafted polymeric films, facilitating surface activation and controlling over the radical grafting polymerization at the interface by "grafting from" and "grafting to" approaches. In this way, unique changes of solid (polymeric) surface properties can be achieved providing biocompatibility, wettability, antibacterial properties, etc.

We report on using FPPs based on a new peroxide monomer, N-(tert-butylperoxymethyl) acrylamide, and acrylamide or octylmethacrylate for modification (activation) of planar (polyolefin) or dispersed (silica particles) substrate. Two interconnected processes, the cross-linking of FPP macromolecules and their grafting to polyolefin surface, govern the peroxidation of the polymer substrate. Ellipsometry, atomic force microscopy (AFM) and contact angle measurements were applied to show that FPP composition and polyolefin reactivity determine the mechanism of peroxidation and properties of grafted polyperoxide films. Polyolefin surfaces peroxidized with FPPs were used for further covalent grafting of macromolecules with polar functional groups – poly(acrylic acid), polyacrylamide, dextran etc.

On FPPs-modified silica particles, radical polymerization of acrylamide resulted in a formation of covalently grafted polyacrylamide (PAAm) macromolecules on silica. By further cross-linking, hydrogel networks containing dispersed particles were developed. As a result, porous hydrogels with reinforced pore "walls" and improved physicochemical properties were formed after leaching out the silica particles.

On FPP-modified planar substrate, hydrogel networks were formed in a presence of cross-linking agent using radical mechanism. As a result, PAAm hydrogels grafted to FPP-modified polypropylene substrates were synthesized by radical polymerization in a presence of N,N'-methylene-bis-acrylamide.

Chapter 2 – Natural polymers, such as polysaccharides and proteins, are materials extensively investigated due to their biocompatibility, biodegradability and non-toxic and non-immunogenic characteristics. Enclosing the biomacromolecules, in a complex structure, these features can be transferred to a biomaterial in order to extend the performance of the device. Basically, the synthesis of bioconjugates, by insertion of natural polymers in a macromolecular network by radical polymerization processes, can be achieved employing two different synthetic approaches. The first method involves the chemical modification of the biomacromolecules to introduce functionality able to undergo radical polymerization reactions. In addition, polysaccharides and proteins, without any functional changes, can take part in graft radical polymerization reactions that involve the heteroatoms of the substrates. Both synthetic approaches allows to prepare biocompatible bioconjugates showing improved physico-chemical and mechanical properties respect to the starting natural species. Furthermore, radical polymerization of biomacromolecules with monomeric species bearing specific functionality, carry out to the synthesis of polymeric network that *undergo* a *phase transition* process in *response* to external stimuli changes (temperature, pH, magnetic and electric field). These findings showed that the radical polymerization techniques, improving the performance of natural polymer, represent an innovative tools for the preparation of macromolecular devices potentially useful in pharmaceutical and biomedical field.

Chapter 3 – Controlled/living radical polymerizations have made great progress in the past two decades based on their advantages over ionic polymerizations as a simple procedure without severe conditions and widely applicable monomers [1]. Examples of the controlled radical polymerizations include the iniferter polymerization [2], the atom transfer radical polymerization (ATRP) using transition metal complexes, such as Cu [3], Ni [4], Co [5], Fe [6], Ru [7], Rh [8], Pd [9], and Re [10], the reversible addition fragmentation chain transfer polymerization (RAFT) [11, 12], the iodide-transfer polymerization [13-15], and the nitroxide-mediated polymerization (NMP) [16, 17]. The primary significance of these controlled/living radical polymerizations lies in the fact that the polymerizations can produce precisely designed architectures. A vast number of precisely designed architectures have been prepared by these polymerizations. It is no exaggeration to say that the controlled/living radical polymerization techniques have now been necessary for the molecular design and creation of new polymer materials.

Chapter 4 – This chapter is focused on plasma polymerization method for the preparation of alkaline anion-exchange membranes. There are mainly two steps in fabrication of plasma-polymerized alkaline anion-exchange membrane: (1) plasma polymerization of monomer into polymer films; (2) quaternization of plasma polymer into alkaline anion-exchange membrane with quaternary ammonium functional groups. Among them, the plasma polymerization process is the most important step since the percentage of quaternary ammonium functional groups in alkaline anion-exchange membrane is determined by the preservation of functional groups in plasma polymerization. A brief introduction to the subject of plasma polymerization, and general characteristics of plasma polymers, is followed by an examination of recent literature on attempts to synthesize electrolyte membranes for fuel cells. The topics covered include effects of plasma polymerization conditions on the structures and properties of the films, the modern analytical techniques used for characterization of plasma polymer films and physical properties of the plasma-polymerized alkaline anion-exchange membranes. The application of the plasma-polymerized alkaline anion-exchange membranes in alkaline direct alcohol fuel cells is also presented.

Chapter 5 – Recent mechanistic developments of radical polymerization in the field of biomaterials will be reviewed. Carbohydrate-based polymers are used in the field of biomedical applications, such as drug delivery devices, contact lenses, bioseparation, biosensors systems tissue engineering scaffolds, cell culture supports, etc. Advances in this field are particularly relevant to applications in the areas of drug delivery and regenerative medicine etc. Particular emphasis is placed on stimuli-responsive (pH and temperature) polymers, micro/nano gels, membranes, and films. The aim of present review analyses and summarizes recent developments in the field of graft modification of carbohydrates by controlled/living radical polymerization.

Chapter 6 – Design and synthesis of structurally well defined polymer is one of the areas of current technology interest for niche product applications. Major research efforts are focused to move from coordination-addition to advanced radical polymerization methodology to produced new polymeric structures using polar and/or non-polar monomer for different end use applications. Evolution of controlled radical polymerization (CRP) lead to synthesis of defined architecture (star, comb, cyclic, dentritic etc) polymer with controlled molecular weights under mild reaction conditions. In the present chapter, various CRP s [Stable Free Radical Polymerization- SFRP, Atom Transfer Radical Polymerization -ATRP and Degenerative Transfer Process- DT] approaches are discussed for the synthesis of functional polymers. The emphasis is given for incorporation of olefin monomer into functional polymer matrix along with structure – property – performance relationship.

Chapter 7 – Highly efficient Type I photoinitiators (PI) based on the silyl radical chemistry are reviewed here for Free Radical Photopolymerization (FRP) or Free Radical Promoted Cationic Photopolymerization (FRPCP). The ability of these structures to initiate the polymerization of acrylates (FRP) or epoxy (FRPCP) monomers is presented here in detail. These compounds (10,10'-bis 10-phenyl-10H-phenoxasilin Si-1 and 9,9'-dimethyl-9,9'-bis 9H-9-silafluorene Si-2) and poly(methyl phenyl silane) Si-3 directly generate silyl radicals under light irradiation. A low oxygen inhibition in the photopolymerization is outlined. In any case, the silylium cations are the initiating species for the ring opening polymerization. A thermal redox process at RT between these organosilanes and oxidizing agents (silver salts) also generates initiating species for the cationic polymerization of epoxy monomers. The present results evidence the high potential of this new class of (photo)initiators based on organosilanes for the thermal redox or photoinitiated epoxy monomer polymerization.

In: Radical Polymerization: New Developments
Editors: I. O. Paulauskas, L. A. Urbonas, pp. 1-57
ISBN 978-1-62100-406-6

Chapter 1

INTERFACE RADICAL REACTIONS OF FUNCTIONAL POLYPEROXIDES FOR FABRICATION OF THREE-DIMENSIONAL POLYMERIC STRUCTURES

Volodymyr Samaryk[a], Andriy Voronov[b*], Igor Tarnavchyk[a], Sergiy Varvarenko[a], Natalya Nosova[a], Ananiy Kohut[b] and Stanislav Voronov[a]

[a] Department of Organic Chemistry, Lviv Polytechnic National University, 12 Bandera str., Lviv 79013, Ukraine.

[b] Department of Coatings and Polymeric Materials, North Dakota State University, NDSU Dept. 2760, P.O. Box 6050, Fargo, ND 58108-6050.

ABSTRACT

Original way of solid surface activation by covalent grafting of functional polyperoxides (FPPs) and further formation of three-dimensional polymeric structures is described. The FPPs contain simultaneously fragments with a high affinity for polymer and mineral surfaces, and reactive fragments with peroxide groups – di-tertriary, primary-tertriary, peresteric. The FPPs can be deposited on a polymer or mineral surface and further covalently attached to the substrate. Localization of peroxide groups at the interface offers interesting aspects in constructing grafted polymeric films, facilitating surface activation and controlling over the radical grafting polymerization at the interface by "grafting from" and "grafting to" approaches. In this way, unique changes of solid (polymeric) surface properties can be achieved providing biocompatibility, wettability, antibacterial properties, etc.

We report on using FPPs based on a new peroxide monomer, N-(tert-butylperoxymethyl) acrylamide, and acrylamide or octylmethacrylate for modification (activation) of planar (polyolefin) or dispersed (silica particles) substrate. Two

* Corresponding author. Tel.: +1 701 231-9563. Fax: +1 701 231-8439. e-mail: andriy.voronov@ndsu.edu.

interconnected processes, the cross-linking of FPP macromolecules and their grafting to polyolefin surface, govern the peroxidation of the polymer substrate. Ellipsometry, atomic force microscopy (AFM) and contact angle measurements were applied to show that FPP composition and polyolefin reactivity determine the mechanism of peroxidation and properties of grafted polyperoxide films. Polyolefin surfaces peroxidized with FPPs were used for further covalent grafting of macromolecules with polar functional groups – poly(acrylic acid), polyacrylamide, dextran etc.

On FPPs-modified silica particles, radical polymerization of acrylamide resulted in a formation of covalently grafted polyacrylamide (PAAm) macromolecules on silica. By further cross-linking, hydrogel networks containing dispersed particles were developed. As a result, porous hydrogels with reinforced pore "walls" and improved physicochemical properties were formed after leaching out the silica particles.

On FPP-modified planar substrate, hydrogel networks were formed in a presence of cross-linking agent using radical mechanism. As a result, PAAm hydrogels grafted to FPP-modified polypropylene substrates were synthesized by radical polymerization in a presence of N,N'-methylene-bis-acrylamide.

INTRODUCTION

Surface immobilization of macromolecules and the formation of thin polymeric films on solid surfaces is a modern and promising way to improve the properties of polymeric materials [1–3]. Surface modification enhances conductivity, wettability, stability, adhesion, antithrombogenicity, antibacterial properties, etc. of polymeric surfaces without deterioration of the polymer bulk properties [3]. Recently, surface modification of a broad range of polymers including polyethylene [4], poly(methyl methacrylate) [5], polytetrafluoroethylene [6], poly(vinyl chloride) [7], etc. has been studied and new smart polymeric materials sensitive to changing external stimuli (pH, temperature, etc.) were produced [8]. Polymer surface modification is conventionally achieved by targeted covalent polymer attachment (grafting), resulting in thin films with interesting properties [2,8,9]. In many cases, radical polymerization is considered a useful approach for polymer grafting on the surface. Radical-mediated polymerization mechanisms have been applied to form thin polymer films on solid substrates to provide coated materials for use in microelectronics, membrane in separation science, (nano)composites with new structural properties and biocompatible materials [10].

To this end, in order to activate the polymer surface and facilitate the initiation of radical polymerization, surface peroxidation can be applied. A number of studies show basic and applied potential value of peroxidation in surface modification [11–15]. Polymeric surfaces were peroxidized using plasma treatment [16–17], UV- [18,19] and γ-irradiation [20,21], oxygenation and ozonation [22,23] and radical initiators [12,24,25], and were then further modified by the grafting of surface films.

Thermal process for the modification of surfaces with thin polymeric films has been described recently using thermosensitive sulfonyl azide groups [26,27]. The method results in surface-attached networks with tailored chemical and physical properties. Variation of the cross-linking density can be regulated through copolymer composition or blending with homopolymer with no sulfonyl azide groups.

As an alternative approach, polyperoxide macromolecules can be used for substrate surface activation [12,28–32]. It is well known that carbochain polymers (including polyethylene, polypropylene, polystyrene, poly(methyl methacrylate), etc.) readily undergo free radical chain transfer and recombination reactions [33]. The prominent feature of polyperoxides is the thermal decomposition of the peroxide groups, generating radicals and macroradicals [34,35]. Generated radicals and macroradicals facilitate the grafting of polyperoxide macromolecules to the substrate surface. Many reports in the literature describe improving the mechanical properties of composites using peroxidized (surface-activated) polymer surfaces to fabricate new polymeric materials, including "core–shell" latexes [36,37], highly filled composites [31,38], based on polymeric blends composites [31,38–40].

Radical copolymerization of monomers containing functional groups was recently used as a new way to synthesize functional polyperoxide (FPP) macromolecules [38]. The FPP macromolecules consist of fragments possessing a high affinity for polymer (for example, polyolefin) surfaces, and reactive fragments with peroxide groups, simultaneously. FPP macromolecules can be deposited on a polymer substrate (Figure 1) and further covalently attached to the substrate by heating [32,41]. In this way, unique changes of polymer surface properties can be achieved. FPPs were used in peroxidation of rubber [40] and considerably improved strength of bonding between rubber and cord.

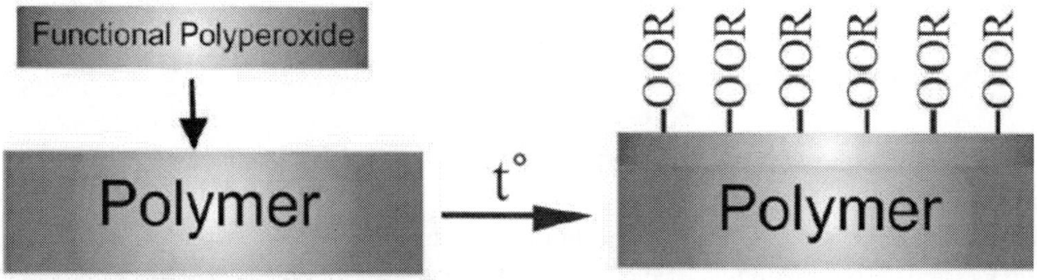

Figure 1. FPP grafting on polymer substrate.

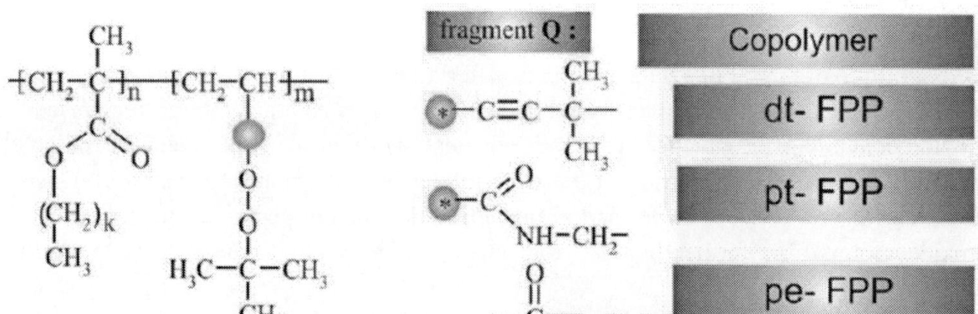

Figure 2. Macromolecular structure of FPPs.

In our recent publications, a mechanism of peroxide group thermolysis, FPP covalent attachment on a polymeric substrate [39,42–44] and further covalent grafting of macromolecules with polar functional groups – poly(acrylic acid), polyacrylamide, dextran etc. to the substrate was discussed. It is assumed that the FPP chemical structure and the nature of the peroxide groups in the FPP are two key parameters determining efficiency of surface peroxidation. In fact, various research groups have studied cross-linking of polypropylene (PP) [45–47], modification of polyethylene (PE) [48,49] in presence of peroxides generating tert-butoxy radicals, and further polymer blends formation. Dokolas and Solomon explained the efficiency of tert-butoxy radicals in the modification of polyolefins, poly(ethylene terephthalate) and poly(methyl methacrylate) by the higher rate of hydrogen detachment in comparison to alkyl and alkoxy primary and secondary radicals [50–52]. Intermolecular interaction between FPP macromolecules and the polymer surface, both at FPP deposition and by elevated temperature is an additional factor that controls polyperoxide grafting. However, FPPs and the polymer substrate should not be thermodynamically compatible, leading to diffusion of the polyperoxide in the polymer bulk and deterioration of surface modification.

In fact, for efficient grafting of FPPs to the polymer surface, the polyperoxide macromolecules should be able to:

- Generate free radicals, for example, methyl $CH_3^{\cdot}$, tert-butoxyl $(CH_3)_3CO^{\cdot}$, or phenyl $C_6H_5^{\cdot}$, actively participating in chain transfer reactions. The temperature and rate of the peroxide groups thermolysis must be suitable for polymer surface modification.
- Interact with the polymer substrate (FPPs must contain fragments ensuring physical bonding to the substrate).

In addition, the FPP macromolecules must be thermodynamically incompatible with the polymer surface.

In this work, we report on a concept of FPP film physical interaction with the polymer surface and its compatibility. FPPs with di-tertiary, dt-FPP, primary-tertiary, pt-FPP and perester, pe-FPP, groups were used in this study (Figure 2). It is assumed that the presence of hydrophobic alkyl fragment groups (acrylate) in FPPs ensures the ability of the peroxide macromolecules to be deposited by physical adsorption onto the polymer substrate. Model design of FPP macromolecules structure prediction and their ability to form reactive film on different polymeric surface substrates is considered in this work too. Thermal behavior and further grafting of three FPPs with structures differing by the nature of the peroxide groups on the polyolefin substrate were investigated in details. At elevated temperatures, the presence of peroxide-containing fragments facilitates covalent grafting of FPPs on the polyolefin (following thermal decomposition of peroxide groups and free radical generation). Covalent attachment of the FPP occurs by partial decomposition of the peroxide groups. Remaining peroxide groups may further be used for activation the polymer substrate in free radical substrate reactions (Figure 1).

In such a way, polyacrylamide (PAAm)-based hydrogels covalently bonded to a PP substrate have been synthesized using two different approaches for forming the hydrogel network. In the first, a grafted hydrogel network is formed by radical polymerization of the acrylamide (AAm) on the peroxidized PP substrate (“grafting from”) in the presence of N,N’-methylene-bis-acrylamide (bis-AAm) (a cross-linking agent) and potassium persulfate (an additional initiator in bulk). In the second approach, a grafted hydrogel network is synthesized by: (i) grafting the PAAm macromolecules from the peroxidized PP substrate and (ii) following intermolecular condensation of the surface-grafted PAAm and adding to the reactive bulk PAAm and poly-N-(hydroxymethyl) acrylamide (used as a cross-linking macromolecule), resulting in the formation of a hydrogel network grafted to the PP substrate.

Besides the peroxidation of the polymeric substrate, FPPs have been used for activation of dispersed mineral surfaces for synthesis of porous hydrogels. Hydrophilic FPP based on acrylamide and N-(*tert*-butylperoxymethyl)acrylamide was applied for the peroxidation of dispersed silica and following initiation of grafting polymerization of acrylamide. The grafting process results in a thin PAAm film chemically attached on the silica surface. The grafted film improves the dispersibility of silica particles in the PAAm hydrogel. After the leaching of silica from the hydrogel, grafted PAAm chains form a "wall" for each of the pore in the hydrogel network. The presence of the pore "walls" significantly enhances the mechanical strength of the porous hydrogels.

EXPERIMENTAL

Materials

Three different functional polyperoxides, poly(2-*tert*-butylperoxy-2-methyl-5-hexene-3-yne-co-octyl methacrylate) (dt-FPP) with 6.5, 13.8 and 19.9 mol.% of peroxide monomer units; poly{N-[(*tert*-butylperoxy)methyl]acrylamide-co-octyl methacrylate} (pt-FPP) with 6.3 and 12.1 mol.% of peroxide monomer units and poly(*tert*-butyl permethacrylate-co-octyl methacrylate) (pe-FPP) with 7.2, 20.1 and 50 mol.% of peroxide monomer units were synthesized as described elsewhere [32,38].

Grafting of FPPs to a Polyolefin Surface

The cleaned surfaces of the polyolefin substrates (PP, PE, and copolymer of PE and PP, TPO) were peroxidized by covalent grafting of the FPP films according to the technique described elsewhere [43]. Briefly, 0.1 mL of a 4 wt. % solution of a corresponding FPP in toluene was spin-coated at 2000 rpm for 1 min onto the polyolefin surface [53]. The spin-coated polyolefin substrates were kept at a certain temperature (80–120°C) in a sealed box under argon. Temperature and duration of each grafting process are presented in a Section 3. Next, the substrates were extracted with acetone in a Soxhlet apparatus for 2 h in order to remove the ungrafted FPPs, and dried to a constant weight in vacuum. Surface energy of the FPP-modified substrates was determined by water and diiodomethane contact angle measurements [54].

Surface coverage (degree of surface modification) was evaluated by the Cassie equation [54] using contact angle measurements data.

Theoretical thickness of the grafted FPP monomolecular film was calculated using the molecular mechanics MM + module and PM3 semi-empirical calculation methods in the HyperChem™ 6.03 molecular modeling program package. Energy minimization was performed using Fletcher–Reeves and Polak–Ribiere algorithms of semi-empirical optimization followed by molecular mechanics optimization with the steepest descent algorithm.

Formation of Hydrogels Grafted to Polypropylene Substrate by Radical Polymerization

The peroxidized PP substrates were placed in glass ampules filled with an aqueous solution containing acrylamide (2–14 wt. %), N,N'-methylene-bis-acrylamide (0.01–1.4 wt. %), and potassium persulfate (0.001–0.015 wt.%). The pH of the solution was adjusted to 7.0 by adding a diluted NH_4OH solution. To remove air from the reaction mixture, the ampoule contents were vacuumed and purged with argon three times. The ampoules were sealed, hold at 80°C for 14 h, and cooled. The PP substrates with grafted hydrogels were rinsed with water for 5 h.

Formation of Grafted Hydrogels by Condensation Mechanism

Stage 1. Grafting of PAAm to the Surface of PP Substrates

The peroxidized PP substrates were placed in glass ampoules filled with an aqueous solution of acrylamide (8 wt. %). To remove air from the reaction mixture, the ampoule content was vacuumed and purged with argon three times. The ampoules were sealed, held at 80°C for 15 h, and cooled. The PP substrates with grafted PAAm were rinsed with water for 5 h and dried to constant weight in vacuum.

Stage 2. Cross-Linking

The PAAm-modified PP substrates were fixed in parallel with the modified sides facing each other, at a distance of 10 mm between them. The space between the substrates was filled with an aqueous solution of PAAm (M_W =130 × 10^3 g/mol, 4 wt.%) and poly-N-(hydroxymethyl)acrylamide (M_W=1100 × 10^3 g/mol, 4 wt.%) with pH 2–3 adjusted by adding 5 N H_2SO_4. The composition was heated at 70°C for 3 h. The hydrogel was rinsed with water to remove sulfuric acid.

Formation of Porous Hydrogels

Stage 1. Peroxidation of Silica Particles

Silica powder (40 g) and a 1.5% aqueous solution of poly[acrylamide-co-N-(*tert*-butylperoxymethyl)acrylamide], the FPP, (140 mL) were agitated at room temperature for 24 h. Silica was isolated by centrifugation, washed three times with distilled water, and dried under reduced pressure until a constant weight.

Stage 2. Grafting of PAAm

A 0.33M aqueous solution of acrylamide (160 mL) and the peroxidized silica particles (40 g) were charged into a glass pressure vessel. The vessel was degassed with argon, closed up, and placed into an oil bath. The polymerization was carried out at 95°C for 2 h. After grafting, silica was placed into a Soxhlet apparatus and extracted with distilled water for 16 h to remove nongrafted PAAm from the grafted PAAm layer. Then, the silica was dried under reduced pressure until a constant weight.

Stage 3. Crosslinking of Hydrogels

To an aqueous solution containing 4% PAAm and 4% poly-N-(hydroxymethyl)acryl-amide, PAAm-modified silica particles and 5N H_2SO_4 (pH 3–4) were added. The mixture was thoroughly agitated for 30 min and kept at 55°C for 4 h.

Stage 4. Leaching out Silica Particles from Hydrogel

Hydrogel filled with the silica particles was immersed in 3% aq HF at room temperature for 48 h at 1:1.25 molar ratio of SiO_2 : HF. The hydrogel was washed five times for 6 h with distilled water (to pH = 6) to remove the excess of hydrofluoric acid.

RESULTS AND DISCUSSION

Prediction of Interfacial Interaction between FPP Film and Polymeric Substrate Surface

The radical reactions of peroxide groups' thermolysis and macroradicals' recombination in condensed phase can be depicted as shown in Figure 3.

Homolysis of the peroxide groups (reaction 1) results in a low-molecular radical (II) and macroradical (I). Low-molecular radicals II participate in chain transfer reactions, illustrated by the reactions 2 and 3. In the course of these reactions additional macroradicals are generated. The interesting fact about the reaction 3 is that formation of the macroradical from the peroxide fragment triggers decomposition of the peroxide group referred to as induced decomposition. Besides, *tert*-butoxyl radical II, as a result of β-decomposition, forms a methyl radical (reaction 4). Thus, these reactions yield both low-molecular radicals and macroradicals. Obviously, thermolysis of peroxide groups in a condensed phase allows accumulation of macroradicals of various nature leading to their recombination (5 and 6). As a result, cross-linking of FPP macromolecules occurs. Undoubtedly, recombination of low-molecular radicals is also possible, but its progress is unlikely due to the low concentration, since most low-molecular radicals are involved in chain transfer reaction.

Following reactions proceed as FPP grafting occurs.

The process of grafting FPP to the polymer substrate occurs due to chain transfer and recombination reactions involving most carbochain polymers (Figure 4). On the other hand, effective surface peroxidation depends on the structure of the polymeric substrate. The dependence of the polymer surface peroxidation degree by dt-FPP and pt-FPP on peroxide groups conversion is shown in Figure 5A. High degree of peroxidation was achieved for a certain conversion of dt-FPP peroxide groups (Figure 5A). Figure 5B features the dependence of the PET surface modification degree on peroxide groups conversion in comparison to the degree of PP surface modification. In the case of PET, the surface modification degree is significantly lower. Similar data are obtained for FPPs used for the modification of poly(phenylene oxide) and nylon 6-6 surfaces (data not shown).

The initial segment of graph 1 in Figure 5b points to the fact that FPP grafting does occur. At the same time, simultaneously with the grafting, a side process lowering efficiency of the surface modification was observed. This process was identified to be a diffusion of the macromolecules from the modifier layer to the bulk of polymeric substrate. It can be assumed that partial compatibility of the modifier macromolecules with the substrate surface macromolecules was observed. The term "partial compatibility" is used here in the same sense it is described in [55], as "thermodynamic compatibility of the polymer mixture containing a small amount of one of the components." In the considered system, this condition is fulfilled, as the quantity deposited on a polymer substrate modifier is very small.

FPP homolysis:

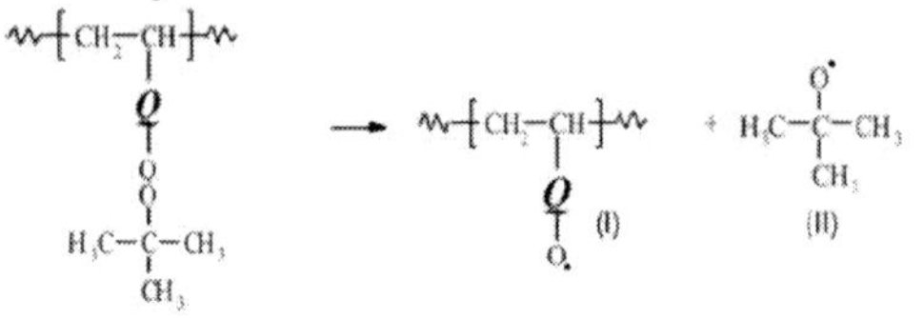

(1)

Chain transfer on *α*-carbon atom of octyl methacrylate fragment and *tert*-butyl peroxide fragment:

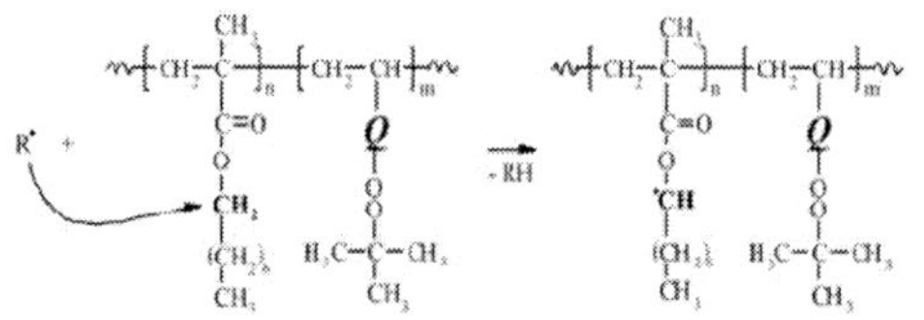

(2)

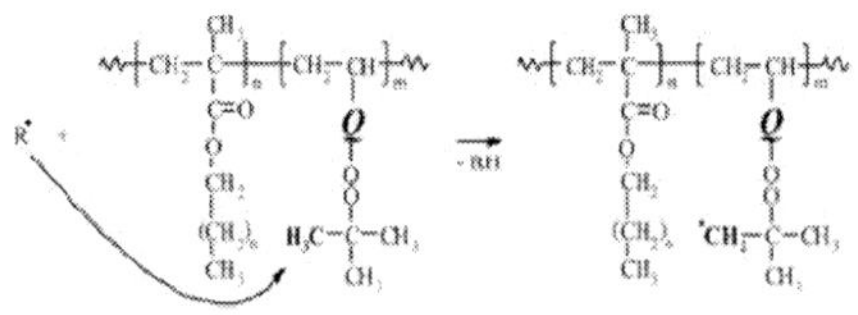

(3)

β-decomposition of *tert*-butoxyl radical:

(4)

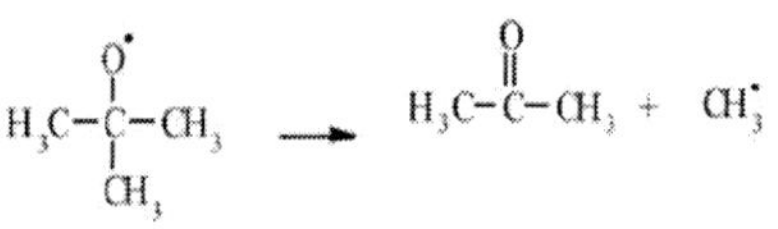

Macroradicals' recombination, formation of cross-linking covalent bonds between FPP macroradicals (CCB).

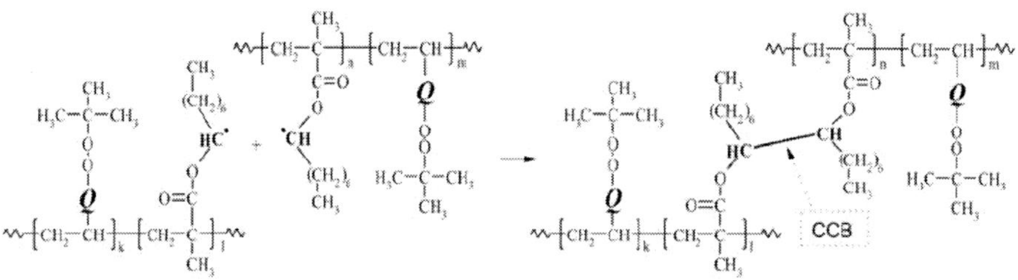

(5)

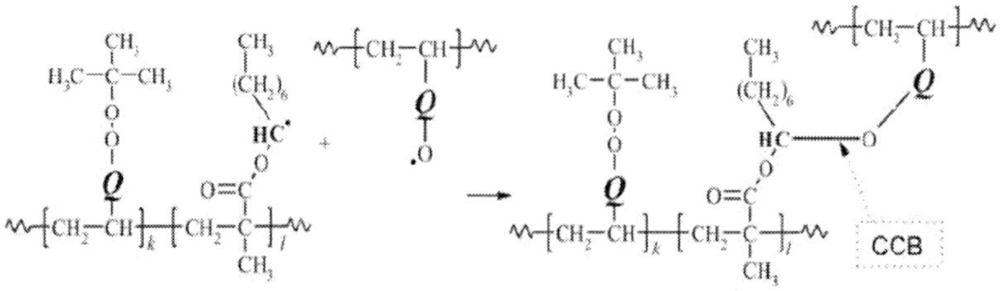

(6)

Figure 3. Thermolysis of FPP peroxide groups and macroradicals' recombination. Chain transfer on surface macromolecules and subsequent formation of macroradicals.

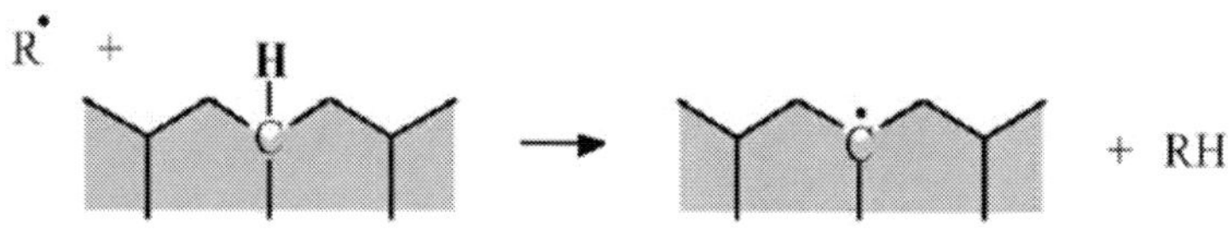

(7)

Recombination between FPP macroradicals and macroradicals of PP surface, formation of covalent bonds between FPP macroradicals and polymer surface macroradicals (ICB):

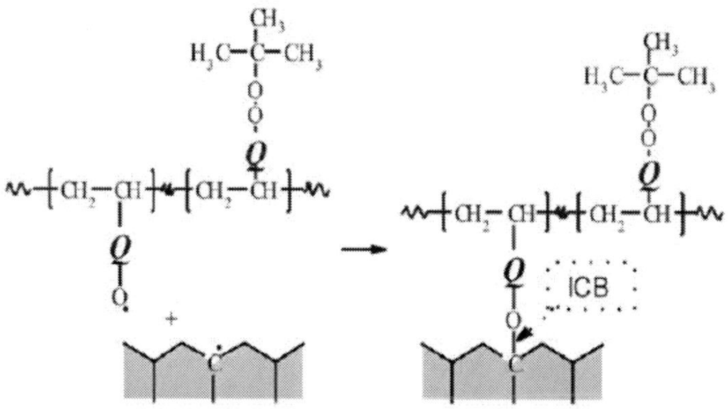

(8)

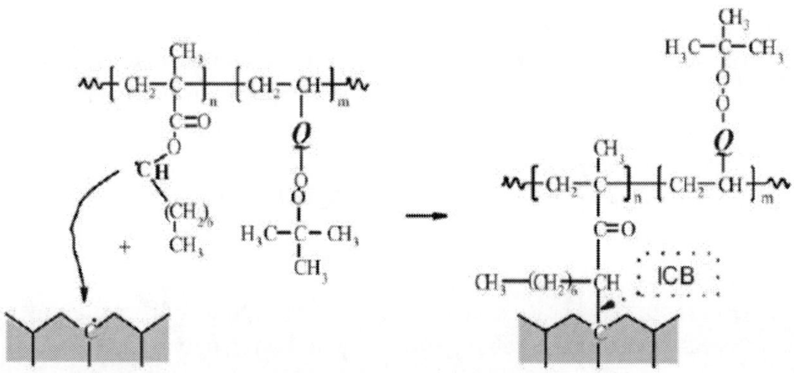

(9)

Figure 4. Grafting of FPP macromolecules to PP surface.

Based on the obtained experimental data, we can formulate the requirements to the structure of the surface modifier to ensure its covalent grafting to the polymeric surface. They are as follows:

1. Macromolecules of the modifier must incorporate peroxide groups for ensuring the formation of covalent bonds between the modifier and substrate.
2. he structure of the modifier macromolecules must ensure physical interaction with the substrate surface, and provide an effective chain transfer reaction with macromolecules of the substrate.

3. The modifying polymer should be thermodynamically incompatible with the polymer of the substrate to prevent the diffusion of modifier macromolecules into the substrate bulk.

The first requirement is fulfilled by the fact that the FPP copolymer contains peroxide groups. In particular, it has been noticed that the peroxide groups in FPP with tert-butoxyl fragments are very efficient in chain transfer reactions, including those onto PP and PE substrates [50,51].

To evaluate how FPP structure determines its interphase interaction with the substrate (the 2nd requirement), Young's equation can be combined with the Good-Girifalco equation [54] to express the interphase interaction criterion (L_{12}) as follows:

$$L_{12} = 2\Phi \cdot \left(\frac{\gamma_1}{\gamma_2}\right)^{1/2}, \text{ where } \Phi \cong \frac{4(V_1 \cdot V_2)^{1/3}}{\left(V_1^{1/3} + V_2^{1/3}\right)^2} \qquad (1)$$

Where γ_i – surface free energy, V_i – molar volume of macromolecules, index 1 refers to the surface, and index 2 refers to the modifier.

The physical content of this criterion is defined as a ratio of reverse work of adhesion (determined by the Dupré ratio) and the modifier surface's free energy. Additionally, this criterion is equal to the cosine of the hypothetical contact angle of the modifier on the polymer surface offset by a figure of one. Apparently, if the value of L_{12} is close to zero, it can be assumed that cohesive forces of interaction between the macromolecules prevail, while interphase interaction forces are insignificant. If $L_{12} = 1$, adhesive and cohesive forces are balanced, and if $L_{12} > 1$, adhesive interaction between the modifier macromolecules and those of the surface is predominated. Interphase interaction between the modifier macromolecules and macromolecules of the substrate (physical bonding) is enabled if $L_{12} > 1$.

To build a prognostic scheme, the unknown from literature, γ_i and V_i values can be experimentally determined, but they can be also obtained from calculations. The calculation algorithm was presented in [56,57]. The L_{12} values for the interaction of FPPs with polymeric substrate surfaces were calculated using table values γ_1 and V_1 for polymeric surfaces and calculated values γ_2 and V_2 for FPP macromolecules. The calculated characteristic dependencies of L_{12} are shown in Figure 6.

Figure 6a shows the described criteria dependence on the alkyl fragment length (k – number of methylene groups) in methacrylate units and Figure 6b indicates the dependence of L_{12} on the ditertiary peroxide groups content in the copolymer at k = 7. It can be concluded that almost all studied FPPs interact considerably with a substrate through reactions involving alkyl radicals. It is in agreement with the results described elsewhere [56]. The optimum of alkyl radical length (k = 6-8) is in a good agreement with experimentally determined FPP grafting efficiency (Figure 7).

The value of the L_{12} is higher for PET than for PP (Figure 6b) and the efficiency of surface modification is expected to be higher too. However, it contradicts to the data given in Figure 5b. This contradiction can be explained by the partial polymers compatibility caused by the increase of interfacial forces, that should be taken into the consideration.

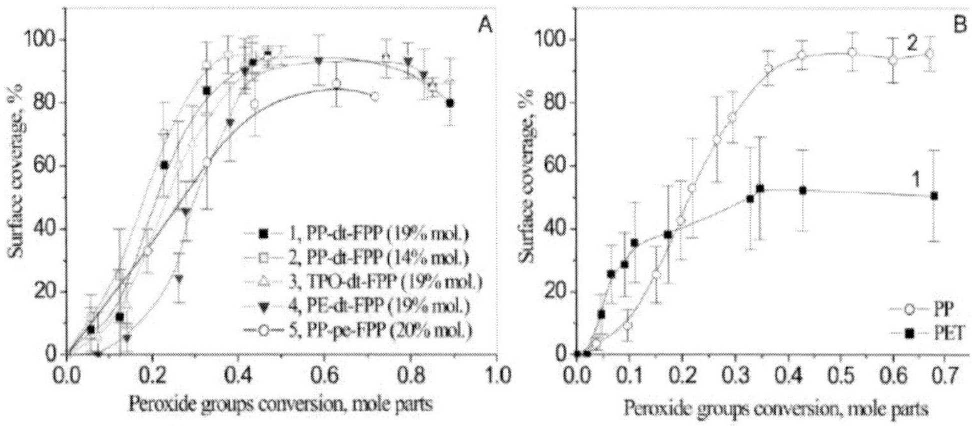

Figure 5. Dependence of polymeric (PP,PE,TPO) surface coverage on peroxide groups conversion: A) modification by dt-FPP at 110°C (1-4), by pe-FPP at 80°C (5); B) modification by dt-FPP (PET (1), PP (2), T = 100 °C).

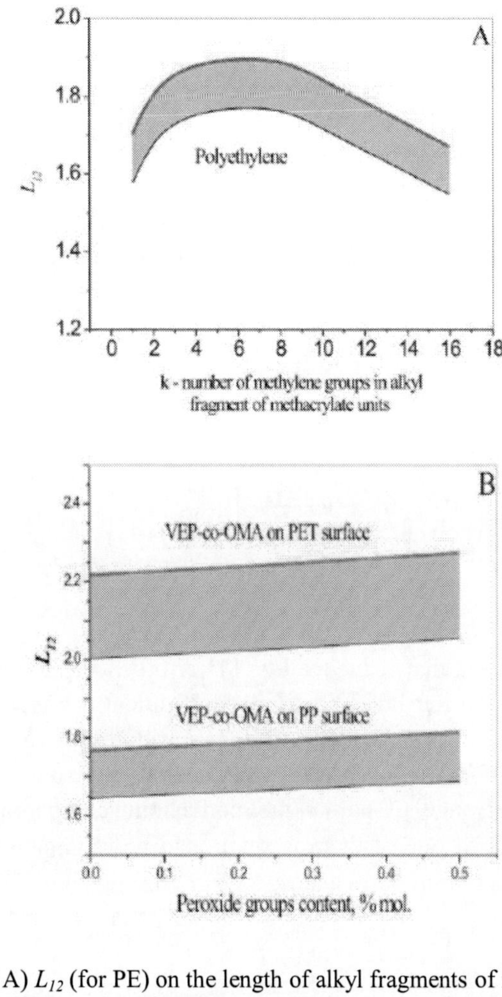

Figure 6. Dependence of A) L_{12} (for PE) on the length of alkyl fragments of methacrylates; B) L_{12} (for dt-FPP (k = 7) on PP and PET) on the remaining peroxide groups content.

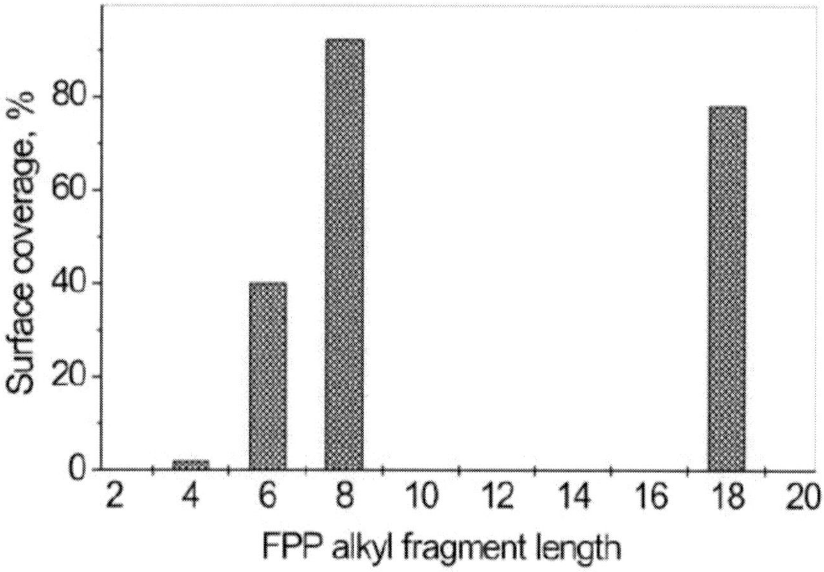

Figure 7. Surface coverage dependence on length of alkyl fragments in dt-FPP (grafting during 62 h, 120°C).

Thermodynamic compatibility of the FPP macromolecules and those of the surface can be estimated using Askadskii's scheme [57]. The model of partial compatibility proposed by Askadskii was tested on a number of polymer mixtures. It is based on the concept of partial thermodynamic compatibility when polymer 1 can be dissolved in a small amount of polymer 2 and vice versa.

$$\mu_1 = \frac{\delta_1^2}{\delta_2^2} < 1.374 \cdot \Phi \cdot \left(\Phi - \sqrt{\Phi^2 - 1 + \frac{\gamma_{12}}{\gamma_2}} \right) \tag{11}$$

$$\mu_2 = \frac{\delta_2^2}{\delta_1^2} < 1.374 \cdot \Phi \cdot \left(\Phi - \sqrt{\Phi^2 - 1 + \frac{\gamma_{12}}{\gamma_1}} \right) \tag{12}$$

where $\gamma_{12} = \gamma_1 + \gamma_2 - 2 \cdot \Phi \cdot (\gamma_1 \cdot \gamma_2)^{1/2}$

In our study, this means that if inequality (11) is fulfilled, the polymer of the substrate can be dissolved in the modifier layer. Similarly, fulfillment of inequality (12) means that a small amount of modifier is dissolved in the substrate polymer. This suggests that macromolecules of the modifier layer can penetrate into the substrate bulk. According to Askadskii's scheme, the range of partial thermodynamic compatibility of two polymers, assuming that the amount of one of them is small, can be estimated using inequalities (11) and (12). The calculations of criteria μ_1 and μ_2 for FPPs are given in Table 1.

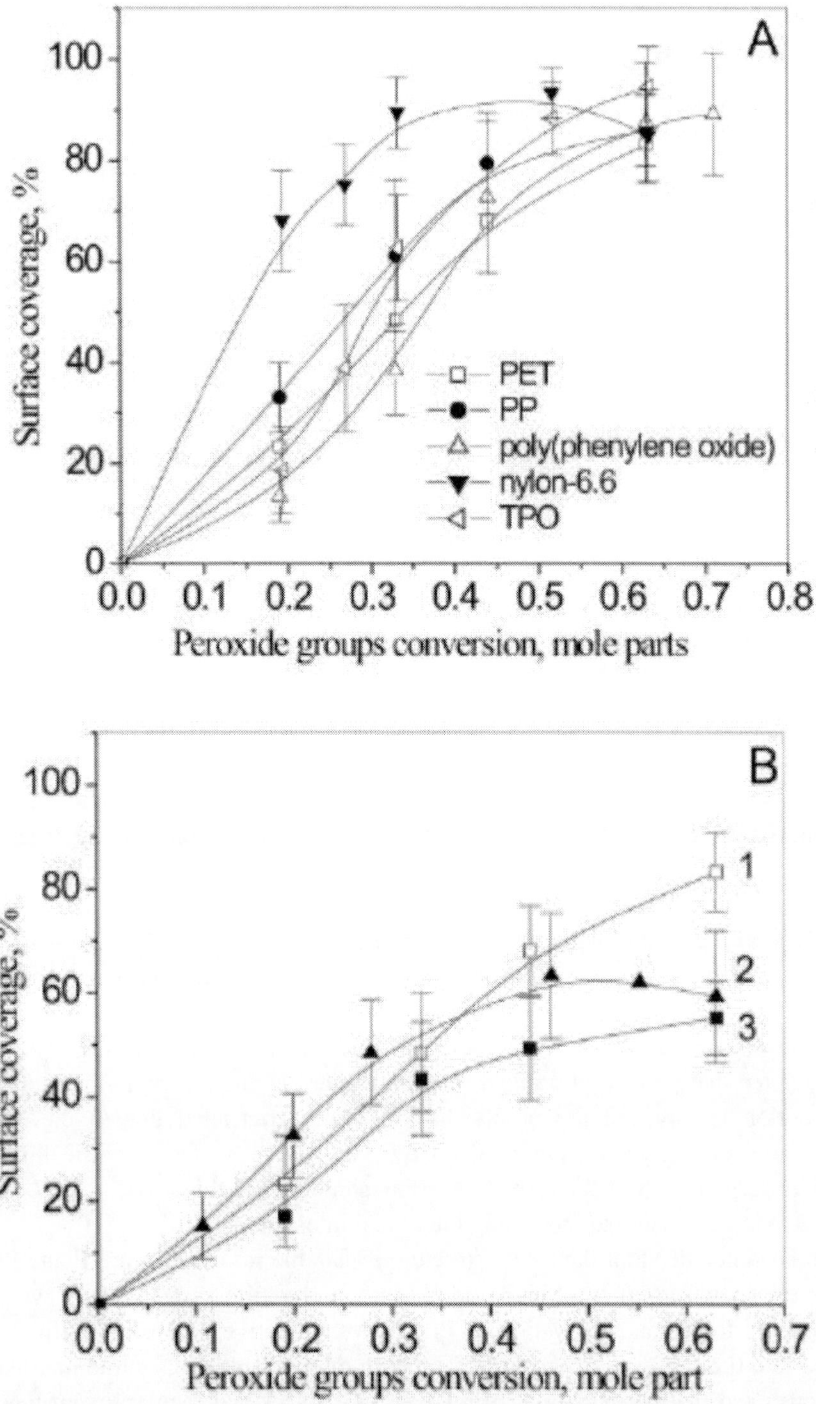

Figure 8. Grafting of pe-FPP to different polymeric substrates (A) (peroxide groups content 20% mol.,* T = 80°C; and modification of PET surface by pe-FPP film from hexane (1,2), benzene (3) (B). * except curve 2 Figure 8B (peroxide groups content 6% mol.).

The data in Table 1 show that for positive values of μ_1, polymer substrate macromolecules can penetrate into the modifier layer. Evidently, this process should not be considered as destructive for the modification, because it leads to an increasing probability of the formation of covalent bonds at the interphase. At the same time, if μ_2 is positive, diffusion of FPP macromolecules into the substrate bulk is thermodynamically possible. This process can hurt modification, due to possible penetration of the FPP macromolecules into the substrate bulk.

Table 1. Solutions of Equations (11) and (12) for μ_1 and μ_2 criteria

Peroxide group / Surface	Ditertriary		Primary-tertriary		Peresteric	
	μ_1	μ_2	μ_1	μ_2	μ_1	μ_2
PP	+	-	+	-	+	-
PE	+/-	-	+	-	+	-
PET	+/-	+	+/-	+	+/-	+
Nylon-6,6	-	+	-	+	-	+
PMMA	+/-	+				

For μ_1: "+" - thermodynamically driven penetration of polymer substrate macromolecules into the modifier layer; "-" - substrate molecules penetration in modifier layer is not possible; "+/-" - depends on the modifier structure.

For μ_2: "+" - thermodynamically driven penetration of FPP macromolecules into the polymer substrate; "-"- FPP macromolecules penetration in polymer substrate is not possible; "+/-" - depends on the modifier structure.

The possible penetration of bulk substrate macromolecules into the modifier bulk does not decrease the grafting efficiency and apparently can even promote the grafting. The latter generally relates mostly to dt-FPP and pt-FPP having thermostable peroxide groups. dt-FPP and pt-FPP grafting occurs within a temperature range of 90–130°C over 6 – 40 hours. For pe-FPP, the modification proceeds during 2–3 hours within the temperature range of 70–80°C. One can obtain high surface coverage on PET and nylon-6,6 surfaces using such modifiers inspite of low efficiency predictions according to the μ_2 criteria (Figure 8a). It should be noted that pe-FPP grafting is possible under conditions where the rate of polymer interpenetration is low and the process is kinetically controlled due to the low thermal stability of the perester group (Figure 8b). However, pe-FPP surface modification efficiency is lower in case of FPP film formation from solution. In this case, reciprocal polymer diffusion rate increases due to increasing interfacial interactions [54].

It can be concluded that the FPPs' structure is suitable for modifying PE and PP surfaces and FPPs are shown to be an effective modifier of polyolefin surfaces. However, using the FPPs modifiers for surface activation of PET is less effective because modifier could have penetrated into the substrate prior to the formation of covalent bonds at the interphase. Thus, it is demonstrated that peroxidation of low-energy surfaces and regularities of the formation of FPP coatings on polymer substrates depends on the nature of the substrate, FPP structure, temperature, and heating time. It is, thus, possible to predict the FPP film formation by prognostic estimations made on the basis of the FPP macromolecule structure and L_{12} values. FPP macromolecules must be thermodynamically incompatible with the polymeric substrate.

Table 2. Thermal behavior of FPP macromolecules

FPP	Peroxide groups, mol %	M_n, g/mol $\cdot 10^{-3}$	ω^{***}, nm	Activation parameters****		Conversion at cross-linking, X_{st}^*, %	T, °C			
							80	100	110	120
				A_0, s^{-1}	E_A, kJ/mol		Time, hrs			
dt-FPP	19.9	7.5	2.8	$(2.2\pm0.1)\cdot10^{13}$	140±5	44 - 46	--	300	47	29
	13.8	14	3.5			39 - 42	--	265	82	25
	6.5	17	--			64 - 68	--	--	150	57
pt-FPP	12.1	130	7.4	$(2.3\pm0.1)\cdot10^{20}$	190±5	12 - 14	--	68	12	3.5
	6.3	180	8.3			20 - 23	--	120	25	6,5
pe-FPP	20.1	80	6.2	$(2.8\pm0.5)\cdot10^{12}$	115±5	79 - 83	6	--	--	--
	7.2	80	6.3			-**	--	--	--	--

* Conversion of the peroxide groups at full cross-linking of FPP macromolecules (X_{st}),

** FPP cross-linking at 90-95 % peroxide group conversion was not fully achieved,

*** ω is the estimated thickness of the monomolecular polyperoxide layer in a condensed phase averaged by mole volume of FPP macromolecule,

**** Peroxide group conversion (determined by liquid chromatography analysis of FPP decomposition product[59]) was used to calculate activation parameters.

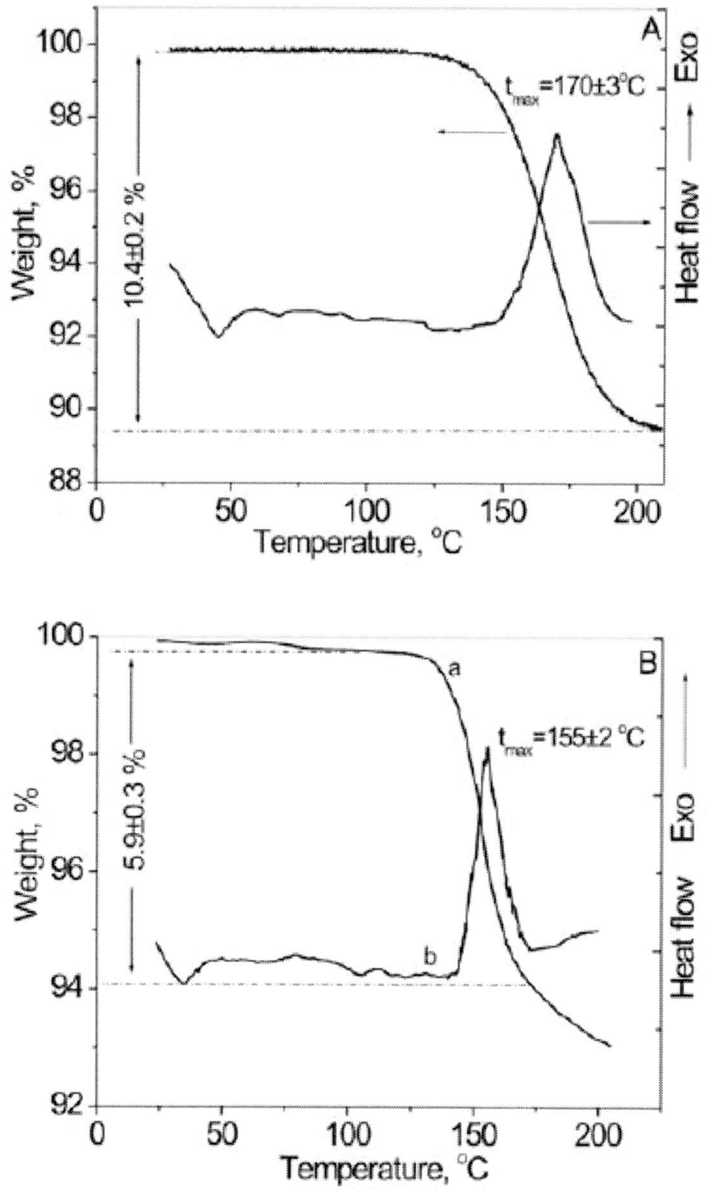

Figure 9. Thermal analysis of dt-FPP (19.9 mol.% of peroxide-containing fragments, A) and pt-FPP (6.3 mol.% of peroxide-containing fragments, B).

Thermal Decomposition of FPPs

Figure 9 shows the thermal analysis (TGA and DSC) for selected dt-FPP and pt-FPP indicating an effect of the peroxide groups nature on FPP thermal stability. The lower thermal stability of the perester group in the FPPs, when compared to di-tertiary and primary-tertiary peroxide groups is known from our previous findings [32,58,59].

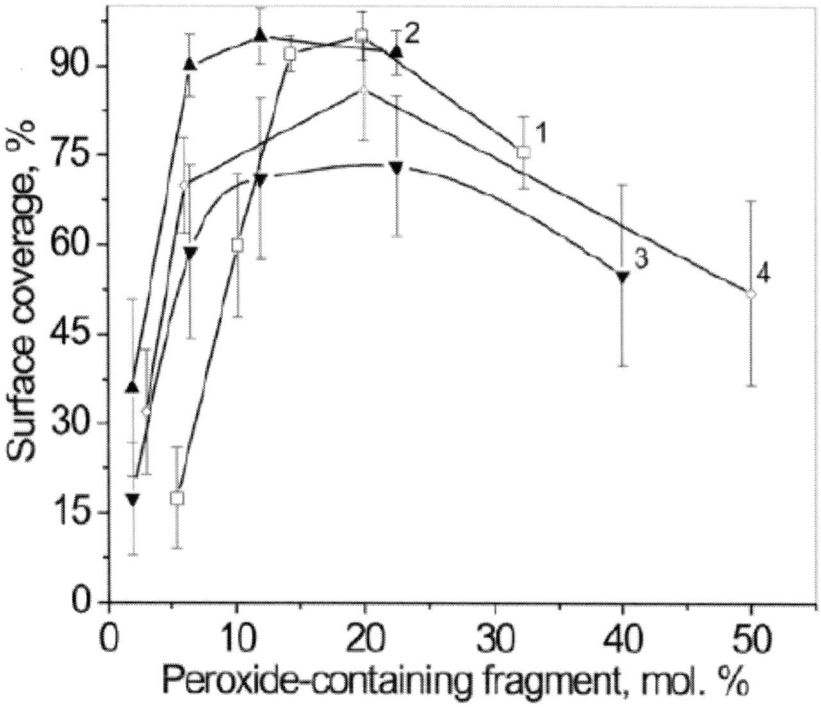

Figure 10. Dependence of surface coverage on nature and content of peroxide groups in FPP (dt-FPP at 110°C – 1, pt-FPP at 100°C – 2, pt-FPP at 130°C – 3, pe-FPP at 80°C – 4).

In the present study, higher thermal stability of di-tertiary peroxide group in comparison to primary-tertiary was found. Using thermal analysis, activation parameters for peroxide group thermolysis were determined (Table 2).

Thermolysis of FPP peroxide groups is accompanied by cross-linking of the polyperoxide macromolecules [44]. To complete thermal behavior analysis, gravimetric measurements were used to determine the degree of FPP cross-linking upon decomposition of the peroxide groups. The cross-linked FPP macromolecules are not soluble in organic solvents. Thus, the degree of cross-linking was measured upon extraction of a sample with acetone [58] and recording of the insoluble fraction in the FPP sample after thermal decomposition. When this fraction approaches 97–100%, full cross-linking of FPP macromolecules is assumed (Table 1).

Remarkably, no temperature effect on X_{st} (peroxide groups conversion at full FPP cross-linking) was found. This is explained by the fact that when intermolecular covalent bonds are formed at a given temperature (upon peroxide group thermolysis and following recombination reaction), the number of these bonds is determined only by number of decomposed peroxide groups.

Effect of FPP Structure on Surface Coverage

The length of the alkyl chain and peroxide group content in FPP structure (k and m in Figure 2) are considered key factors in determining the FPP grafting on the PP substrate. The effect of alkyl fragment length in dt-FPP was shown above (Figure 7).

Figure 10 shows the effect of peroxide group nature and content in FPP on surface coverage. For four different FPPs, the maximum in surface coverage was found to depend on the content of peroxide-containing fragments and the nature of the peroxide group in FPP. Interestingly, the content of peroxide fragments in FPP corresponding to the maximum in surface coverage depends only slightly on temperature, but differs significantly with the nature of the peroxide group. The highest surface coverage corresponds to structures with 14–20 mol.% of peroxide fragments for grafted dt-FPP. For pt-FPP, a concentration of 6–12 mol.% of peroxide fragments is required for maximum surface coverage. Finally, pe-FPP should contain 18–22 mol.% to achieve maximum surface coverage of PP.

In practice, however, an optimal ratio between cross-linking FPP macromolecules and grafting FPP macromolecules is required for high surface coverage. Increasing peroxide group content in FPP facilitates primarily cross-linking and, only indirectly, affects grafting. Excessive cross-linking will obviously deteriorate grafting.

To ensure experimental parameters for efficient surface coverage with grafted FPP macromolecules, we monitored the effect of peroxide group conversion and decomposition rate at various temperatures on surface coverage for pt-FPP macromolecules (Figure 11) and dt-FPP macromolecules (Figure 12).

For the pt-FPP grafting, the highest surface coverage, exceeding 90%, was achieved at a peroxide group conversion of 10–15 mol.% and a rate of peroxide group thermal decomposition of $0.8 \times 10^{-5} - 2.2 \times 10^{-5}$ mol/(l·s) (Figure 11,12). The required decomposition rate can be ensured by varying the peroxide group content in FPP and the grafting temperature.

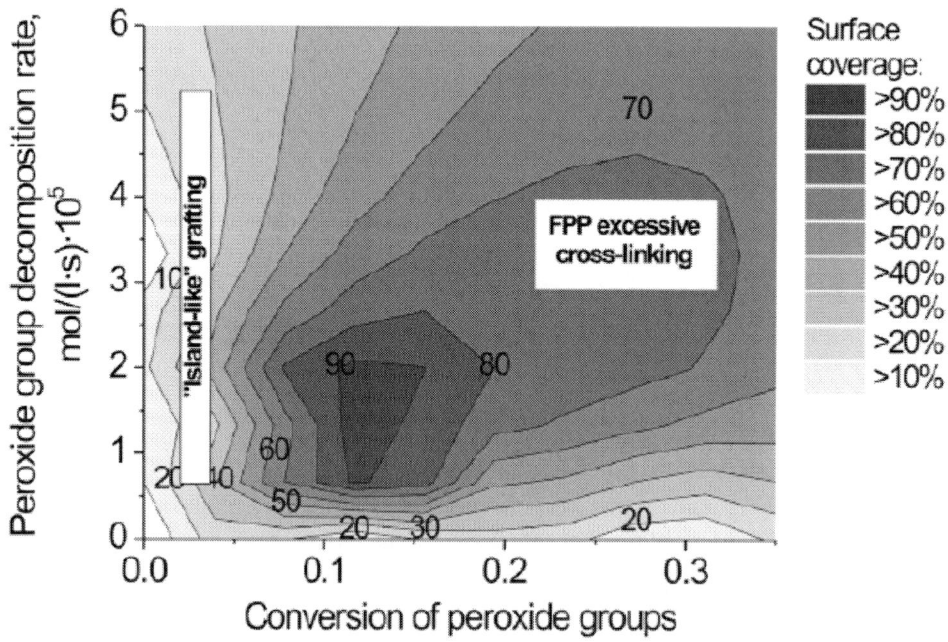

Figure 11. Dependence of surface coverage on peroxide groups conversion of pt-FPP at various temperatures.

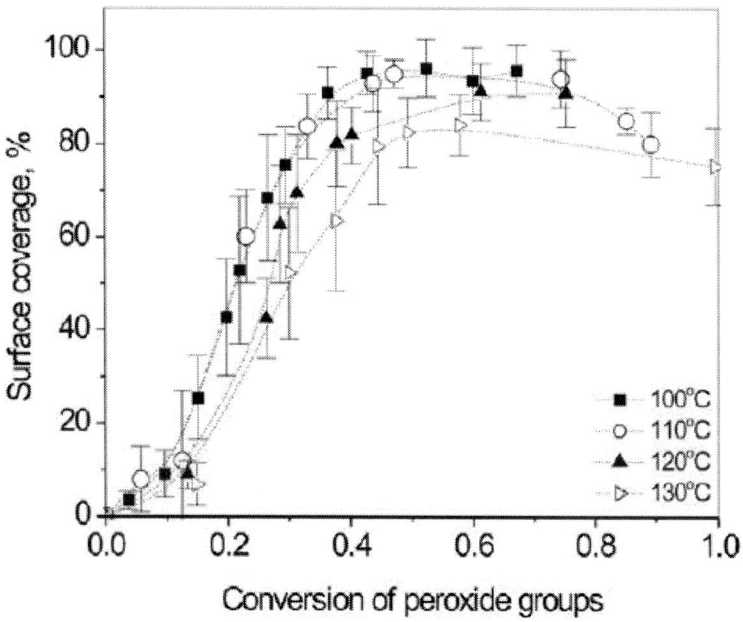

Figure 12. Dependence of surface coverage on peroxide groups conversion and thermal decomposition rate of dt-FPP.

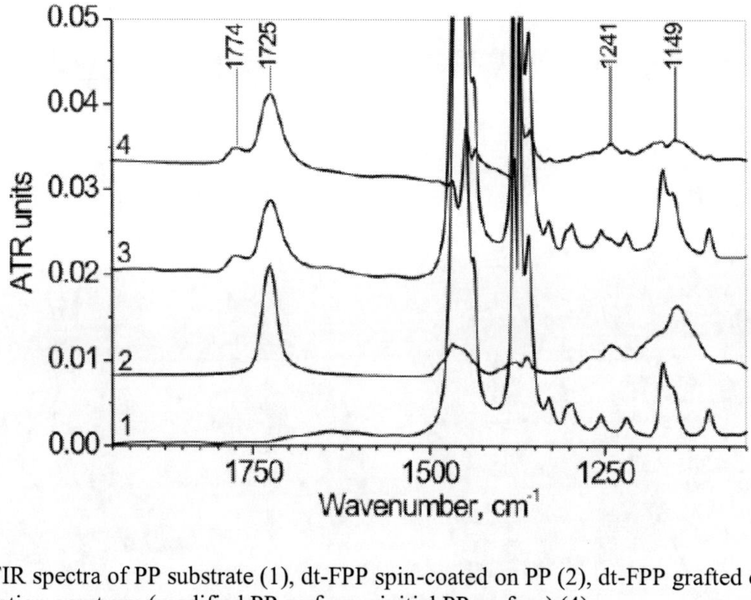

Figure 13. FTIR spectra of PP substrate (1), dt-FPP spin-coated on PP (2), dt-FPP grafted on PP (3). The compensation spectrum (modified PP surface – initial PP surface) (4).

The effect of grafting temperature confirms that FPP macromolecule cross-linking is crucial in polyperoxide grafting. Beyond the optimal decomposition rate, the grafting either does not proceed or is suppressed by cross-linking. The decreasing surface coverage at temperatures higher than 120°C can obviously also be explained by the excessive thermal decomposition rate of the peroxide groups in FPP (Figure 12).

FPP Grafting on PP Surface

Successful grafting of FPP macromolecules on the PP substrate was confirmed by FTIR spectroscopy (Figure 13). Adsorption bands at 1725 cm^{-1} (carbonyl group), 1241 cm^{-1} and 1149 cm^{-1} (ester groups) were observed in the compensation spectrum (*plot 4*).

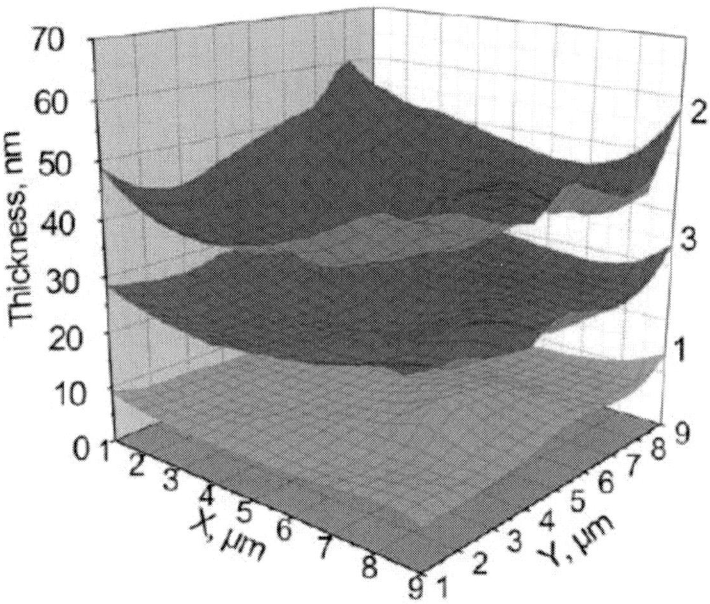

Figure 14. Ellipsometric mapping of: (1) PP-coated silica, (2) spin-coated film of dt-FPP on PP-coated silica (3) grafted dt-FPP on PP-coated silica (after extraction of non-grafted dt-FPP).

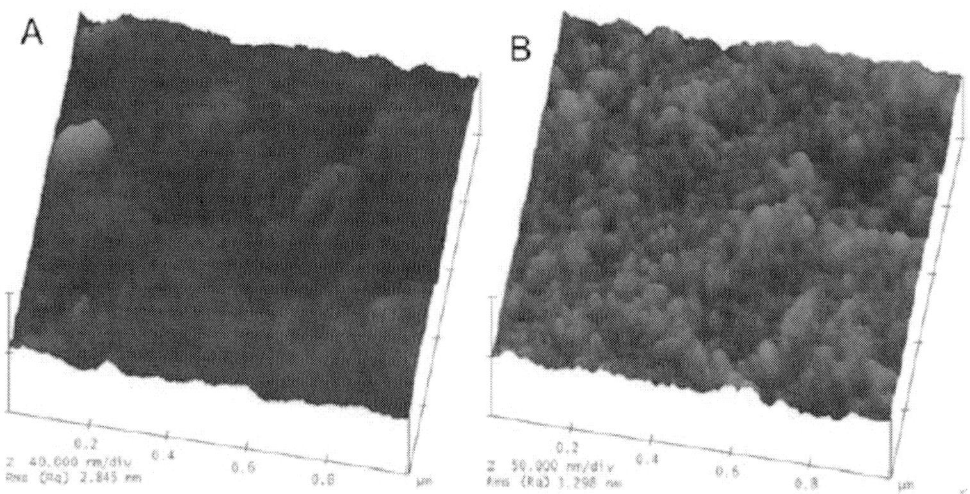

Figure 15. AFM of FPP-modified PP substrate before (A) and after dt-FPP grafting (B) (grey scale is 40 nm for (A) and 50 nm for (B)).

These bands are typical for the initial FPP (plot 2) and are not observed in the spectrum of PP (*plot 1*) that confirms FPP grafting on PP. The absorption band at 1774 cm^{-1} on the compensation spectra is apparently due to the carbonyl group resulting from FPP cross-linking. No correspondent band was detected on the spectrum of the nongrafted FPP (*plot 2*).

To further confirm grafting and to characterize the properties of the polyperoxide-attached films on the PP substrate, ellipsometric mapping of the grafted dt-FPP macromolecules was performed on model PP films attached to a silica substrate (Figure 14) using a well established process in our lab [44]. Plot 1 in Figure 14 represents the model PP film on silica. Ellipsometric mapping of the spin-coated dt-FPP on the PP film is shown in plot 2. Plot 3 corresponds to the mapping image after grafting dt-FPP and following extraction (to remove non-grafted polyperoxide macromolecules).

Using the HyperChem™ 6.03, the theoretical thickness of the grafted dt-FPP monolayer was calculated (7 nm) indicating that the experimental thickness of the grafted polyperoxide film after extraction significantly exceeds the theoretical value. It is assumed (taking into account that non-grafted dt-FPP macromolecules are thoroughly removed from the film by extraction) that the recorded ellipsometric thickness of the grafted polyperoxide on PP (about 20 ± 2 nm) is significantly higher due to cross-linking of dt-FPP macromolecules resulting from free radical reactions, recombination and chain transfer.

As a result of dt-FPP cross-linking during grafting, surface roughness increases. Typical for fractal structures, morphology without distinct ordering was identified in an atomic force microscopy study (Figure 15). FTIR spectroscopy and ellipsometric mapping data for the modified model PP are in good agreement with AFM measurements performed on modified commercially available PP substrates.

It was expected that grafting of polyperoxide would result in changes in the surface energy of the modified PP substrate. To this end, a contact angle study was performed on the PP surface modified with three different selected FPPs – dt-FPP with 19.9 mol.% peroxide-containing fragments, pe-FPP with 50 mol.% peroxide-containing fragments and pt-FPP with 6.5 mol.% peroxide-containing fragments.

Although surface energy changes slightly upon polyperoxide grafting, there is a significant difference in its polar component λ^h_S (hydrogen bonding component) depending of FPP composition. It makes possible to confirm and follow grafted FPP film formation by changing the water contact angle and polar component of the surface energy (Table 3).

Table 3. Contact angles and calculated surface energy of PP and FPP-modified PP substrates

Surface	Modification conditions		Contact angle		Surface energy, mN/m			Surface coverage, %
	T,°C	Time, hr	H_2O,°	CH_2I_2,°	λ_S^d	λ_S^h	λ_S	
PP			106	55.3	31.2	0.2	31.3	-
PP-pt-FPP*	110	8	84	50.0	31.1	3.6	34.7	92 ± 3
PP-pe-FPP*	80	1.5	72	55.7	23.4	11.8	35.2	83 ± 4
PP-dt-FPP*	110	100	91	54.1	30.8	1.6	32.4	95 ± 4

* content of peroxide-containing fragments: 19.9 mol. % (dt-FPP), 50 mol. % (pe-FPP), 6.5 mol. % (pt-FPP).

Figure 16 shows the ellipsometric mapping of grafted dt-FPP films on PP under different experimental conditions. dt-FPP consisting of 13.8 mol.% of peroxide-containing fragment was used in experiments 1, 2 and 4, and 19.9 mol.% in experiment 3. The ellipsometric thickness of the spin-coated dt-FPP before heating was 10 ± 1.8 nm in experiments 1 and 2, and 40 ± 2.6 nm and 135 ± 4.5 nm in experiments 3 and 4, respectively.

According to ellipsometric measurements, the average thickness of the grafted dt-FPP is 0.5 ± 0.05 nm in experiment 1, 3.7 ± 0.2 nm in 2, 17.5 ± 0.8 nm in 3 and 64 ± 5 nm in experiment 4. We chose corresponding to initial stages dt-FPP-modified substrates 1 and 2 (Figure 16) for AFM study. The AFM images (Figure 17,18) reveal that dt-FPP films are not continuous and consist of individual island-like cross-linked constituents. Based on recorded AFM images and ellipsometric analysis, the size of the dt-FPP "islands" and their number at initial stages of grafting were assessed and these data are presented in Table 4.

The degree of surface coverage increases with an increasing conversion of peroxide groups. It obviously happens due to an increasing number of "islands" and their growing size on the substrate. We assume that the growing size of individual "islands" is a prime factor for increasing surface coverage. The assumption is based on the fact that the number of "islands" increases only 1.2–1.3 times in comparison to growing in 8–9 times number of dt-FPP macromolecules constituting "islands". This can be explained by the formation of two different covalent bonds. Presumably, the number of covalent bonds between FPP macroradicals and macroradicals of PP surface, GB is smaller than the number of covalent bonds between macromolecules of polyperoxide, CB.

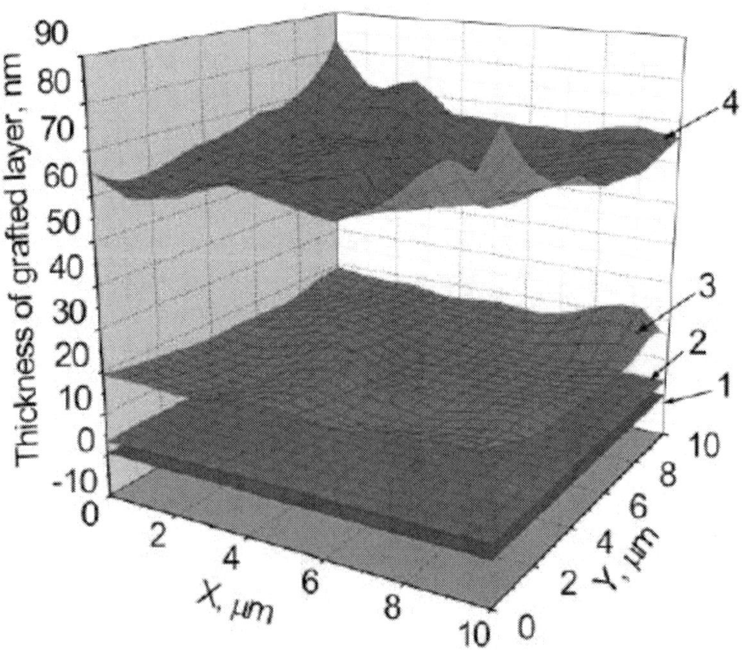

Figure 16. Ellipsometric mapping of dt-FPP grafting at different experimental conditions: (1) at heating time 5 h, and temperature 120°C, (2) – 10 h, 120°C, (3) – 62 h, 110°C, (4) – 24 h, 120°C.

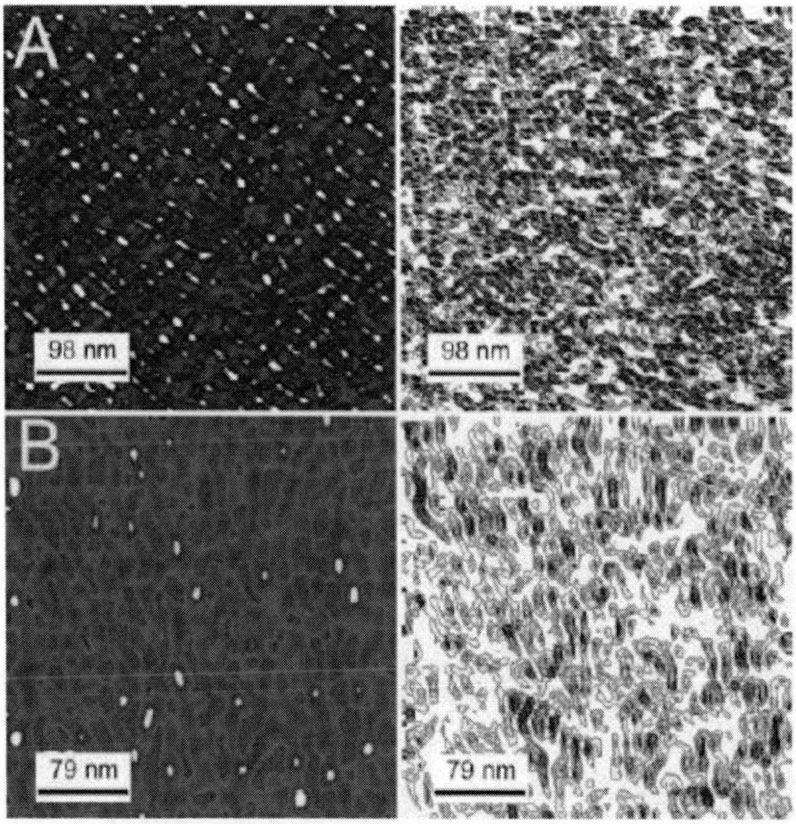

Figure 17. AFM topography (left) and phase contrast (right) of dt-FPP modified PP substrate (A – surface 2 in Figure 16, B – surface 1 in Figure 16).

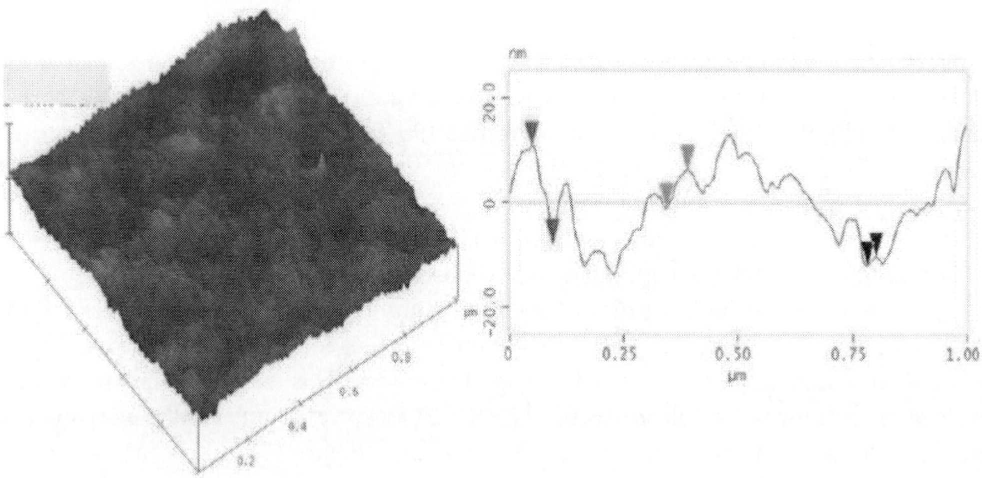

Figure 18. AFM morphology (A) and cross-section (B) of grafted dt-FPP on PP substrate (grafted film corresponds to surface 4 in Figure 16).

Table 4. Initial stage of dt-FPP grafting (quantitative data based on ellipsometry and AFM)

Time, h	Peroxide group conversion, %	Grafted "island" characteristics				Grafted film thickness, nm		Surface coverage, %
		Number per μm^2	Average thickness, nm	Average area, nm^2	Number of dt-FPP cross-linked macromolecules in grafted "island"	Ellipsometry	AFM	
5	9.5	520±25	1.4±0.2	500±80	20 – 22	0.5±0.1	0.7±0.2	23±3
10	18.0	625±30	5.2±0.3	920±120	170 - 180	3.5±0.2	5.2±0.3	56±4

From the data in Table 4, it can be seen that about 20 CB bonds are formed relative to the formation of 1 GB bond. A dt-FPP macromolecule (molecular weight 6500 g/mol, content of peroxide-containing fragment 19.9 mol.%) consists, on average, of 8 peroxide groups. At all estimations performed, the number of peroxide groups required to form 1 GB exceeds the peroxide group content of a single dt-FPP macromolecule. Taking the latter into account, it is statistically unlikely that a single dt-FPP macromolecule will be grafted to the PP surface. Thus, the polyperoxide layer will be formed as a result of the grafting of cross-linked FPPs.

An increasing number of peroxide groups per cross-linked FPP "island" enhances probability of grafting upon the cross-linking and "island" formation. The latter fact may explain why grafting of dt-FPP was not recorded in ellipsometric analysis and AFM measurements below conversion of peroxide groups 4–7%. Nevertheless, when conversion approaches 8–10%, cross-linked FPP macromolecules make the first move to form grafted "islands" (Figure 18).

The "island" morphology disappears with further grafting of dt-FPP macromolecules, when conversion of the peroxide groups increases. Grafting and growing "islands" lead to continuous FPP film and high surface coverage (90–96%) (surface 3 in Figure 16, film thickness 17–22 ± 1.5 nm). There are no more individual FPP macromolecules in the reaction, and continuous FPP film is formed at X_{st}. Finally, grafted film is formed at the expense of cross-linked structures (surface 4 in Figure 16). At that stage, well-developed surface morphology with fractal formations was observed.

The size of the fractal formations for selected grafted dt-FPP films correlates with the size of "islands" indicating that the grafted film is in fact formed by the amalgamation of "islands" during increasing conversion of peroxide groups. This may result in a morphology for the final film formed on the model PP surface that looks very similar to the morphology of the FPP nanolayer formed on the surface of industrial PP (Figure 15).

Mathematical Model of FPP Grafting

According to the mechanism proposed in the work, polyperoxide film grafting to polymer surface proceeds due to recombination of the polymer surface macroradicals and the FPP ones. The process is initiated by thermolysis of peroxide groups of the polyperoxide layer deposited on the polymer surface. Macroradicals from polymer surface macromolecules are generated by the reaction of chain transfer of low-molecular radicals resulting from thermolysis. At the same time, recombination of FPP macroradicals leads to the cross-linking of the polyperoxide layer. Thus, cross recombination results in formation of interphase covalent bond (ICB) with FPP fixed to the polymer surface (Figure 3). FPP macroradicals recombination causes formation of cross-linking covalent bond between FPP macroradicals (CCB) (Figure 4). Cross-linking of the polyperoxide layer leads to increase in the polyperoxide's molecular weight. The notion "cross-linked FPP macromolecule" (CM) is introduced. The number of primary macromolecules of FPP included into CM is regarded as the CM size (*z*). Macroradicals' recombination, however, is not the only way of their consumption. Therefore, probabilities of ICB and CCB formation are fairly low, and the process of their formation can be considered as probable. In this case, there can be introduced the estimate of the average number of peroxide groups decomposition of which results in formation of one CCB (n_{ef}, estimated to fall into the range 2.5-8, depending on the polyperoxide nature) and one ICB (n_{sm}, estimated to be no less than 200). The number of peroxide groups (n_0) in one polyperoxide macromolecule is 7-80 depending on the polyperoxide type. Given such relations between these values, it can be asserted that probability of formation of at least one CCB by an individual FPP macromolecule ($n_{ef} < n_0$) is quite high. This is verified experimentally – most FPP get cross-linked at thermolysis of their peroxide groups. On the other hand, probability of ICB formation by an individual FPP macromolecule for $n_{sm} >> n_0$ is insignificant. However, it is shown in the work that the number of peroxide groups per CM rises with increase in *z* if $n_{ef} < n_0$. This results in increased probability of ICB formation. Thus, by the proposed mechanism grafted polyperoxide film is mainly formed not due to the grafting of individual macromolecules of FPP, but through the grafting of CM. Thereat, the notion of the size of $CM_{Z=L}$ for which probability of CM grafting is close to unit is introduced. In other words, CM for which *z* is higher than or equal to *L* ($CM_{Z>L}$) and which contact the surface directly are considered grafted to the surface. It is clear that in the course of formation of such $CM_{Z>L}$ no less than n_{sm} peroxide groups decompose.

Given the above, the statistical model of the process is founded on the hypothesis that probability of CM formation is proportional to the number of peroxide groups in the objects (primary non-cross-linked polyperoxide molecules and CLM with a lower cross-linking degree) forming CM and the number of these objects. Taking into account rules for constructing a complex random event using a set of dependent and independent simple events, the hypothesis for *z*-sized CM assumes Eq. (13):

$$P_z = n_0 \cdot X \cdot (i + j) \cdot n_i \cdot n_j \qquad z = i + j \tag{13}$$

where P_z is a factor proportional to the probability of *z*-sized CM formation, n_i, n_j are numbers of *i*-sized and *j*-sized CM respectively, and *X* is a conversion of peroxide groups.

The factor proportional to including *z*-sized CM into larger CM can be expressed by Eq. (14):

$$R_z = n_0 \cdot X \cdot n_z \cdot (z \cdot N + N_0) \tag{14}$$

where N is number of macromolecules (initial FPP and CM) at a given point of time, and N_0 is the number of FPP deposited macromolecules per surface area.

In order to transform the derived factors into the probability of CM changing number, each factor can be divided by the sum of all possible interactions. This normalizing factor is defined by Eq. (15):

$$\sum_{z=2}^{N_0} P_z = 2 \cdot n_0 \cdot X \cdot N \cdot N_0$$

$$\sum_{z=1}^{N_0-1} R_z = 2 \cdot n_0 \cdot X \cdot N \cdot N_0 \tag{15}$$

Dividing Eq. (13) and (14) by their respective normalizing factors gives the probability for the formation (cP_z) and consumption (cR_z) of z-sized CM:

$$cP_z = \frac{z \cdot \sum_{i=1}^{z-1} n_i \cdot n_{z-i}}{2 \cdot N \cdot N_0}; \quad cR_z = \frac{1}{2} \cdot n_z \cdot \left(\frac{z}{N_0} + \frac{N_0}{N} \right) \tag{16}$$

Changing number of z-sized CM for the time *[k-1]* to *[k]*, where the total number of CM drops by $\Delta N^{[k]} = N^{[k-1]} - N^{[k]}$ can be determined by balance Eq. (17):

$$\Delta n_z^{[k]} = n_z^{[k]} - n_z^{[k-1]} = -\left(cP_z^{[k-1]} - 2 \cdot cR_z^{[k-1]} \right) \Delta N^{[k]} \tag{17}$$

Combining Eqs. (16) and (17) and converting the absolute number of macromolecules (discrete value) to the mol. number of the macromolecules (continuous value) results in a system of differential Eqs. (18):

$$\frac{dm_z}{d\xi} = -\frac{1}{2} \frac{z \cdot \left(\sum_{i=1}^{z-1} m_i \cdot m_{z-i} \right)}{M_0 \cdot \xi} + m_z \cdot \left(z + \frac{1}{\xi} \right) \qquad z = 1 \ldots N_0 \tag{18}$$

Where M_0 is the FPP (mol. number) deposited onto the surface, and ξ is a dimensionless reaction coordinate defined by Eq. (19):

$$\xi = {}^{N}\!/_{N_0} \tag{19}$$

The system integration with the boundary conditions $m_1 = M_0$ at $\xi = 1$ and $m_z = 0$ at $\xi = 1$ for $z = 2 \div N_0$ leads to Eq. (20):

$$m_z(\xi) = M_0 \cdot \xi \cdot \frac{z^{(z-2)} \cdot (1-\xi)^{(z-1)}}{(z-1)!} \cdot e^{-z(1-\xi)} \qquad z = 1 \dots N_0 \tag{20}$$

Eq. (20) describes the accumulating CMs of all sizes at all reaction stages. However, it has a practical value only for initial FPP molecules, when $z = 1$, because of the difference in size of CM. Thus, Eq. (21) enables tracing a consumption of initial FPP macromolecules.

$$m_1(\xi) = M_0 \cdot \xi \cdot e^{-(1-\xi)} \tag{21}$$

Applying a Stirling equation, Eq. (20) can be simplified and is applicable as an integrand:

$$m_z(\xi) = 2 \cdot \sqrt{{}^{2}\!/_{\pi}} \cdot \xi \cdot M_0 \cdot e^{z \cdot \xi} \cdot (1-\xi)^{z-1} \cdot z^{{}^{2}\!/_{3}} \qquad z = 1 \dots N_0 \tag{22}$$

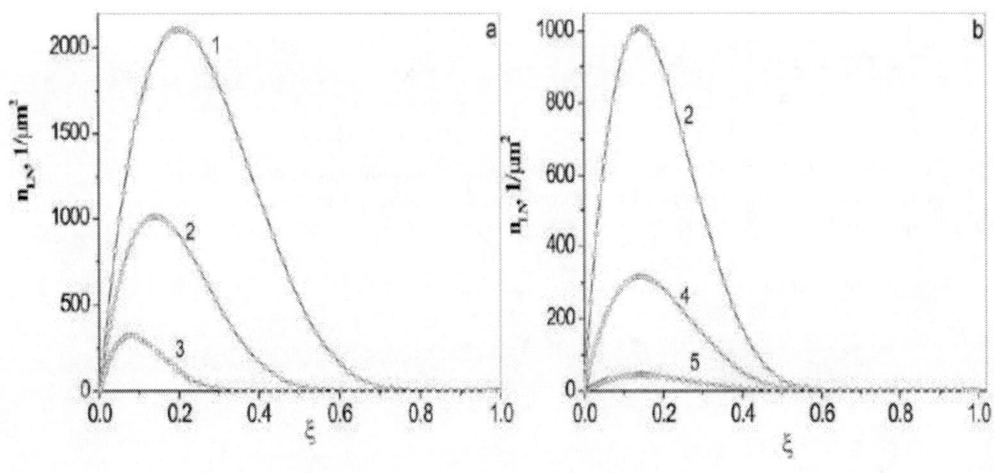

Figure 19. Calculated number of $CM_{z>L}$ vs. the model's internal coordinate ξ: at various L values (1 - 10, 2 - 20, 3 - 60); b) at various numbers of FPP, macromolecules on the substrate (2 - $1.5 \cdot 10^{-19}$, 4 - $4.7 \cdot 10^{-20}$, 5 - $6.8 \cdot$ 10-21 mol/μm^2) (calculations done for dt-FPP, molecular weight 6,800).

Using Eq. (22), a number of CM with a size exceeding L ($CM_{z>L}$) can be estimated (Figure 19):

$$\mathbf{n}_{LN} = \int_{L}^{M_0 \cdot N_A} \mathbf{m}_z(\xi)\, dz \tag{23}$$

It is seen that the proposed model neatly predicts that $CM_{z>L}$ number. ξ_p determines an interval $\xi_p < \xi < 1$, where number of $CM_{z>L}$ is small. Experimentally, no FPP grafting to the substrate was verified within $\xi_p < \xi < 1$.

According to in-built mechanism of the model, surface modification occurs for $0 < \xi < \xi_p$. ξ_p is determined by the L value. Cross-linking decreases the total number of CM. Decreasing n_{LN} after the maximum on Figure 19 can be explained by depletion of $CM_{z<L}$, (source for $CM_{z>L}$ formation), combined with decreasing total number of CM.

The number of CM with a size smaller than ($CM_{z<L}$) and the number of FPP initial macromolecules can be determined by a material balance Eq. (24):

$$\mathbf{n}_{1L} = \xi \cdot \mathbf{M}_0 - \mathbf{n}_{LN} \tag{24}$$

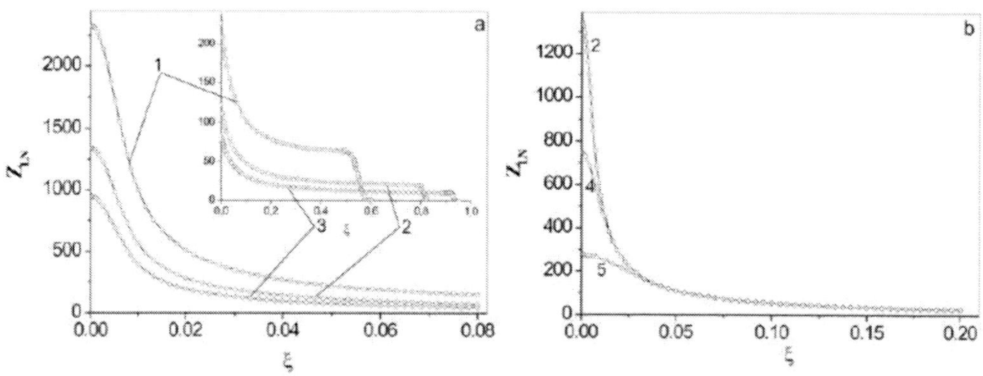

Figure 20. Average size of $CM_{z>L}$ vs. the model's internal coordinate ξ: a) at various L values (1 - 10, 2 - 20, 3 - 60); b) at various numbers of FPP macromolecules on the substrate (2 - $1.5 \cdot 10^{-19}$, 4 - $4.7 \cdot 10^{-20}$, 5 - $6.8 \cdot 10^{-21}$ mol/μm^2).

The current average $CM_{z>L}$ and $CM_{z<L}$ sizes can be evaluated by the statistical Eqs. (25),

$$\mathbf{z}_{LN} = \int_{L}^{M_0 \cdot N_A} \left(\mathbf{m}_z(\xi) \cdot \mathbf{z}\right) dz \Big/ \mathbf{n}_{LN}$$

$$\mathbf{z}_{1N} = \int_{1}^{L} \left(\mathbf{m}_z(\xi) \cdot \mathbf{z}\right) dz \Big/ \mathbf{n}_{1L} \tag{25}$$

Calculated data on average $CM_{z>L}$ size (Figure 20) illustrates three stages of FPP grafting. The 1st stage corresponds to accumulation of $CM_{z<L}$ in FPP film without grafting to the substrate ($\xi_p < \xi < 1$). At the 2nd stage, $CM_{z>L}$ is formed, ensuring FPP grafting to the surface due to FPP initial macromolecules. Here, the average size value z_{LN} increases slightly and "island-like" morphology is experimentally shown. The 3rd stage begins when number of initial FPP macromolecules is considerably depleted. The evidence for that is cross-linking and the rapid increase of CM average size (Figure 20), as well as a decrease of the $CM_{z>L}$ number (Figure 19). When ξ approaches 0, the model predicts a particular z_{LN} value instead of infinity (that would make prediction of grafted film's thickness impossible). The fraction of $CM_{z>L}$ can be estimated by Eq.26:

$$\upsilon = {}^{\mathbf{n}_{LN}}\!\Big/_{\xi\cdot \mathbf{M}_0} \qquad (26)$$

From Eqs. (24-26) based on regularities of hypergeometric distribution, one can estimate an average $CM_{z>L}$ number in the FPP film in contact with a substrate:

$$\widehat{\mathbf{n}}_{LN} = \mathbf{N}_h \cdot \frac{\upsilon}{(1-\upsilon)\mathbf{z}_{1L} + \upsilon \cdot \mathbf{z}_{LN}} \qquad (27)$$

Where $\widehat{\mathbf{n}}_{LN}$ is the number of $CM_{z>L}$ in the film, N_h is total number of FPP macromolecules in the film with a thickness z_{LN}.

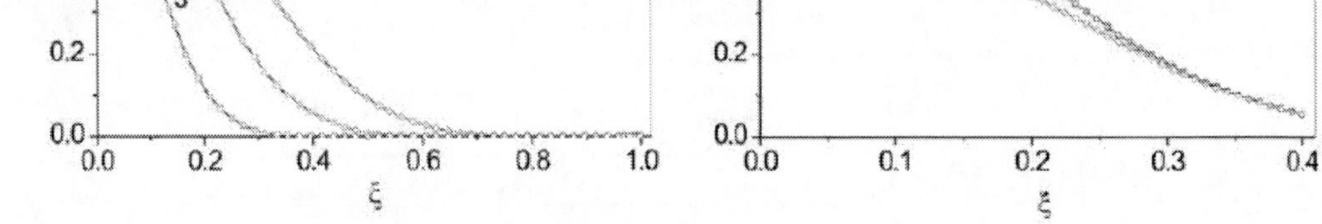

Figure 21. Surface coverage degree vs. ξ value: a) at various L values (1-10, 2-20, 3-60), b) at various amount of FPP on the substrate (4 - $1.1\cdot10^{-22}$, 5 - $1.1\cdot10^{-21}$, 6 – $6.8\cdot10^{-21}$, 7 - $4.7\cdot10^{-20}$, 8 - $1.5\cdot10^{-19}$ mol/μm^2).

Using Eq. (27) , instantaneous surface coverage can be calculated:

$$S_m = \widehat{n}_{LN} \cdot z_{LN} / N_h \tag{28}$$

Combining Eqs. (27) and (28), one can calculate surface coverage:

$$S_m = \frac{erf(\sqrt{\Theta}\sqrt{N_0}) - erf(\sqrt{\Theta}\sqrt{L})}{erf(\sqrt{\Theta}\sqrt{N_0}) - erf(\sqrt{\Theta})} \quad \Theta = -\xi - \ln(1-\xi) \quad N_0 = M_0 \cdot N_A \tag{29}$$

The data showing surface coverage vs. ξ value dependence are presented in Figure 21. To interpret experimental results adequately, one should consider that the model implies dependence of surface coverage on the amount of FPP initially deposited onto the surface (for example, by spin-coating). To this end, when less than $1.1 \cdot 10^{-21}$ mol/µm^2 FPP was used, high surface coverage was not achieved and "island-like" morphology was formed. High surface coverage is expected at $4.7 \cdot 10^{-20}$ mol/µm^2 of FPP. Here, the dependence of surface coverage on the amount of deposited FPP macromolecules becomes negligible, which is in good agreement with the experimental data.

Taking into consideration the properties of the error function $(\lim_{x \to \infty}(erf(x)) = 1)$, Eq. (29) can be simplified and used in experiments with a significant amount of FPP:

$$S_m = \frac{1 - erf(\sqrt{\Theta}\sqrt{L})}{1 - \xi} \tag{30}$$

The thickness of the grafted film can be expressed using the thickness of the FPP film, z_{LN} value and surface coverage:

$$h = \omega \cdot \sqrt[3]{z_{LN}} \cdot S_m \tag{31}$$

Combining Eqs. (30) and (31) results in Eq. (32):

$$h = \varpi \cdot S_m \cdot \sqrt[5]{\frac{\pi \cdot L}{\Theta}} \cdot \sqrt[3]{\frac{N_0}{2}} \cdot \sqrt[3]{\frac{D}{\sqrt{LN_0} \cdot (1-\xi)^{N_0} \cdot e^{N_0 \xi} - N_0 \cdot e^{-L\Theta} - \sqrt{\pi L \Theta} \cdot D}}$$

$$D = erf\left(\sqrt{\Theta} \cdot \sqrt{N_0}\right) - erf\left(\sqrt{\Theta} \cdot \sqrt{L}\right) \tag{32}$$

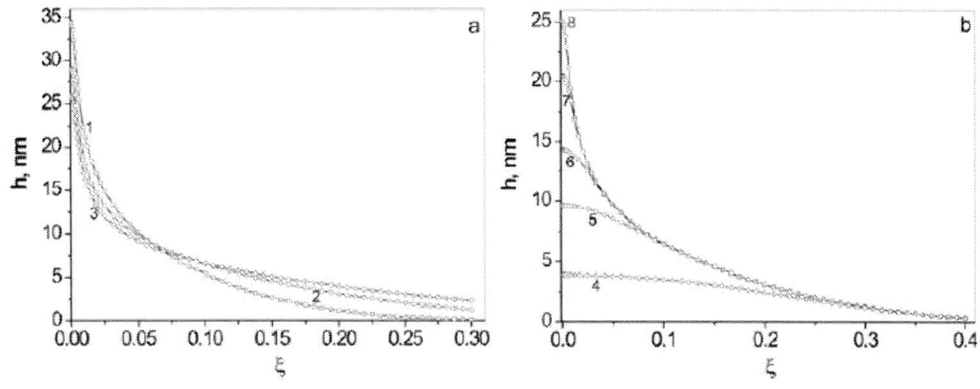

Figure 22. Calculated dependence of grafted film thickness on the model's internal coordinate ξ: a) at various L (1-60, 2-20, 3-10); b) at various FPP quantities on the surface (4 - $1.1\cdot10^{-23}$, 5 -$1.1\cdot10^{-22}$, 6 - $1.1\cdot10^{-21}$, 7 - $4.7\cdot10^{-20}$, 8 - $1.5\cdot10^{-19}$ mol/μm^2).

By definition, ξ changes from 0 (at an initial moment) and tends to 1 at cross-linking of deposited macromolecules during grafting. The relationship between an intrinsic coordinate of the model and the peroxide group conversion can be expressed by following empirical equation:

$$\xi = A \cdot \ln\left(\frac{X_{st}}{X + X_{st}/e^{1/A}}\right) \tag{33}$$

Where X is the peroxide group conversion in the final FPP film, X_{st} - the peroxide group conversion when full cross-linking occurs, A - a coefficient relating cross-linking rate and thermal decomposition rate, and characterizing the nature of the peroxide group within a range of the given model. A describes the ability of FPP macromolecules to cross-link.

Combining Eqs. 29 and 33, one can determine X_{st} and ω (Table 2) from experimental data (Figure 11, 12). For dt-FPP grafting, L correlates to the number of grafted "islands" (Table 5) at initial stages of FPP grafting.

The latter fact verifies the model parameters showing that cross-linking of 17-22 FPP single macromolecules is required for grafting to the PP substrate. Increasing temperature increases this number to 55. As a result, the grafting efficiency drops. This is confirmed by experimental data presented in Figure 11 and 12. The parameter L is increasing in a range of PP > TPO > PE. This is in good agreement with polyolefin reactivity in a chain transfer reactions. PP is the most reactive because of a considerable number of tertiary carbon atoms. The number of reactive centers in TPO is much lower and PE possesses even fewer.

Using model parameters, one can estimate an average thickness of grafted FPP film by Eq. 32. This is in good agreement with an experimental thickness measured by ellipsometry (Table 4). Deviations in calculated thickness are mainly due to deviations in determining X_{st} and the peroxide group conversion for completed grafting of the FPP film. Inaccuracy of calculating ω does not cause considerable deviation in the evaluation of the thickness.

Table 5. FPP characteristics and conditions of peroxidation of various polymer surfaces, estimated values of mathematical model parameters for the formation of grafted polyperoxide layer and characteristics of polyperoxide layers grafted to polymer surface

Polymer surface	Peroxide group content in FPP			Temperature, °C	Time, h	Peroxide group conversion, %	Parameters of the model*				Parameters of the grafted layer			
	Mole fraction	Number of groups per macromolecule	X_{st}				A	L	$\chi_k \cdot 10^3$	n_{ef}	L_k^{**} at χ_k	Sm , %	Thickness of the grafted layer, nm	Number of peroxide groups $\cdot 10^5$ mol/m^2
dt-FPP														
PP	19.9	7.5	0.45 ± 0.03	100	395 ÷ 305	42 ÷ 45	0.25 ± 0.03	19 ± 2	8 ÷ 10	4 ± 0.2	520 ÷ 750	96 ± 4	18 ÷ 25	1.2 ÷ 1.6
				110	43 ÷ 46			20 ± 2			550 ÷ 800	96 ± 4	18 ÷ 23	1.1 ÷ 1.4
				120	29 ÷ 31			35 ± 3			850 ÷ 1000	92 ± 6	22 ÷ 28	1.4 ÷ 1.8
				130	10 ÷ 11			55 ± 3			1050 ÷ 1300	84 ± 6	24 ÷ 29	1.5 ÷ 1.8
	13.8	9.8	0.40 ± 0.03	110	79 ÷ 80	37 ÷ 40	0.22 ± 0.03	18 ± 2		5.2 ± 0.3	450 ÷ 600	94 ± 4	20 ÷ 28	1.0 ÷ 1.4
TPO	19.9	7.5	0.45 ± 0.03		43 ÷ 46	42 ÷ 45	0.25 ± 0.03	32 ± 2		4 ± 0.2	650 ÷ 950	95 ± 3	25 ÷ 30	1.6 ÷ 2.0
PEt					43 ÷ 46			48 ± 2			850 ÷ 1050	93 ± 5	28 ÷ 32	1.8 ÷ 2.1
pt-FPP														
PP	12.1	79.5	0.13 ± 0.02	110	12.5 ÷ 13	12 ÷ 13	0.17 ± 0.007	9 ± 1	8 ÷ 10	7.1 ± 0.4	350 ÷ 500	95 ± 5	40 ÷ 50	2.4 ÷ 3.0

PP	6.3	57.3	0.22 ± 0.02	110	24 ÷ 25	20 ÷ 22	0.15 ± 0.005	13 ± 2		7.8 ± 0.3	500 ÷ 650	95 ± 3	50 ÷ 65	1.4 ÷ 1.8
PP	12.1	79.5	0.13 ± 0.02	120	12.5 ÷ 13	12 ÷ 13	0.17 ± 0.007	15 ± 2		7.1 ± 0.4	500 ÷ 600	92 ± 3	40 ÷ 55	2.4 ÷ 3.3
pe-FPP														
PP	20	81.2	0.65 ± 0.05	80	3.0 ± 0.2	62 ÷ 65	0.16 ± 0.05	28 ± 2	6 ÷ 8	2.5±0.1	660 ÷ 850	90 ± 6	50 ÷ 65	2.0 ÷ 2.7
TPO								32 ± 3			725 ÷ 920	95 ± 5	60 ÷ 70	2.5 ÷ 2.9

* Optimized parameters of mathematical model.

** Grafted "island" size at the moment of complete polyperoxide layer formation.

Table 6. Surface characteristics of peroxidized polymeric surfaces modified by grafted macromolecules

Surface type	Conditions of surface modification			Contact angle		Components and total surface free energy, mN/m			Degree of modif., %**
	Solvent	T, °C	Time, hr.	H_2O, °	CH_2I_2, °	λ_S^d	λ_S^h	λ_S	
Non-modified PP				106	55.3	31.2	0.2	31.3	-
Non-modified PET				81/65	24	44.1	2.4	46.5	-
Non-modified nylon-6.6				86/67	50	32	5.5	37.5	-
Non-modified polyphenylene oxide				75	27	40.4	5.4	45.8	-
VEP-co-OMA									
VEP-co-OMA (19.9 % mole) continuous coating				90.2	55.1	29.1	2.4	31.5	-
PP modif. with VEP-co-OMA (19.9 %)		110	100	91.0	54.1	30.8	1.6	32.4	95
PP-VO-PHEM*	acetone	110	120	67.0	48.0	26.1	14.1	40.2	97
PP- VO- PAAm	methanol	110	120	34.5	54.4	19.0	41.5	60.5	100
PP- VO- PAAm	water	110	120	96.1	49.2	34.5	0.5	35.0	18
PP- VO-PVAc	acetone	110	120	74.4	50.2	23.0	29.1	52.2	81

Table 6. Continued

Surface type	Conditions of surface modification			Contact angle		Components and total surface free energy, mN/m			Degree of modif., %**
	Solvent	T, °C	Time, hr.	H_2O, °	CH_2I_2, °	λ_S^d	λ_S^h	λ_S	
PP-VO-PVA		110	120	61.0	46.3	27.3	18.7	46.0	100
PP- VO-PAA	acetone	110	120	52.6	48.2	24.0	27.1	51.1	87
PP- VO-PVP	acetone	110	120	51.9	46.9	26.0	24.0	50.0	100
PP- VO-PMMA	acetone	110	120	72.5	35.6	36.3	7.2	43.5	97
PP- VO-PAEMA	acetone	110	120	49.8	57.0	19.7	30.1	49.8	100
PP- VO-D	water	110	120	68.8	48.7	27.5	12.4	39.9	44
PP- VO-DS	water	110	120	55.6	44.9	27.4	20.8	48.2	58
PP-VO-HP	water	110	120	75.4	58.1	23.8	10.2	34.0	31
PEst-co-OMA									
PEst-co-OMA (6 % mole) continuous coating				90.0	54,5	28.9	2.4	31.3	-
PP modified with PEst-co-OMA (6 %)		80	1.5	93.6	56.6	28.9	1.6	30.5	80
PP- PEst - PAAm	water	80	3.0	96.1	52.0	32.6	0.7	33.3	18
PEst-co-OMA (50 % mole) continuous coating				64.4	59.0	21.1	18.3	39.4	-
PP modified with PEst-co-OMA (50 %)		80	1.5	72.0	55.7	23.4	11.8	35.2	83
PP- PEst -D	water	80	3.0	61.4	51.2	25.0	18.2	43.2	68
PP- PEst -DS	water	80	3.0	49.3	43.2	27.2	25.1	52.3	75
PET modified with PEst -co-OMA (50%)		80	1.5	65.0	60.0	19.4	19.9	39.3	99
PET- PEst -PAA	acetone	80	3.0	42.0	63.5	15.2	39.5	54.7	100
PET- PEst -D	water	80	3,0	43.9	47.4	24.0	31.0	55.0	89
PET- PEst -DS	water	80	3.0	29.9	42.7	24.6	39.3	63.9	91
N 6,6 modified with PEst-co-OMA (50 %)		80	1.5	67.0	85.0	16.5	19.0	35.5	94

N.6,6 – PEst -PAA	acetone	80	3,0	52.0	88.0	5.4	45.8	51.2	92
PFO modified with PEst-co-OMA (50%)		80	1.5	66.2	61.0	23.0	18.4	41.4	96
PFO – PEst -PAA	acetone	80	3.0	46.0	63.5	15.5	37.1	52.6	99
PO-co-OMA									
PO-co-OMA (6.3 % mole) continuous coating				82.0	51.0	29.8	5.0	34.8	-
PP modified with PO-co-OMA (6,3 %)		110	8.0	84.0	50.0	31.1	3.62	34.73	92
PP-PO- PAAm	water	80	15.0	46.2	43.1	26.7	27.5	54.2	90
PO-co-OMA (12 % mole) continuous coating				88.0	50.0	31.9	2.5	34.4	-
PP modified with PO-co-OMA (12 %)		110	8.0	89.6	51.0	31.7	2.1	33.8	92
PP- PO- PAAm	water	80	12.0	44.2	43.5	26.2	29.2	55.4	91
Surface type	Conditions of surface modification			Contact angle		Components and total surface free energy, mN/m			Degree of modif., %**
	Solvent	T, °C	Time, hr.	H_2O, °	CH_2I_2, °	λ_S^d	λ_S^h	λ_S	
PP- PO-PHEM	water	80	8.0	69.8	52.0	26.3	12.3	38.6	92
PP- PO-PHEM	methanol	80	8.0	89.5	50.2	32.1	2.1	34.2	42
PP- PO-PHEM	methanol	100	8.0	67.9	55.6	23.8	14.6	38.4	96

* PP-VO – PP modified with VEP-co-OMA; PP-Pest – PP modified with PEst-co-OMA; PP-PO – PP modified with PO-co-OMA; PHEM – poly(hydroxyethyl methacrylate); PAAm – polyacrylamide; PAA – poly(acrylic acid); PVAc – poly(vinyl acetate); PVA – poly (vinyl alcohol); PVP – poly(vinyl pyridine); PMMA – poly (methyl methacrylate); PAEMA – poly(N-(2-aminoethyl)methyl acrylamide); D – dextran; DS – dextran sulphate; HP – heparin;

** The accuracy of estimation of modification degree is 3 - 6%.

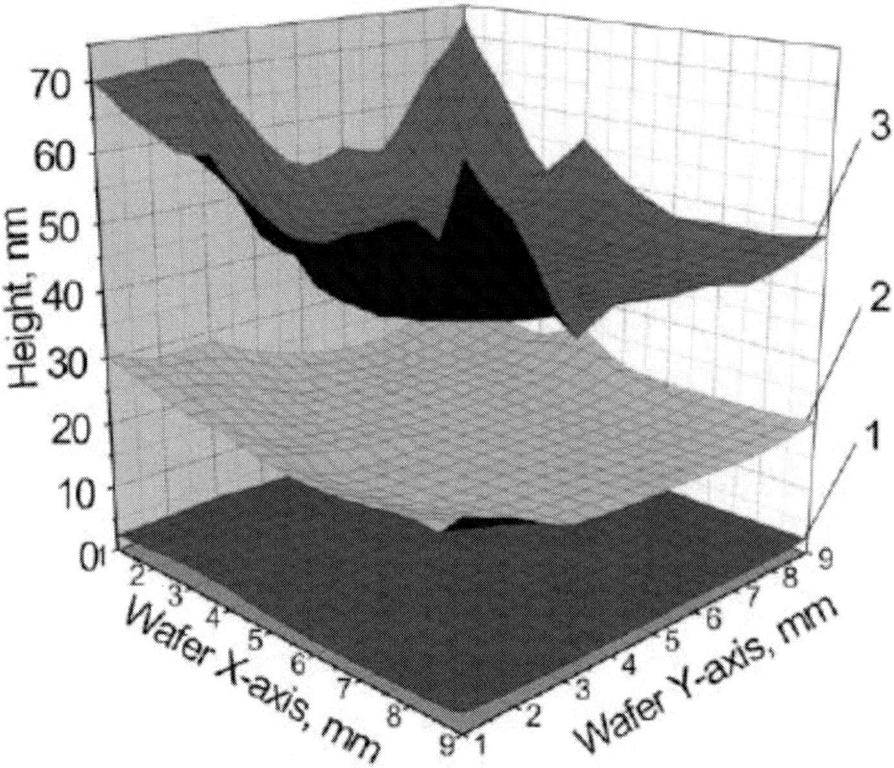

Figure 23. Ellipsometric analysis of a poly(acrylic acid) layer (3) grafted to PP surface (1) modified with a grafted dt-FPP nanolayer (2) in methanol at 110°C for 95 h.

Hence, resulting data show that the grafted FPP layer is formed from "islands" of cross-linked FPP macromolecules. The structural unit of each "island" comprises 500 - 1000 single FPP macromolecules. These structures ensure a high degree of surface coverage (above 90%).

The thickness of the FPP grafted film varies between 20 and 60 nm and is mainly determined by the nature of the peroxide group in FPP. It is possible to localize 0.01 to 0.03 mmol/m^2 of reactive peroxide groups of PP substrate. By following thermal decomposition, peroxide groups initiate free radical polymerization reactions facilitating grafting processes.

Modification of the Peroxidized Polymer Surface with Functional Monomers and Polysaccharides

Chemical functionalization of a polymeric substrate, as a rule, results in increasing surface energy. Peroxide groups of the grafted FPPs on a polymeric substrate can be used for surface hydrophilization and formation of coatings providing tailored properties. Both "grafting to" and "grafting from" approaches can be applied to fabricate grafted coatings.

Table 6 shows the contact angles and calculated free energy of peroxidized polymeric substrates modified by grafting hydrophilic macromolecules, compared to non-modified surfaces. Using a "grafting from" approach, a higher surface modification degree (80 ÷ 95%) is observed in comparison to a "grafting to" approach (40 ÷ 91%). Using dt-FPP-activated substrates results in a higher modification degree in non-aqueous media, whereas its efficiency in an aqueous solution is lower. To obtain a high degree of surface modification in water, grafting pe-FPP or pt-FPP is required. As it is shown, grafted layers of water-soluble polymers significantly reduces the water contact angle by increasing the polar component of the free energy of the surface. As shown in Table 6, the surface energy of PP surface can be controlled by graft-polymerization of a functional monomer or grafting of a polymer using preliminary peroxidized surface. The latter is especially important for polyolefin materials whose reactivity is very low because of the absence of any functional groups in their structure.

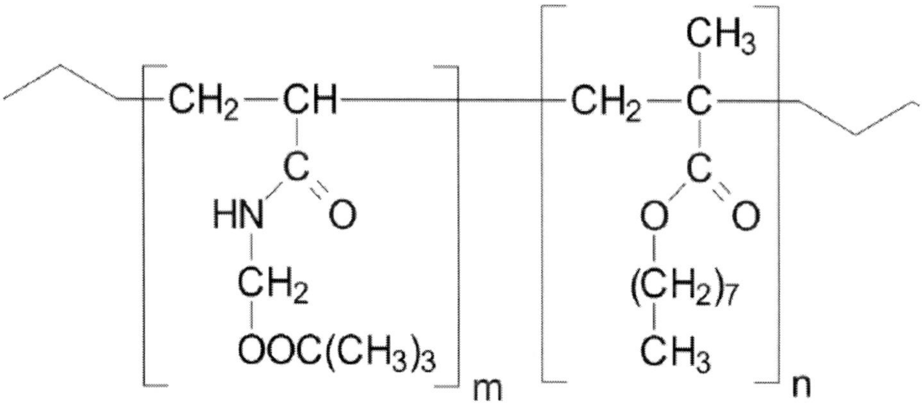

Figure 24. Poly-N-[(*tert*-butylperoxy)-methyl]acrylamide-co-(octyl methacrylate), m:n = 1:6.

Peroxidized PP surface has been shown to initiate radical graft-polymerization from the surface (Table 6) upon heating in methanol or aqueous solution of acrylic acid. The presence of a layer of grafted poly(acrylic acid) macromolecules on the polypropylene surface has been confirmed by increase in polar constituent (λ_S^h) of free surface energy indicating that the hydrophilization of hydrophobic PP surface occurs.

Figure 23 shows ellipsometric mapping of the peroxidized PP surface with a covalently bound layer of poly(acrylic acid). To the cross-linked dt-FPP layer of 18 ± 2 nm, a 25 ± 3 nm film of poly(acrylic acid) has been grafted.

Covalent Grafting of Paam-Based Hydrogels to a PP Substrate by Radical Polymerization

The chemical structure of the FPP used for the peroxidation (activation) of the PP surface is shown in Figure 24.

To synthesize a hydrogel network using radical reactions, the polymerization of AAm from the peroxidized PP substrate has been initiated in the presence of bis-AAm as a cross-linking agent. It was expected that the addition of a low-molecular cross-linking agent would result in the formation of a PAAm hydrogel network covalently bonded to the PP surface. The scheme for radical polymerization of acrylamide in the presence of bis-AAm and hydrogel network grafting is shown in Figure 25.

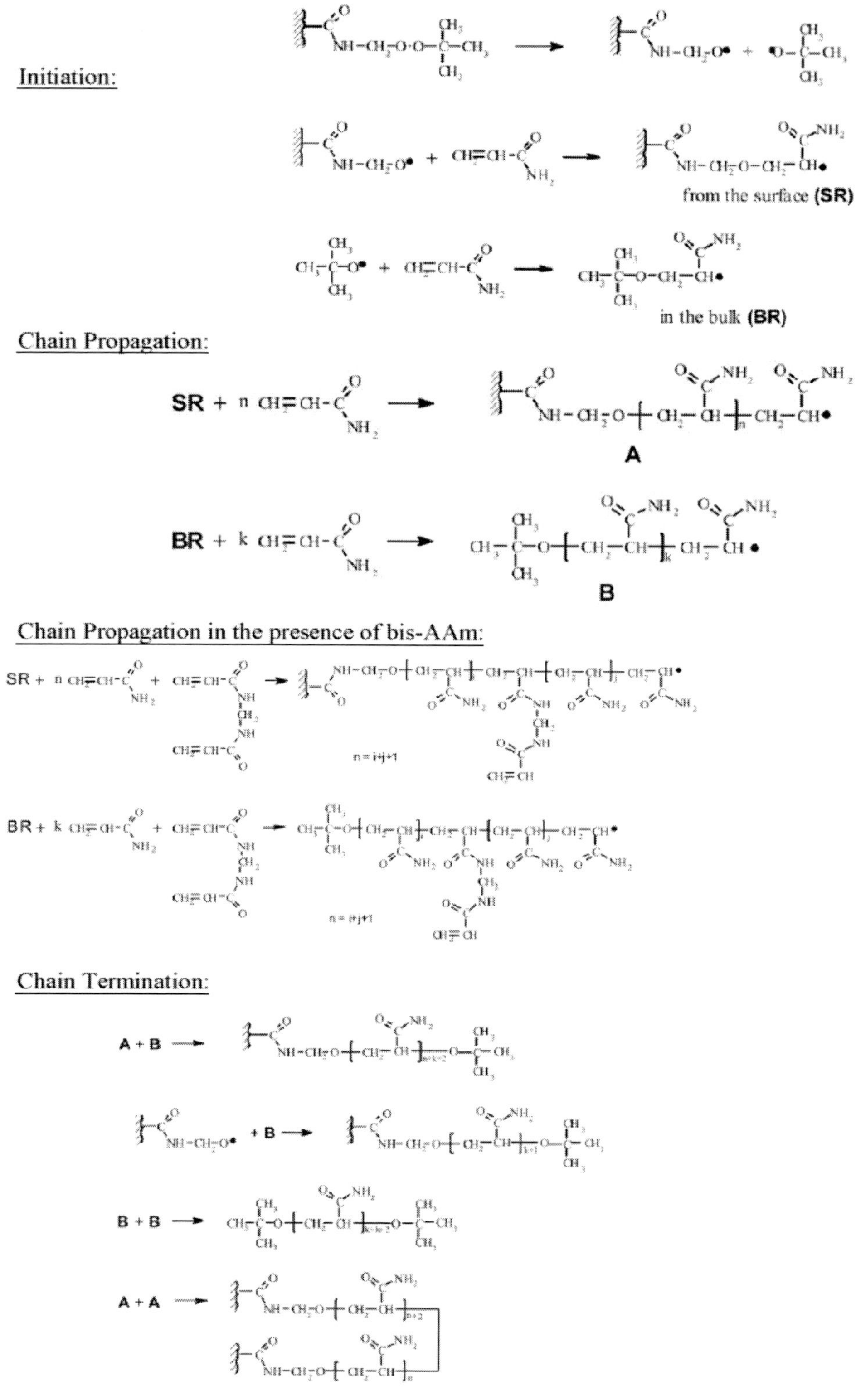

Figure 25. PAAm grafting on peroxidized PP surface by "grafting-from" approach.

It was assumed that, as a result of polymerization, the unsaturated fragments would be incorporated into macromolecules, thus forming the cross-links shown in Figure 26. However, no grafted PAAm-based hydrogel network was detected, even when a broad concentration range of bis-AAm (0.1–0.6 wt. %) and AAm (4–8 wt.%) was used for polymerization in aqueous medium in the presence of the peroxidized PP substrate. Obviously, the peroxide groups localized on the PP surface generate radicals that ensure only grafting of PAAm macromolecules (as it was, in fact, later confirmed when intermolecular condensation was applied). However, number of localized groups is not sufficient for the cross-linking reaction and formation of the grafted PAAm network. To enhance the formation of radical active centers, a water-soluble initiator of radical polymerization, potassium persulfate (PPS), was added. Formation of ultrathin cross-linked polymeric films in a presence of free initiator has been previously reported [60]. The formation of a grafted hydrogel network in the presence of a radical initiator in bulk was monitored by water contact angle measurements. Indeed, the addition of PPS provided more consistency in the formation of a grafted PAAm hydrogel network.

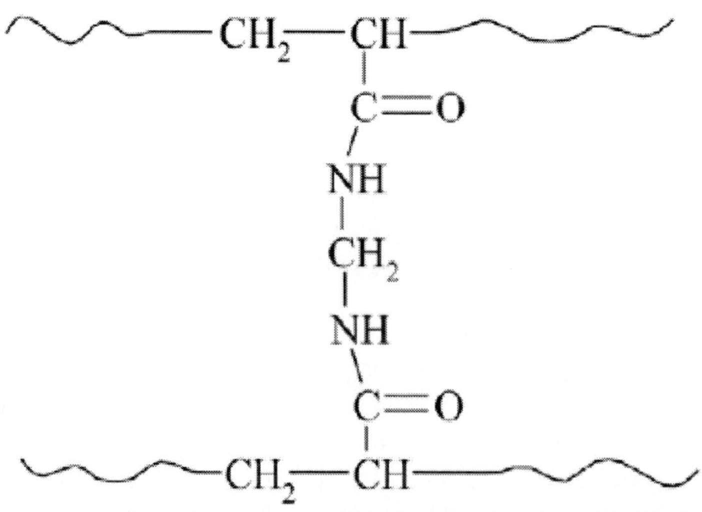

Figure 26. Structure of the cross-links formed in radical polymerization of AAm in the presence of bis-AAm.

The appearance of the grafted hydrogel on the PP substrate and the effect of the PPS concentration and the monomers ratio on the water contact angle measured on the modified PP surface are shown in Figure 27. Both the PPS concentration and the monomers ratio have an impact on the wettability of the PP substrate after polymerization. Upon polymerization, the hydrophobic PP surface becomes hydrophilic, indicating the attachment of more hydrophilic PAAm chains to the substrate. Interestingly, once the concentration of PPS exceeds 7.5×10^{-3} wt.%, a rapid increase in the water contact angle was observed for both monomer ratios used in the experiments. Obviously, this can be explained by fact that, at a PPS concentration higher than 7.5×10^{-3} wt.%, the rate of polymerization in bulk dramatically exceeds the rate of polymerization initiated from the surface at this concentration, thus preventing grafting.

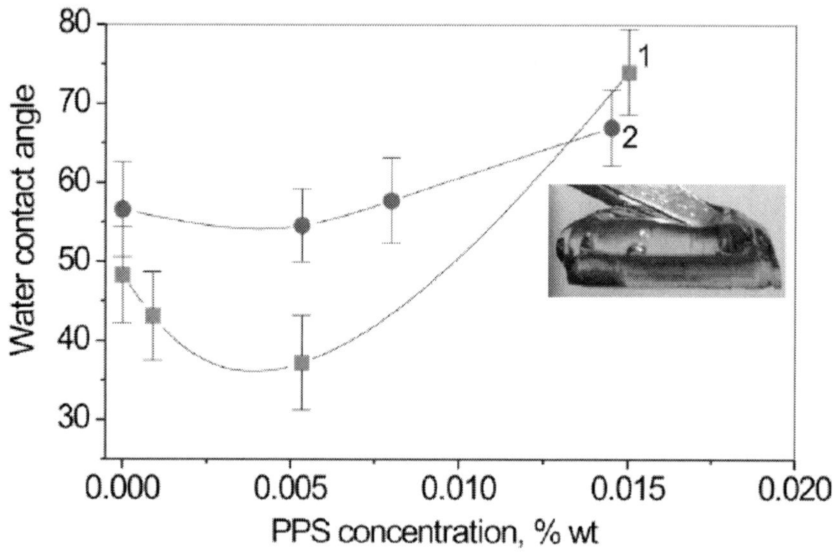

Figure 27. Water contact angle change with [PSS] at different monomer ratios: (1) [bis-AAm] = 0.34 wt.%, [AAm] = 7.5 wt.%; (2) [bis-AAm] = 0.25 wt.%, [AAm] = 7.5 wt.%. Grafted hydrogel appearance in polymerization of [bis-AAm] = 0.25 wt.%, [AAm] = 7.5 wt.% in the presence of [PPS] = 0.005 wt.% is shown in the photograph.

The chosen experimental conditions ensure the formation of a hydrogel with a height of 14–25 mm in the swollen state (Figure 27, photograph). To verify the covalent bonding of the hydrogel network and the PP substrate, the grafted hydrogel was extensively removed from the surface.

After drying, the substrate with grafted hydrogel residues was immersed for 2–3 min in the aqueous solution of Brilliant Black BN dye and rinsed in distilled water. The residues were not seen visually. However, once the dried sample was immersed in the dye solution, almost immediately dark spots became visible, indicating adsorption of the dye on covalently bonded hydrogel residues. The hydrophilic molecules of brilliant black were not otherwise adsorbed on the highly non-polar PP substrate. Thus, the appearance of dark spots indicates the presence of covalently bonded PAAm hydrogel residues on the PP substrate (Figure 28 B).

Experimental series at varying concentrations of AAm and bis-AAm showed that the water contact angle recorded on the residues of the hydrogel networks depended on the monomers' concentration and the ratio used in the synthesis of the grafted hydrogel (Figure 28 A). Most hydrophilic substrates were identified for hydrogels made of AAm at concentrations ranging from 8.5 to 14 wt. %, and the ratio of bis-AAm/AAm at 0.05–0.08.

An alternative way to ensure the formation of covalent bonds between hydrogels and PP substrates is to measure the water contact angles on dried PP samples containing hydrogel residues after removal. These angles have been compared to contact angles on the PP surface modified by grafting PAAm macromolecules (without bis-AAm cross-linking). Plot 2 in Figure 29 shows that the water contact angle measured on the dried PP surface containing grafted hydrogel residues gradually decreased in time and reached 22°after 4 min. In turn, the water contact angle of the surface-modified without cross-linking substrate remained constant at 40° after a very small decrease (Figure 29, plot 1). This value of the water contact angle on the PAAm-modified without cross-linking PP is in good agreement with data observed earlier [54].

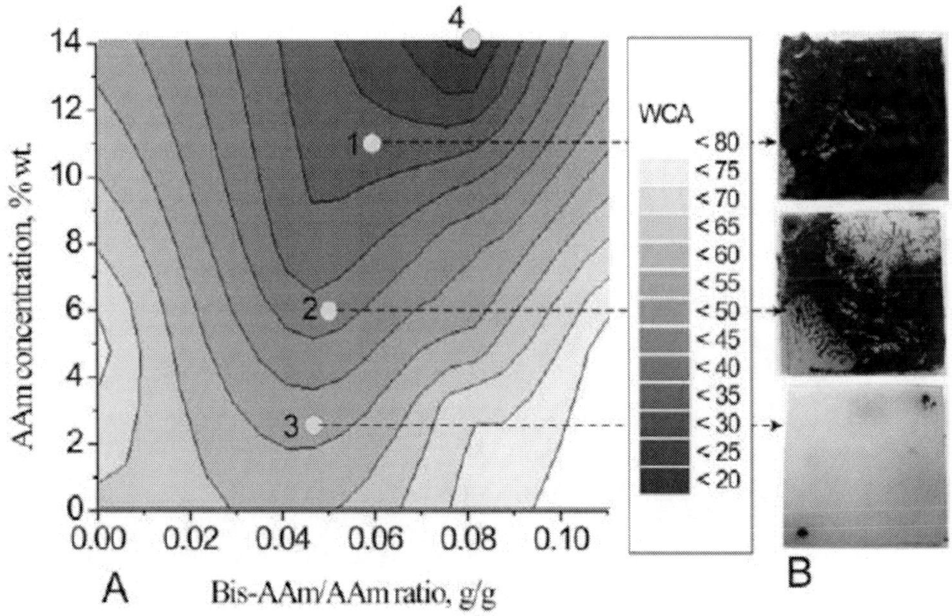

Figure 28. Effect of [AAm] and the bis-AAm/AAm ratio on the water contact angle of PP after the grafted hydrogel was removed, rinsed, and dried (A), grafted hydrogel residues on surfaces after immersion in Brilliant Black BN solution and rinsing in distilled water (B).

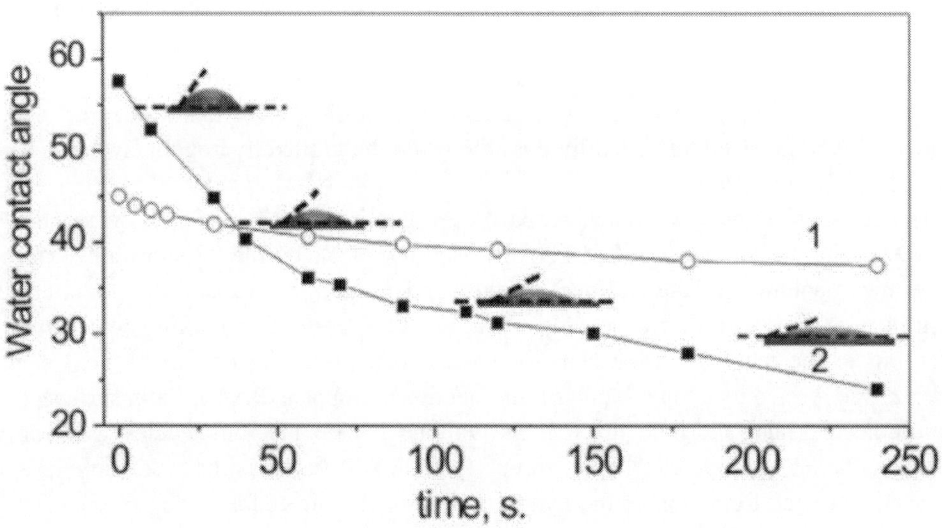

Figure 29. Water contact angle on PAAm-modified PP substrate: (1) covalently grafted PAAm without cross-linking, (2) grafted hydrogel residues (corresponding to experiment 4 in Figure 28 A).

The decreasing contact angle (plot 2) indicates that the hydrogel residues that remain after the extensive cleaning and extraction undergo swelling by contact with wetting water. The swelling results in an enhanced substrate wetting and a decreasing contact angle. In fact, this indicates the presence of covalently bonded hydrogel residues and confirms the successful grafting of the PAAm network to the PP substrate by radical polymerization.

Using radical polymerization, a grafted hydrogel was formed on the surface of PP mesh (Figure 30). This sample was first swollen in water, then dried and reswollen once again in a 1% Brilliant Black BN aqueous solution. The photographs show the appearance of the grafted hydrogel network after each of the experimental procedures. The dried grafted hydrogel reswelled and changed color upon placement in the dye solution. Grafted hydrogel (Figure 30 B) perfectly reswelled after being dried and possessed an ability to immobilize hydrophilic dye molecules on the surface of the hydrophobic PP mesh.

Figure 30. Grafted hydrogel network (15 wt.% of gel-forming polymers) covalently bonded to PP mesh: swollen (A), dried (B), reswollen in dye solution (C).

Covalent Grafting of Paam-Based Hydrogels Top Substrate by Intermolecular Condensation

Although the successful grafting of the PAAm hydrogel network to the PP substrate by radical polymerization has been carried out, the approach described above has been found to be effective only for a narrow concentration range of gel-forming components (additional initiator and monomers). When concentration of the additional initiator is less than 0.002%, the polymerization in reactive volume is obviously too slow to support grafting reactions. By high concentration of the additional initiator (>0.008%), in turn, high polymerization rate in the reactive volume prevents grafting. In addition, control of cross-linking degree, and thus degree of hydrogel swelling, is hardly possible when the grafted hydrogel network is formed by radical polymerization.

Hence, an alternative approach based on the intermolecular condensation reaction was applied for the formation of grafted PAAm-based hydrogels on the PP substrate surface. To ensure the attachment of the hydrogel by the condensation mechanism, the peroxidized PP substrate was first modified by covalent grafting of PAAm. The grafting was carried out according to the technique [58,61] in a separate experimental step. In the next step, the network was formed by condensation of the functional groups in PAAm (both surface-grafted macromolecules and added to the reactive volume PAAm macromolecules) with poly-N-hydroxymethylacrylamide functional groups (pHMAAm) (used as the cross-linking agent). The scheme of the formation of the hydrogel network by the condensation mechanism [62–64] is shown in Figure 31.

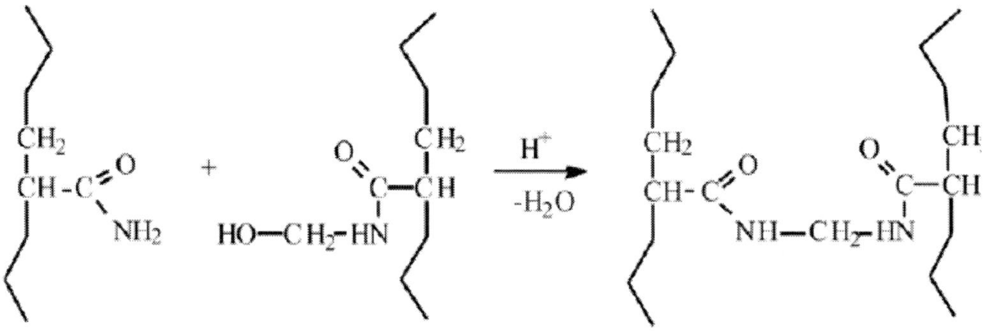

Figure 31. Formation of the hydrogel network by intermolecular condensation between the functional groups of PAAm and pHMAAm.

The grafting of PAAm to the PP surface (separate first reaction step) was confirmed in FTIR spectroscopic and contact angle measurements. There is an appreciable difference among the spectra of non-modified polypropylene surface (substrate) and the PAAm-modified polypropylene shown in Figure 32. As compared to the spectrum of PP substrate, the following new characteristic absorption bands appear in the PAAm-modified PP surface spectrum: a broad absorption band in the range 3500–3100 cm^{-1} resulting from the N–H stretch oscillation, and an intensive band at 1670 cm^{-1} corresponding to the C=O stretch oscillation. The bands clearly indicate the presence of a grafted PAAm on the substrate. The adsorption bands at 2950, 2910, 2840, 1460, and 1380 cm^{-1}, found in the both spectra, are attributed to the antisymmetric and symmetric stretching (first three bands) and deformation (last two bands) of PP alkyl groups.

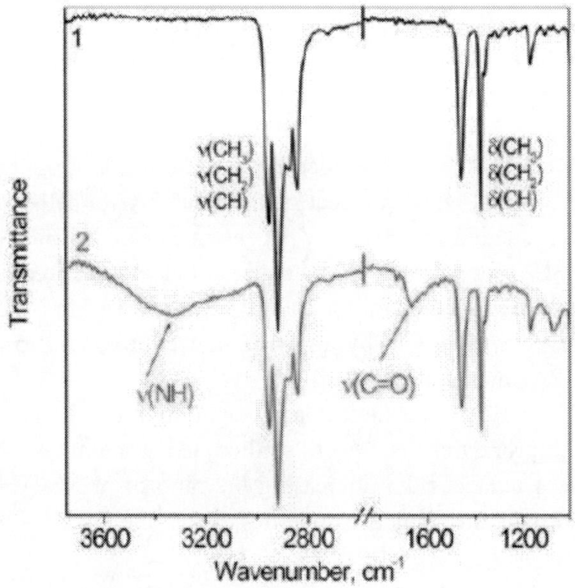

Figure 32. FTIR spectrum of the grafted PAAm-modified PP substrate.

Water and diiodomethane contact angles recorded on nonmodified PP, peroxidized PP, and grafted PAAm-modified PP surfaces and calculated surface energy values [54] are shown in Table 7. Changes in the contact angles were observed after both reaction steps. The changes confirm the successful peroxidation and grafting of PAAm on the PP substrate. The recorded data indicate the changing surface energy of the polymer substrate from hydrophobic to hydrophilic, upon PAAm grafting. Interestingly, that significant change in free surface energy after the PAAm grafting resulted mainly from increasing polar constituent λ_s^h, whereas the disperse energy constituent λ_s^d was rather less sensitive to the surface modification.

Table 7. Contact angles and calculated free surface energy values on the non-modified, peroxidized, and grafted PAAM-modified PP surface

no.	Surface	Surface characteristics				
		Contact angle Θ, °		Free surface energy, mN/m		
		H_2O	CH_2I_2	λ_s^d	λ_s^h	λ_s
1	PP	106	64	31.2	0.19	31.39
2	PP +FPP	82	52	29.6	4.3	33.9
3	PP + FPP + PAAm	39	53	20.4	37.3	57. 7

Before the hydrogel grafting on the PP substrate by the condensation mechanism, formation of the (non-grafted) network was studied using PAAm ($M_W = 130 \times 10^3$ g/mol) and pHMAAm (at total 3 wt.% macromolecules concentration, at ratio 1:1 mol of functional groups) at 70°C and different pH values. One purpose of this experiment was to verify cross-linking and find the optimal pH value and cross-linking time to ensure hydrogel network formation.

The data in Figure 33A indicate considerably faster cross-linking by decreasing the pH of the reactive volume. To this end, an effect of the total PAAm/pHMAAm concentration on cross-linking has been studied at pH 2.0, when the most rapid formation of the hydrogel is observed. Figure 33B presents the dependence of the time required for gel-forming components to be fully cross-linked on the total PAAm/pHMAAm concentration. Full cross-linking can be achieved within 3 h, by applying concentrations of 5–8 wt.% (1:1 ratio of functional groups in PAAm and pHMAAm).

The effect of the pHMAAm molecular weight and the PAAm/pHMAAm ratio on hydrogel mechanical properties has been studied using complex modulus of elasticity measurements. Table 8 indicates that increasing the ratio of pHMAAm/PAAm results in an enhancement of hydrogel mechanical properties, once the maximum degree of swelling decreases (samples 1,4,7).

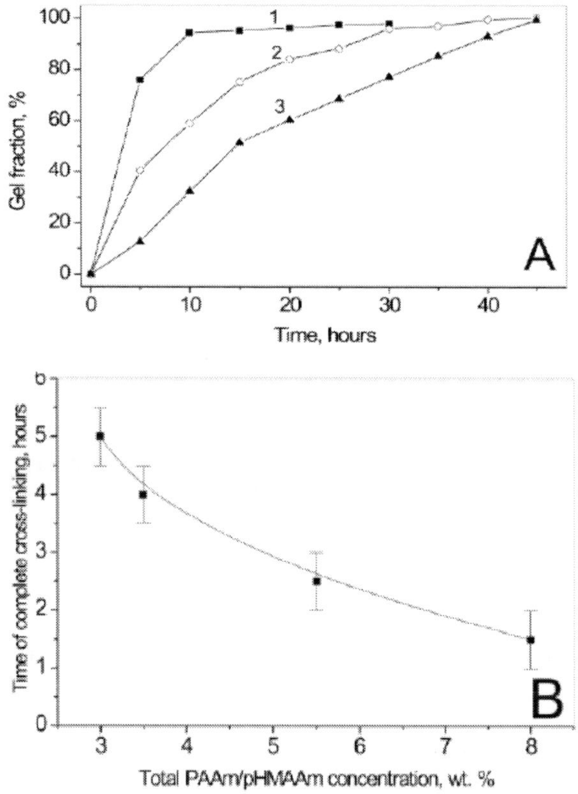

Figure 33. (A) Gel-fraction volume change with cross-linking time at pH: 1.95 (1), 2.6 (2), 3.0 (3). (B) Cross-linking time (required for the gel-fraction volume to exceed 97%) vs. total wt. % concentration of macromolecules in the reactive volume (ratio 1:1 mol of functional groups at pH 2).

Table 8. Formation and properties of the PAAm/pHMAAm non-grafted hydrogel network

No	Reaction conditions			Molecular weight of pHMAAm, $\times 10^{-3}$ g/mol	Maximum swelling in water, %	Complex modulus of elasticity G^*, Pa
	Concentration, wt. %		Ratio PAAm/ pHMAAm (mol.)			
	PAAm	pHMAAm				
1	4	5.6	1.0	1800	445	1904
2	4	4.0	0.71	180	1125	334
3	4	4.0	0.71	1100	1371	916
4	4	4.0	0.71	1800	577	736
5	4	2.4	0.43	180	1106	146
6	4	2.4	0.43	1100	2138	279
7	4	2.4	0.43	1800	1866	233
8	4	0.8	0.14	180	1829	Destroyed
9	4	0.8	0.14	1100	1989	Destroyed
10	4	0.8	0.14	1800	--	Destroyed

The highest elasticity modulus was observed (at the 1:1 mol ratio of functional groups in gel-forming polymers and pHMAAm molecular weight 1100 × 10^3 g/mol) with the total concentration of polymers 6–8 wt.%. Nevertheless, the presented data show that, to ensure optimal gel properties (higher strength and maximum swelling), a molar ratio of 0.4–0.7 of the functional groups in gel-forming polymers is required.

Figure 34 presents a storage modulus (G') for hydrogel networks corresponding to compositions 5,6,7 in Table 8 in dependence on the loading rate (ω). No significant changes in the modulus upon changing the loading rate were observed. The latter indicates that the samples were three-dimensional polymer networks [65], and cross-linking occurred for each of three pHMAAm samples with different molecular weights.

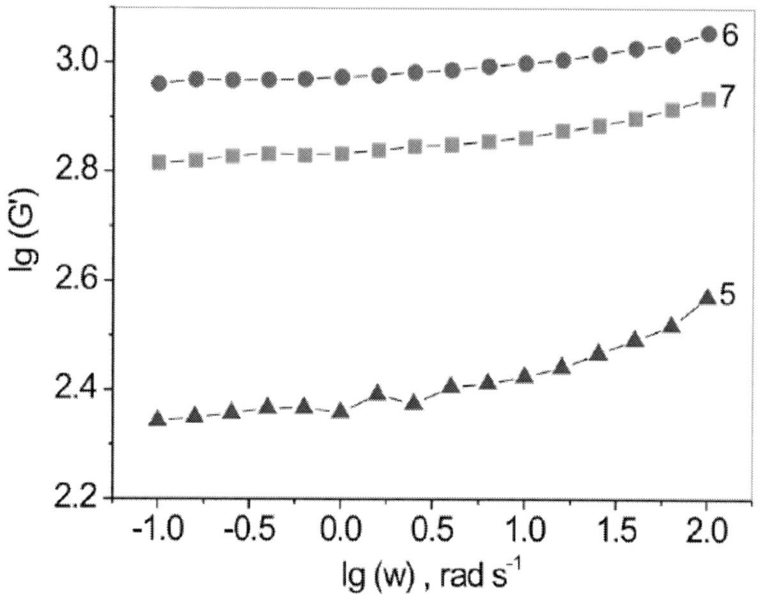

Figure 34. Storage modulus change with loading rate for non-grafted hydrogel samples (5,6,7 correspond to compositions shown in Table 8).

The composition number 3 (Table 8) has been chosen for an experiment on the covalent attachment of the hydrogel network to the PP substrate by the proposed intermolecular condensation mechanism. To confirm covalent grafting of the hydrogel on the PP substrate, a tensile failure test was carried out for the hydrogel covalently bonded to PP and for the hydrogel formed in the presence of the non-modified PP substrate.

To prepare the samples for mechanical testing, the hydrogels were synthesized between two PP substrates, modified and nonmodified by grafted PAAm (stage II in the proposed approach). No hydrogel network residues (adhesion failure) were observed for the hydrogel formed between the non-modified PP substrates (Figure 35 A). In turn, when the PAAm-modified substrates were chosen, cohesion failure occurred (Figure 35 B), clearly indicating that grafted hydrogel residues remained. The tensile strain values at failure were low for the samples made of the non-modified substrate, whereas the values recorded for the failure of the PAAm-modified surfaces were within the 30–50 kPa range. This is in a good agreement with the compressive failure strain of the gel compositions presented in the literature (40–80 kPa) [66].

Figure 35. Failure of hydrogel samples at mechanical loading.

Covalent grafting of the hydrogel on PP by the condensation mechanism has been well shown by the water contact angles data. The angles were measured after the mechanical and extractive removal of the residual grafted network from the PP (the samples after mechanical testing were extensively cleaned and then used for the water contact angle measurements).

Figure 36 (plot 2) indicates that the water contact angle on the dried PP surface containing grafted hydrogel residues gradually decreased in time. This is similar to the wettability of the residues from the network made by radical polymerization. The decreasing contact angle confirmed the swelling of the grafted hydrogel residues upon their contact with wetting water. In fact, the angle confirmed successful grafting of the PAAm network to the PP substrate by intermolecular condensation.

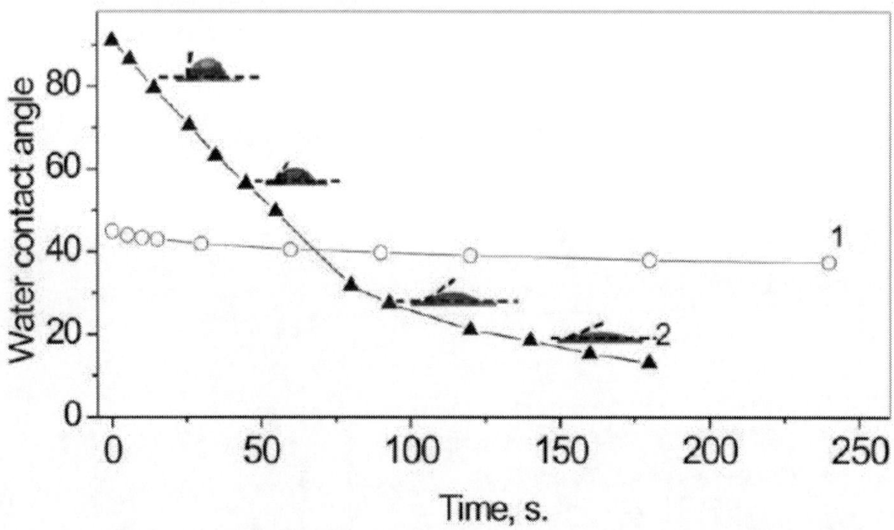

Figure 36. Water contact angles change with time for: (1) grafted PAAm-modified PP substrates, (2) grafted hydrogel residues after the mechanical testing and following extraction.

Covalent Grafting of Paam to Dispersed Silica Surface and Formation of Porous Hydrogels with a Regular Pore Distribution

The chemical structure of the hydrophilic FPP for the peroxidation (activation) of the dispersed silica surface is shown in Figure 37. The FPP's acrylamide fragments provide the affinity to the dispersed mineral surface serving physical adsorption of FPP on the surface of dispersed silica particles. On the other hand, the peroxide groups of the FPP are able to initiate free-radical processes, on the surface, at elevated temperatures. As we have shown above, pt-FPP, developed from the peroxide monomer of the same chemical structure, successfully initiated free radical polymerization on the polyolefins surface.

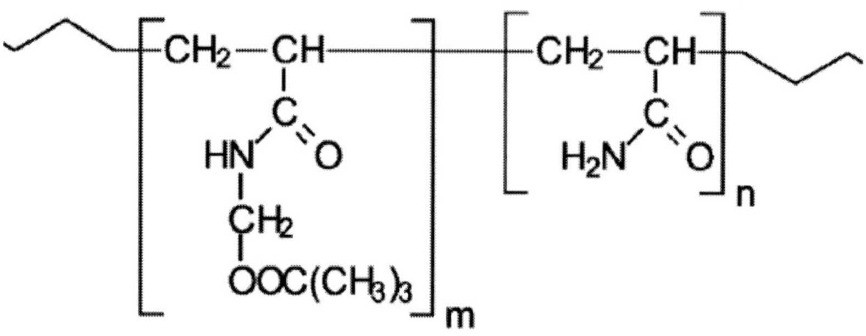

Figure 37. Poly[acrylamide-co-N-(*tert*-butylperoxymethyl)acrylamide], m:n = 9.6 : 90.4.

The porous PAAm hydrogels have been prepared by a procedure schematically represented in Figure 38. In the first step (peroxidation), a thin layer of FPP has been adsorbed on the silica particles. Then, acrylamide has been polymerized on the peroxidized particulate surface, resulting in a grafted film on the silica particles. The peroxide groups immobilized in the first step on the dispersed silica initiate free radical polymerization of acrylamide, on the silica surface, under the given polymerization conditions.

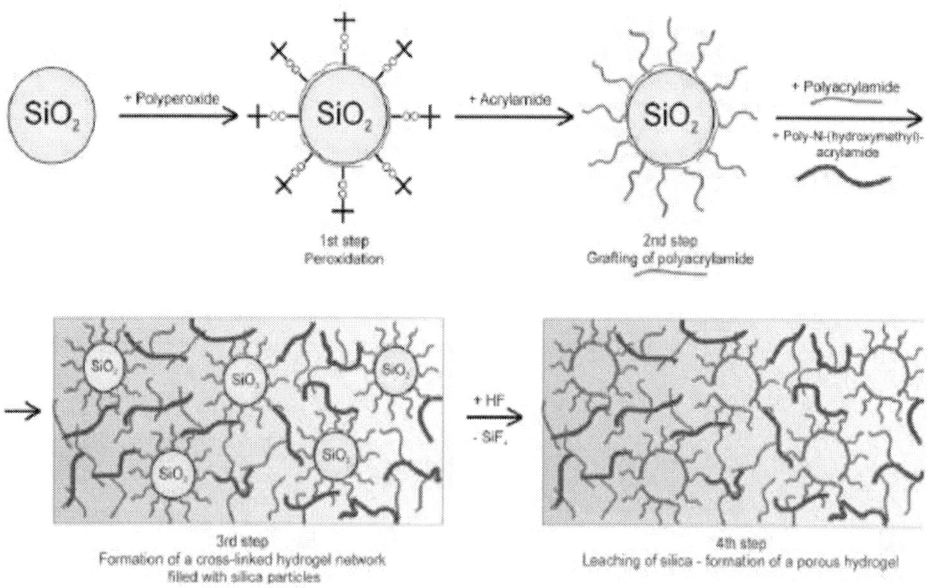

Figure 38. Formation of porous hydrogels.

The hydrogel network filled with the modified silica particles has been formed in the next step (crosslinking), and finally, silica particles have been leached out with hydrofluoric acid to give a porous hydrogel (leaching out silica).

The formation of the hydrogels (crosslinking) has been carried out by the interaction of the PAAm chains grafted to silica surface with an aqueous solution of a mixture PHMAAm used as a structurizing agent and polyacrylamide. A three-dimensional crosslinked hydrogel network has been developed due to the interaction of NH_2 groups of polyacrylamide with the hydroxyl groups of poly-N-(hydroxymethyl)acrylamide (Figure 31,38). We expect that PHMAAm does not interact only with 'free' polyacrylamide macromolecules in the aqueous solution. Simultaneously, the structurizing agent reacts with the grafted silica surface polyacrylamide chains, and forms covalent bonds between the grafted PAAm and the 'free' PAAm from the aqueous solution. Subsequent removal of the dispersed silica, by treating the hydrogel with hydrofluoric acid, results in the formation of a porous hydrogel. In fact, the interaction between silica and HF is a common reaction, resulting in the formation of volatile SiF_4. We assume, that the pore "walls" are formed by the polyacrylamide, which has been previously attached to the silica particles improve the mechanical properties of the porous hydrogel.

FPP has been adsorbed on the silica particulate surface from an aqueous solution. It can be assumed that, the adsorption of FPP macrochains onto silica particles is due to the van der Waals forces, as well as the formation of hydrogen bonds between the amide groups present in the FPP macromolecules and the hydroxyl groups that are usually present on the silica surface.

Thermogravimetric analysis has been used to estimate the amount of the polyperoxide adsorbed on the dispersed silica surface.

Figure 39 shows a TGA diagram of the silica modified with FPP. The sample shows an appreciable weight loss, mainly in the region between 150 and 500°C. According to TGA analysis, the content of the FPP adsorbed from water on the silica surface is 0.78 wt % silica, and this value can be assumed to be the equilibrium adsorption of the FPP on the dispersed SiO_2, from the 1.5% aqueous solution of the FPP. The sample shows two distinct weight loss regions like the bulk polyperoxide. The first turning point ($T \approx 185°C$) correlates well with that of the bulk FPP (data not shown) confirming the presence of the peroxide groups immobilized on the silica surface.

To carry out the grafting of PAAm, the peroxidized silica particles have been immersed in an aqueous solution of acrylamide. The peroxide groups immobilized on the silica surface decompose at 95°C and initiate the polymerization of acrylamide. As a result, a PAAm film grafted on the surface of dispersed silica has been formed. Both FT-IR spectroscopy and thermogravimetric analysis confirmed the grafting of polyacrylamide on the silica surface. There is an appreciable difference between the spectra of bare silica and the silica with grafted polyacrylamide.

Figure 40 shows an FT-IR spectrum of the PAAm-grafted silica. As compared to the spectrum of bare silica, new characteristic absorption bands appear in the PAAm-modified silica spectrum: (i) a band at 1670 cm^{-1} attributed to the valence oscillation of the C=O bond in the amide groups (amide I); a band at 1610 cm^{-1} resulting from deformation of the N–H bond (amide II); and (iii) a broad broken band in the range of 3550–3050 cm^{-1} attributed to the N–H stretch oscillations. These data clearly show the presence of a grafted PAAm film on the surface of the silica particles.

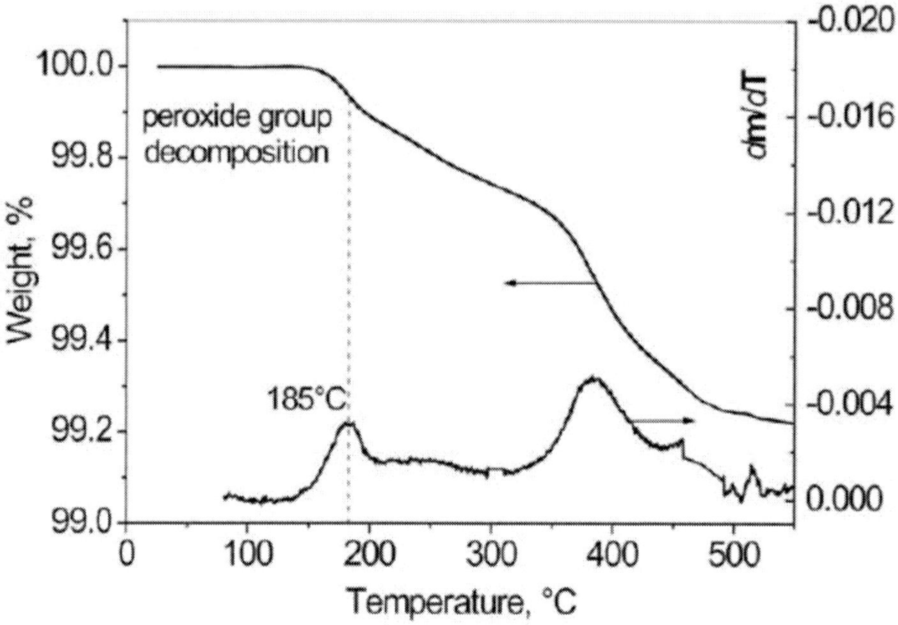

Figure 39. Thermogravimetry data for silica modified with poly[acrylamide-co-N-(tert-butylperoxymethyl)acrylamide] FPP.

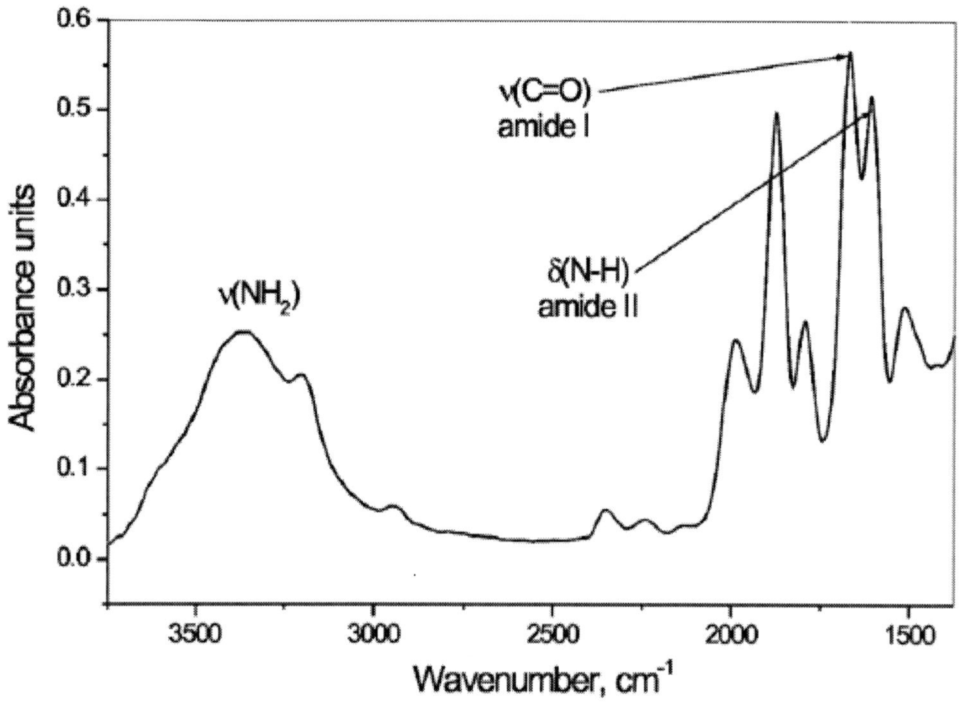

Figure 40. FT-IR spectrum of silica particles modified with PAAm.

The amount of PAAm grafted on the SiO_2 particle surface has been estimated from the thermogravimetric analysis data. Figure 41 shows the TGA diagrams of the nonmodified SiO_2 (a), silica modified with the FPP (b), and silica modified with tethered polyacrylamide macrochains (c). The PAAm-modified sample shows an appreciable weight loss in the region between 250 and 550°C. According to the TGA analysis, the total weight loss for silica modified with PAAm is about 2.35% relative to the initial SiO_2 powder. The percentage of the grafted PAAm has been calculated as 1.57%, after subtracting the contribution of FPP decomposition from the total weight loss.

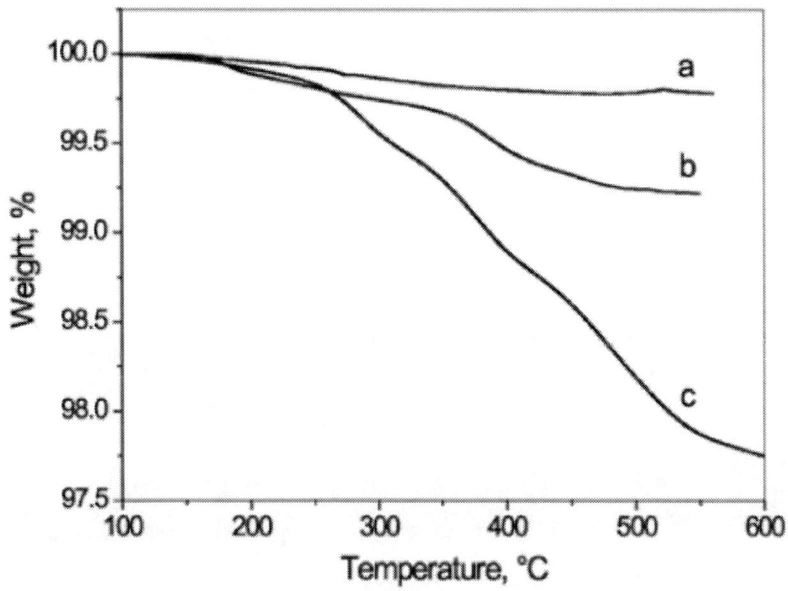

Figure 41. Thermogravimetry diagrams for silica: (a) bare, (b) modified with FPP, and (c) with grafted PAAm chains.

For the development of hydrogels filled with silica particles, the latter have been dispersed in an aqueous solution of a mixture of PAAm and PHMAAm, before the crosslinking reaction. When the hydrogel composition has been crosslinked, the resulting hydrogels have been immersed into an aqueous solution of hydrofluoric acid to remove the silica particles and to develop the pores.

We have studied an effect of the grafted silica PAAm chains on the morphology and the physicomechanical properties of porous hydrogels formed with the PAAm-modified silica particles and compared it with that formed using nonmodified SiO_2.

Figures 42, 43 present scanning electron micrographs (SEM) of samples broken in liquid nitrogen and dried hydrogel samples filled with the grafted PAAm-modified and nonmodified silica particles. It is clearly seen that more regular distribution of the modified silica particles has been observed in the hydrogel composition when compared with nonmodified particles. Furthermore, cohesion failure of the bulk matrix without any stripping of the surface of the filler particles has been detected, when the samples were imaged using BSE detector (as different contrast) (Figure 42 c).

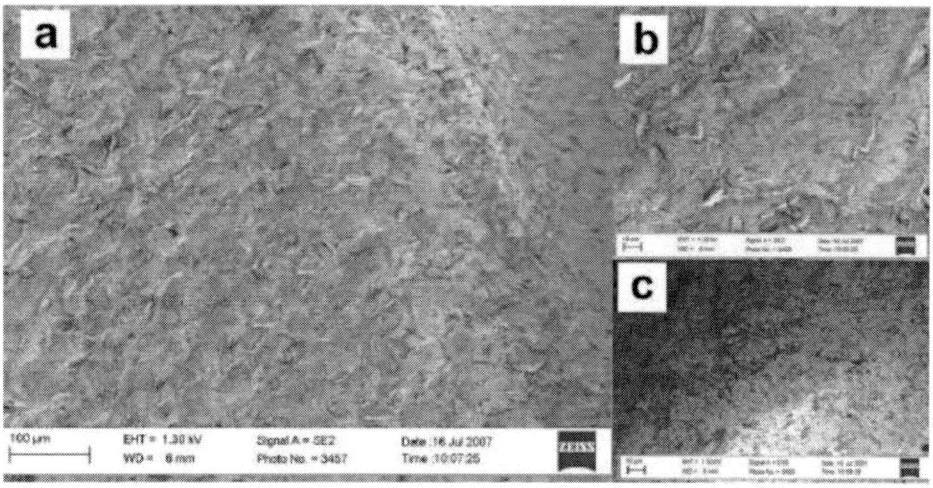

Figure 42. SEM micrographs of hydrogels filled with modified silica (a, b). The micrograph (c) is imaged at different (grayscale) contrast using BSE detector.

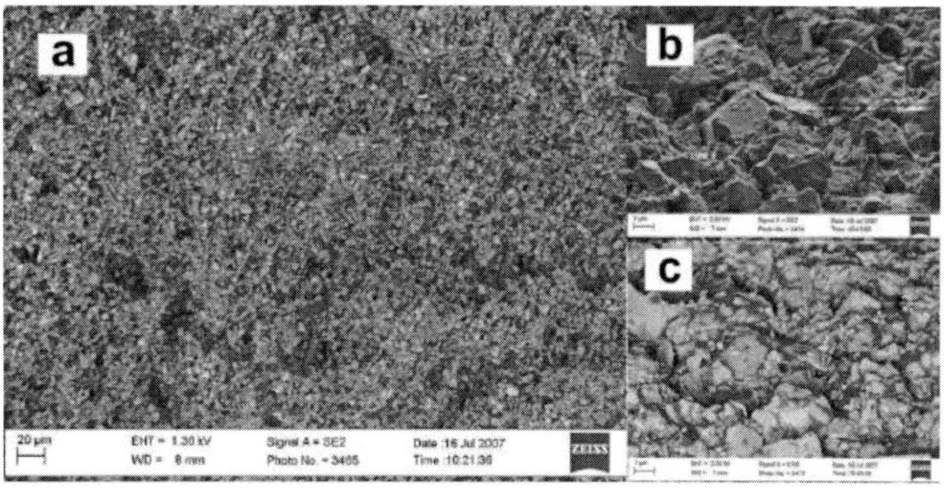

Figure 43. SEM micrographs of hydrogels filled with nonmodified silica (a, b). The micrograph (c) is imaged at different (grayscale) contrast using BSE detector.

On the contrary, a micrograph in Figure 43 b shows appreciable agglomeration of the particles, and adhesion failure on the filler–matrix interface has been observed, when the samples were measured using BSE detector (Figure 43 c). These data demonstrate that the grafted PAAm macromolecules are involved in the formation of a crosslinked polymeric hydrogel composition, because of their covalent bonding to a hydrogel network.

The modification of the silica surface has not changed the ability of the silica particles to leach out from a filled hydrogel. Both modified and nonmodified silica have been successfully leached out from hydrogels using an HF aqueous solution. SEM micrographs of the broken surfaces of porous hydrogels, after the removal of silica, (Figure 44) have shown that silica has been completely dissolved by the treatment of filled hydrogels with hydrofluoric acid.

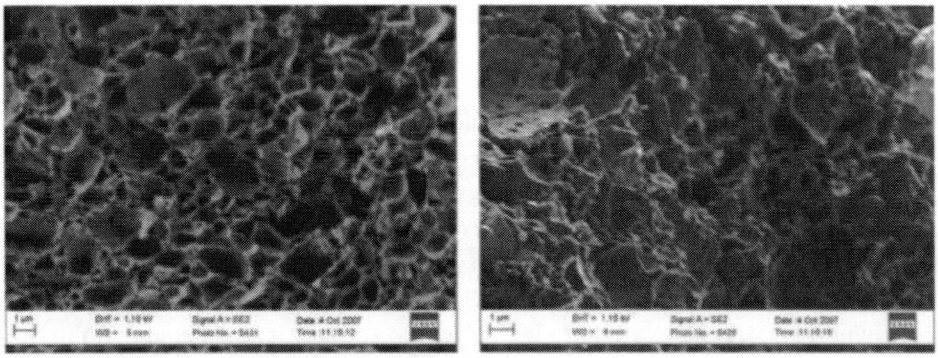

Figure 44. SEM micrographs of porous hydrogels formed after leaching out silica particles: (a) modified with polyacrylamide and (b) nonmodified.

More regular distribution of the modified SiO_2 particles has resulted in the formation of hydrogels with a regular pore distribution. An important feature of the porous hydrogel formed with the PAAm-modified silica particles is the small amount of pore containing "wall" defects when compared with the hydrogel formed with the nonmodified SiO_2 (Figure 44).

The data in Table 9 show that the formation of porous hydrogels with reinforced pore "walls" results in a 1.5–3 times increase in the mechanical strength. The maximum swelling degree and swelling rate retain almost the same as that of the hydrogels formed with the nonmodified silica.

Table 9. Characteristics of porous hydrogels

Amount of SiO_2, %	SiO_2	Swelling rate, $g_{water}/(g_{polym} \times sec)$	Maximum swelling degree, %	Complex modulus of elasticity G^*, Pa
25	non-modified	0.015	7520	1800
	PAAm-modified	0.013	7020	3400
50	non-modified	0.043	11730	1500
	PAAm-modified	0.045	10540	2100
65	non-modified	0.052	13150	600
	PAAm-modified	0.048	11400	1600

The swelling rate of the nonporous hydrogel formed under the same conditions has been found to be as $7.0 \times 10^{-5} g_{water}/(g_{polym} \times s)$, and its maximum swelling degree as 1160%. Hence, the formation of pores in the polymeric hydrogels significantly improves their swelling behavior.

CONCLUSION

We described peroxidation of polymeric substrate surface (PP, PE, TPO) by localization of FPP macromolecules at the interface and further covalent grafting of FPP as a concept for polymer surface modification. Ellipsometry, atomic force microscopy, and contact angle measurements were used to show that FPP chemical structure and polyolefin reactivity determine the mechanism of peroxidation and the properties of the grafted polyperoxide films. Formation of the grafted FPP film occurs by two interconnected processes, cross-linking of FPP macromolecules and their grafting to the polyolefin surface. We found reaction conditions when high degree of surface modification is achieved for grafted FPP film on the polymer surface. The grafted fractal (grafted “island”) consists of about 500–1000 cross-linked FPP macromolecules. This finding was described by a mathematical model characterizing polyperoxide grafting to the polyolefin surface. It was found that thickness of the grafted film (20–50 nm) depends on the polyperoxide chemical structure and molecular weight. As a result of peroxidation, 0.01–0.03 mmol/m^2 peroxide groups can be covalently attached to the polyolefin surface. These groups are further able to initiate radical polymerization on the polyolefin surface.

For the first time, PAAm-based hydrogels covalently bonded to a peroxidized polymer substrate have been synthesized. Peroxide groups in the grafted FPP initiate radical polymerization from the substrate. Two different approaches for hydrogel formation are proposed. First, a grafted PAAm hydrogel was formed by radical polymerization of AAm initiated from the peroxidized substrate in the presence of N,N’-methylenebis-acrylamide (cross-linking agent) and potassium persulfate (additional initiator of radical polymerization in bulk). Alternatively, covalent attachment of the PAAm hydrogel was carried out in a two-stage process: (i) grafting the PAAm from the peroxidized substrate and (ii) cross-linking the grafted PAAM and added to the reactive volume PAAm by the intermolecular condensation reaction with poly-N-(hydroxymethyl)acrylamide (used as a crosslinking agent). The alternative mechanism was found to be more advantageous for the formation of covalently bonded hydrogels on the PP substrate, while formation of the grafted network by radical polymerization occurs at a limited concentration range of the monomer and cross-linking agent.

Using peroxidized silica paricles as a porogen, a new and a facile strategy to the development of porous PAAm hydrogels with enhanced mechanical properties and regular pore distribution have been elaborated. This methodology involves the formation of porous hydrogel network by leaching out chemically modified silica particles. To improve the dispersibility of silica in the hydrogel composition, the silica particles have been modified with the chemically attached PAAm chains. The grafting polymerization of AAm on the silica surface has been initiated by FPP. The grafted PAAm chains form pore walls after leaching out the silica particles, thereby reinforcing the hydrogel composition and improving the mechanical strength of the hydrogel.

In summary, localization of FPP peroxide groups at the interface offers interesting aspects in constructing grafted polymeric films, facilitating surface activation and control over the radical grafting polymerization at the interface by “grafting from” and “grafting to” approaches.

REFERENCES

[1] I. Luzinov, S. Minko, V. Tsukruk, V. *Prog, Polym. Sci.* 29 (2004) 635–698.

[2] K. Kato, E. Uchida, Y. Uyama, Y. Ikada, *Prog. Polym. Sci.* 28 (2003) 209–259.

[3] A.S. Hoffman, *Macrom. Symp.* 101 (1996) 443–454.

[4] D.E. Bergbreiter, J.G. Franchina, K. Kabza, Macromolecules 32 (1999) 4993–5007.

[5] M. Tirrell, E. Korroli, M. Biesalski, *Surf. Sci.* 500 (2002) 61–83.

[6] N. Inagaki, S. Tasaka, Y. Goto, J. *Appl. Polym. Sci.* 66 (1997) 77–84.

[7] Y. Babukutty, R. Prat, K. Endo, M. Kogoma, S. Okazaki, M. Kodama, *Langmuir.* 15 (1999) 7055–7062.

[8] S. Minko (Ed.), Responsive Polymer Materials: Design and Applications, Blackwell Publishing, Ames, 2006.

[9] Y. Ikada, *Biomaterials.* 15 (1994) 725–736.

[10] A. Pomogailo, V. Kestelman, *Metallopolym. Nanocompos.,* Springer-Verlag, Berlin, 2005.

[11] D.J. Angier, In Book Chemical Reactions of Polymers, second ed., Interscience Publishers, New York, 1964.

[12] A.H. Hogt, in: J. Meijer, J. Jelenoc (Eds.), Reactive Modifiers for Polymers, Blackie Academic and Professional, London, 1997.

[13] S. Minko, G. Gafiychuk, A. Sidorenko, S. Voronov, *Macromolecules.* 32 (1999) 4525–4532.

[14] S. Minko, A. Sidorenko, M. Stamm, G. Gafiychuk, V. Senkovsky, S. Voronov, *Macromolecules.* 32 (1999) 4532–4539.

[15] A. Sidorenko, S. Minko, G. Gafiychuk, S. Voronov, *Macromolecules.* 32 (1999) 4539–4546.

[16] D.S. Wavhal, E.R. Fisher, *Langmuir.* 19 (2003) 79–85.

[17] Y.V. Pan, R.A. Wesley, R. Luginbuhl, D.D. Denton, B.D. Ratner, *Biomacromolecules.* 2 (2001) 32–36.

[18] Y. Ogiwara, M. Kanda, M. Takumi, H. Kubota, *J. Polym. Sci. Polym. Lett.* 19 (1981) 457–462.

[19] Z. Feng, B. Ranby, Angew. *Macromol. Chem.* 196 (1992) 113–125.

[20] V.Ya. Kabanov, *Uspekhi Khimii.* 67 (1998) 861–895.

[21] E. Bucio, G. Cedillo, G. Burillo, T. Ogawa, *Polym. Bull.* 46 (2001) 115–121.

[22] K. Fujimoto, Y. Takebayashi, H. Inoue, Y. Ikada, J. *Polym, Sci. Part A Polym. Chem.* 31 (1993) 1035–1043.

[23] P. Gatenholm, T. Ashida, A. Hoffman, J. *Polym. Sci.* 35 (1997) 1461–1467.

[24] E. Borsig, L. Hrcková, A. Fiedlerová, M. Lazár, M. Rätzsch, A. Hesse, *JMS Pure Appl. Chem.* A35 7 (1998) 1313–1326.

[25] S. Navarre, B. Maillard, J. *Polym, Sci. Part A Polym. Chem.* 28 (2000) 2957–2963.

[26] K. Schuh, O. Prucker, J. Ruhe, *Macromol.* 41 (2008) 9284.

[27] G.K. Raghuraman, K. Schuh, O. Prucker, J. Ruhe, *Langmuir.* (2010) 769.

[28] B. Boutevin, Y. Pietrasanta, M. Robin, *Macromol. Symp.* 57 (1992) 371–381.

[29] S. Voronov, V. Tokarev, V. Datsuk, *Prog. Colloid Polym. Sci.* 101 (1996) 189.

[30] S. Voronov, V. Tokarev, K. Oduola, *J. Appl. Polym. Sci.* 76 (2000) 1217.

[31] S. Voronov, V. Tokarev, V. Datsuk, *J. Appl. Polym. Sci.* 76 (2000) 1228.

[32] V. Samaryk, I. Tarnavchyk, A. Voronov, S. Varvarenko, N. Nosova, A. Kohut, S. Voronov, *Macromolecules.* 42 (2009) 6495–6500.
[33] G. Odian, *Principles of Polymerization,* Wiley, New York, 1981.
[34] K.U. Ingold, *Free-Radical Substitution Reactions,* Wiley, New York, 1971.
[35] V.L. Antonovsky, S.L. Khursan, Physical Chemistry of Organic Peroxides, *Akademkniga,* Moscow, 2003.
[36] S.A. Voronov, V.S. Tokarev, V.V. Datsyuk, H.J. Adler, G. Puschke, A. Pich, *Rep. Nat. Acad. Sci.* Ukraine 2 (1999) 145–149.
[37] H.J. Adler, in: A. Pich, A. Henke, G. Puschke, S. Voronov (Eds.), In *Polymer Colloids,* vol. 801, ACS, Washington, 2002, p. 413.
[38] S. Voronov, Functional Polyperoxides. Theoretical Basis of Their Synthesis and Application in Compounds, in: V. Tokarev, G. Petrovska (Eds.), Lviv Polytechnic State University, Lviv, 1994.
[39] S. Voronov, V. Samaryk, Y. Roiter, J. Pionteck, P. Pötschke, S. Minko, V. Tokarev, S. Varvarenko, N. Nosova, J. *Appl. Polym. Sci.* 96 (2005) 232–242.
[40] S. Voronov, I. Shmurak, V. Puchin, Yu. Lastukhin, V. Tokarev, Ye. Kiselyov, *Rep. Nat. Acad. Sci.* Ukraine 1 (1981) 50–53.
[41] S. Voronov, V. Samaryk, *Chem. Chem. Technol.* 1 (2007) 1–13.
[42] S. Voronov, V. Samaryk, S. Varvarenko, N. Nosova, Yu. Roiter, *Rep. Nat. Acad. Sci.* Ukraine 6 (2002) 147–150.
[43] N. Nosova, Yu. Roiter, V. Samaryk, S. Varvarenko, Yu. Stetsyshyn, S. Minko, M. Stamm, S. Voronov, *Macromol. Symp.* 210 (2004) 339–348.
[44] Yu. Roiter, V. Samaryk, N. Nosova, S. Varvarenko, P. Pötschke, S. Voronov, *Macromol. Symp.* 164 (2001) 377–387.
[45] M. Xanthos, In Reactive Extrusion, Principles and Practice, Hanser, Munich, 1992.
[46] R. Sing, *Prog. Polym. Sci.* 17 (1992) 251–281.
[47] D. Braun, S. Richter, G.P. Hellmann, M. Rätzsch, J. Appl. Polym. Sci. 68 (1998) 2019–2028.
[48] S. Navarre, M. Degueil, B. Maillard, *Polymer.* 42 (2001) 4509–4516.
[49] S. Navarre, B. Maillard, *Eur. Polym. J.* 36 (2000) 2531–2539.
[50] P. Dokolas, M. Looney, S. Musgrave, *Polymer.* 41 (2000) 3137–3145.
[51] P. Dokolas, D. Solomon, *Polymer.* 41 (2000) 3523–3529.
[52] C. Ingold, Theoretical Foundations of Organic Chemistry (Russian translation), *Mir.* Publishers, Moscow, 1973.
[53] D. Jenkins, S. Hudson, *Chem. Rev.* 101 (2001) 3245–3273.
[54] D.W. Van Krevelen, Properties and Chemical Structure of Polymers (Russian translation), Khimia Publishers, Moscow, 1976.
[55] D. Paul, S. Newman, "Polymer blends. In 2 v.", (Russian translation), *Mir,* Moscow 1981.
[56] T. I. Borysova, L. L. Burshtein, V. P. Malynovska et al., *Polym. Sci. A* 1970, 22(2), 621.
[57] A. A. Askadskii, *Russian Chemical Reviews*. 1999, 4, 349.
[58] S.A. Voronov, Conformation of Macromolecules: Termodynamic and Kinetic Demonstrations, in: V.S. Tokarev, V.Y. Samaryk (Eds.), Nova Science Publishers, New York, 2007.

[59] O. Budishevska, I. Dronj, A. Voronov, N. Solomko, A. Kohut, O. Kudina, S. Voronov, *React. Funct. Polym.* 69 (2009) 785–791.

[60] W. Huang, G.L. Baker, M.L. Bruening, *Angew. Chem.* 113 (8) (2001) 1558–1560.

[61] V. Samaryk, S. Varvarenko, I. Tarnavchyk, N. Nosova, N. Puzko, S. Voronov, *Macromol. Symp.* 267 (2008) 113–117.

[62] J. Dutkiewicz, *J. Macromol. Sci.* A20 (9) (1983) 957–965.

[63] C.J. McDonald, R.H. Beaver, *Macromolecules.* 2 (1979) 203–208.

[64] GB Patent 1354349, *Flocculating Agents,* 1971.

[65] A.Ya. Malkin, in: V.A. Kabanov, Encyclopedia Polymerov, vol. 2, *Sovetskaya Encyclopedia,* Moscow, 1974 (in Russian).

[66] V.V. Lopatin, A. Askadskii, Polyacrylamide Hydrogels in Medicine, *Nauchnyi Mir,* Moscow, 2004.

In: Radical Polymerization: New Developments
Editors: I. O. Paulauskas, L. A. Urbonas, pp. 59-96
ISBN 978-1-62100-406-6

Chapter 2

BIOMACROMOLECULES IN RADICAL PROCESSES: INNOVATIVE STRATEGIES FOR THE SYNTHESIS OF BIOMATERIALS

U. Gianfranco Spizzirri**, Francesca Iemma, Manuela Curcio, Ilaria Altimari, and Nevio Picci

Pharmaceutical Science Department, University of Calabria, Rende (CS), Italy.

ABSTRACT

Natural polymers, such as polysaccharides and proteins, are materials extensively investigated due to their biocompatibility, biodegradability and non-toxic and non-immunogenic characteristics. Enclosing the biomacromolecules, in a complex structure, these features can be transferred to a biomaterial in order to extend the performance of the device. Basically, the synthesis of bioconjugates, by insertion of natural polymers in a macromolecular network by radical polymerization processes, can be achieved employing two different synthetic approaches. The first method involves the chemical modification of the biomacromolecules to introduce functionality able to undergo radical polymerization reactions. In addition, polysaccharides and proteins, without any functional changes, can take part in graft radical polymerization reactions that involve the heteroatoms of the substrates. Both synthetic approaches allows to prepare biocompatible bioconjugates showing improved physico-chemical and mechanical properties respect to the starting natural species. Furthermore, radical polymerization of biomacromolecules with monomeric species bearing specific functionality, carry out to the synthesis of polymeric network that *undergo* a *phase transition* process in *response* to external stimuli changes (temperature, pH, magnetic and electric field). These findings showed that the radical polymerization techniques, improving the performance of natural polymer, represent an innovative tools for the preparation of macromolecular devices potentially useful in pharmaceutical and biomedical field.

* Corresponding author: g.spizzirri@unical.it.

1. Introduction

Selecting a suitable polymer is of the utmost importance because it can affect the biological and physicochemical behaviour of a biomedical device such as biocompatibility, drug release rate and targeting. A desirable polymer should be biocompatible, biodegradable, non-toxic, able to control the release rate. Synthetic biodegradable polymers, such as poly-lactic acid (PLA) and its derivative polylactic co-glycolic acid (PLGA) have been frequently used in the preparation of biomedical devices [1]. These polymers can be used as scaffold in tissue engineering or tailored for controlled release of drugs and they are one of the few polymers currently approved for use in humans [2]. However, the stability of some bioactive agents, such as proteins and peptides can be affected during the degradation of this material. When PLA or PLGA are degraded, lactic acid and glycolic acid are produced. The surrounding environment becomes extremely acidic, which could lead to the inactivation of proteins and peptides [3].

Considerable interest in recent years has been shown in the use of natural polymer to produce biomedical devices. Thus, the covalent conjugation of a biodegradable macromolecule, as a protein or polysaccharides, to a radical monomer represents a versatile strategy to produce intelligent biodegradable hydrogels, suitable for pharmaceutical and biomedical applications [4]. The cross-linking of biomacromolecules is somewhat important in pharmaceutical field because it increase the mechanical properties of the hydrogels without to affect, in severe degree, the biodegradability. Basically, the insertion of natural polymers in a macromolecular network by radical polymerization processes, can be achieved employing two different synthetic approaches. The first method involves the chemical modification of the biomacromolecules to introduce functionality able to undergo radical polymerization reactions. In addition, polysaccharides and proteins, without any functional changes, can take part in graft radical polymerization reactions that involve the heteroatoms of the substrates.

2. Synthesis of Biomaterials by Radical Polymerization of Biomacromolecules Derivatives

Various biopolymers were functionalized with various methacrylates, yielding methacrylated derivative of biopolymers. These biomacromonomers were polymerized to prepare biodegradable hydrogels and nanogels for tissue engineering scaffolds and drug delivery carriers (Tables 1 and 2).

2.1. Chitosan

Chitosan (CS), poly[β-(1–4)-2-acetamido-2-deoxy-D-glucopyranose], was obtained by alkaline deacetylation of chitin [5]. Because of its biodegradability, biocompatibility and non-toxicity, it has been used as an anticoagulant, a wound healing accelerator and drug delivery carriers [6, 7]. It is a potentially useful biomaterial for use in the preparation of hydrogels as tissue engineering constructs and drug delivery matrices.

Chitosan can be modified easily to form porous hydrogels for existence of reactive amino group [8].

Table 1. Radical polymerization of protein derivatives

Reactive species	Polymerization	Initiator system	Potential Application	Ref.
MA-BSA, DMAA	W/O emulsion	APS/TEMED	Drug carrier	[73, 74]
MA-BSA, NaMA	W/O emulsion	APS/TEMED	pH-responsive drug delivery	[75]
MA-BSA, NIPAAm and NIPAAm/HEMA	W/O emulsion	APS/TEMED	Thermo-responsive drug delivery	[76, 77]
MA-BSA, NaMA, NIPAAm	W/O emulsion	APS/TEMED	Dual stimuli responsive drug delivery	[78]
A-BSA, AAm, MEBA	Bulk	APS/TEMED	Drug delivery	[79]
MA-GL	O/W emulsion	AIBN	Tissue engineering scaffold	[80, 81]
$PGMA_1$-HGL or $PGMA_2$-HGL, HEMA, PEGDM	Bulk	Irgacure 819	Tissue engineering scaffold	[84]
MA-HGL, NIPAAm, MEBA	W/O emulsion	APS/TEMED	Thermo-responsive drug delivery	[85]

Methacrylated bovine serum albumin (MA-BSA), acrylated bovine serum albumin (A-BSA), methacrylated gelatin (MA-GL), methacrylated gelatin derivatives ($PGMA_1$-HGL), ($PGMA_2$-HGL), methacrylated gelatin hydrolisate (MA-HGL), N,N-dimethylacrylamide (DMAA), sodium methacrylate (NaMA), N-isopropylacrylamide (NIPAAm), hydroxymethacrylate (HEMA), acrylamide (AAm), N,N-methylenebis (acrylamide) (MEBA), poly(ethylene glycol 400) dimethacrylate (PEGDM), ammonium peroxydisulfate (APS), *N,N,N′,N′*-tetramethylethylenediamine (TEMED), 2,2′-azobis(2-methyl propionitrile) (AIBN).

Kim et al. developed and characterized a novel in situ CS-poly(ethylene oxide) hydrogel (PEO-CS) via two steps: 2-carboxyethyl acrylate molecules were grafted to the primary amine functional groups in CS in the first step to provide acrylated chitosan (A-CS), and then Michael type addition reaction was processed between the grafted acrylate end groups and the thiol end groups of the PEO hexa-thiol molecule. Grafting of acrylate molecules to the amine groups in the deacetylated water soluble A-CS was confirmed by observing the new acrylate peaks by the FT-IR and NMR spectra of the A-CS samples, as well as changes in relative viscosities of CS and A-CS. Rheological analyses were performed to characterize the obtained systems, while the biological activities of the hydrogels were evaluated by observing smooth muscle cell behaviors, such as cell adhesion and viability, as well as by measuring the number of adhered cells on their surfaces [9].

UV-curable chitosans derivatives (UV-CS) have been prepared through multistep reaction [10] from chitosan with photosensitive methacrylated aldehydes. *N*-Selective introduction of the side chain to the chitosan was accomplished by reductive *N*-alkylation, using $NaBH_3CN$ via Schiff base. The UV-CS derivatives were inserted in a polymeric network by photopolymerization induced using Irgacure 1000 as initiator, to produce biocompatible materials for biomedical applications.

Therefore, Dong et al. prepared both water-soluble and organic solvent-soluble maleoyl chitosan (NM-CS) by reaction of chitosan with maleic anhydride under different reaction conditions introducing vinyl groups and carboxyl groups onto the $-NH_2$ groups of chitosan [11]. The graft copolymerization of NM-CS and butyl acrylate (BA) in acetic acid aqueous solution was investigated, using the γ-ray of ^{60}Co γ-irradiation method. The NM-CS/BA films have enhanced hydrophobic for their practical use, which can be expected to have broad application for seed coating, antistaling agent of vegetable and fruit, and so on [12]. Copolymers based on water-soluble NM-CS and N-isopropylacrylamide (NIPAAm) were prepared by UV radiation. NM-CS/NIPAAm hydrogels showed a swelling ratio depending on both pH value and temperature of the aqueous solution [13].

Table 2. Radical polymerization of polysaccharide derivatives

Reactive species	Polymerization	Initiator system	Potential Application	Ref.
UV-CS	Bulk	Irgacure 1000	Biomedical implants	[10]
NM-CS, NIPAAm	Bulk	Darocure 1173	Thermo-responsive drug delivery	[13]
NM-CS, NIPAAm	Bulk	Electron beam irradiation	Thermo-responsive drug delivery	[16]
NM-CS, NIPAAm, AAm	Bulk	APS/TEMED	Thermo-responsive drug delivery	[17]
NM-CS, BA	Bulk	$^{60}Co\ \gamma$ -ray source	Seed coating and antistaling agent of vegetable and fruit	[12]
ST-CS	Bulk	Carboxylated camphorquinone	Tissue engineering scaffold	[18]
EGAMA-CS, PEGDMA	Bulk	Irgacure 2959	pH-responsive drug delivery	[19]
MA-GCS	Bulk	Photoinitiator I2959	Tissue engineering, drug delivery	[23]
GMA-DX	Bulk	APS/TEMED	Drug delivery	[26]
MA-DX, MAA	w/o emulsion	APS/TEMED	pH-responsive drug delivery	[30]
MA-DX	w/w emulsion	APS/TEMED	Drug delivery	[31]
MA-DX, Ca^{2+}-Alginate (IPN)	Bulk	Irgacure 2959	Drug delivery	[32]
MA-DX, AEM, FA	Bulk	APS	Antioxidant, drug delivery	[33]
HEMA-DX	w/w emulsion	KPS/TEMED	Drug delivery	[38]
HEMA-DX	Bulk	Irgacure 2959	Drug delivery	[39]
LCHEMA-DX	Bulk	APS/TEMED	Drug delivery	[35]
MA-HA, PEGDMA	Bulk	Irgacure 2959	Tissue engineering	[43, 44]
MA-HA	Bulk	AIBN	Tissue engineering scaffold	[45]
LCMA-HA	Bulk	Irgacure 2959	Growth factor release	[46]
GMA-HA, HEA	Bulk	Photoinitiator VA086	Drug delivery	[47]
MA-HA	Bulk	KPS/TEMED	Tissue engineering, drug delivery	[48]
APMA-HA, PEGDMA, NVP	Bulk	EosinY/triethanolamine	Tissue engineering	[49]
MA-ST, GMA-ST	Bulk	NaPS	pH-responsive drug delivery	[55]
HEMA-HEST	w/w emulsion	Irgacure 2959	Growth factor release	[57]
A-CL, DMAA	Bulk	APS	Antioxidant, drug delivery	[51]
MA-HPCL or SMA-HPCL	w/o emulsion	KPS	Drug delivery	[52]
MA-GA	Bulk	NaPS/TEMED	Drug delivery	[59]
MA-GG	Bulk	Irgacure 2959	Tissue engineering scaffold	[64]
MA-IN	Bulk	APS/TEMED	Colonic drug targeting	[67]
MA-IN, AA	Bulk	APS/TEMED	pH-responsive drug delivery	[69]
GMA-PC	Bulk	NaPS	Coating materials	[71]

UV-curable chitosans (UV-CS), maleoyl chitosan (NM-CS), styrenated chitosan (ST-CS), chitosan–ethylene glycol acrylate methacrylate (EGAMA-CS), glycidyl methacrylate-dextran (GMA-DX), methacrylated dextran (MA-DX), hydroxyethyl methacrylate dextrane (HEMA-DX), dextrane (LCHEMA-DX), methacrylated hyaluronate (MA-HA), lactate methacrylate hyaluronate (LCMA-HA), glycidyl methacryl hyaluronate (GMA-HA), methacrylated galactomannan (GA-MA), methacrylated guar gum(MA-GG), methacrylated inulin (MA-IN), glycidyl methacrylated pectin (GMA-PC), methacrylated starch (MA-ST), glycidyl methacrylated starch (GMA-ST), hydroxymethacrylathydroxyethylstarch, (HEMA-HEST), acrylated cellulose (A-CL), methacrylated hydroxypropyl cellulose (MA-HPCL), spacer methacrylated hydroxypropyl cellulose (SMA-HPCL), N-isopropylacrylamide (NIPAAm), butyl acrylate (BA), acrylamide (AAm), polyethylene glycol diamethacrylate (PEGDMA), methacrylic acid (MAA), amminoethylmethacrylate (AEM), ferulic acid (FA), hydroxyethyl acrylate (HEA), N-vinyl pyrrolidone (NVP), acrylic acid (AA), N,N-dimethylacrylamide (DMAA), Ammonium peroxydisulfate (APS), potassium peroxydisulfate (KPS), sodium peroxydisulfate (NaPS), *N,N,N',N'*-tetramethylethylenediamine (TEMED), 2,2'-azobis(2-methylpropionitrile) (AIBN).

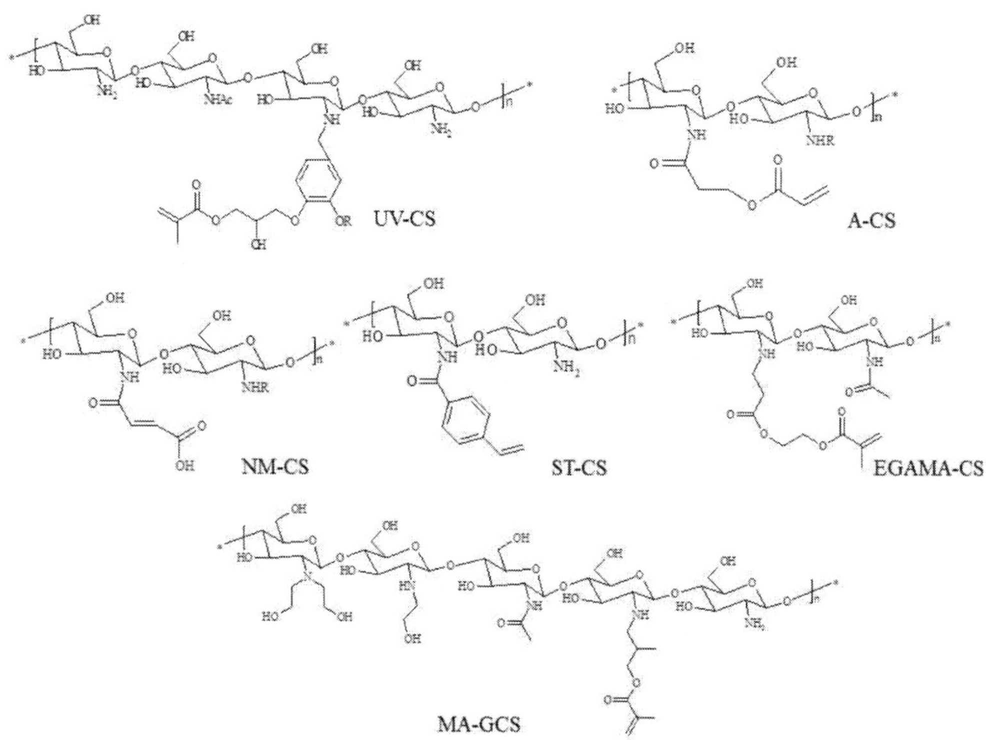

Figure 1.

It is well known that physical, chemical and radiant methods can be applied to prepare polymer blends. Among them, radiation technique is relatively simple for the improvement or modification of polymer materials through cross-linking, grafting, or degradation [14]. Moreover, the product can be free from impurities such as chemical residues from initiators for no catalysts or additives are needed to initiate the reaction in radiation processing. In addition, the degree of crosslinking and grafting can be controlled by the change of radiation absorbed dose [15]. Finally, thermo- and pH-sensitive hydrogels by grafting NIPAAm onto NM-CS, as biodegradable cross-linker, by electron beam irradiation were prepared [16]. Additionally, thermo-sensitive hydrogels based on NM-CS and a series of pNIPAAm and poly(*N*-isopropylacrylamide-co-acrylamide) were prepared, and their lower critical solution temperature (LCST) were determined. By altering the NIPAAm/acrylamide molar ratio of NIPAAm and acrylamide in the hydrogels, the LCST could be increased to 39°C. In addition, the LCST of the hydrogel was significantly influenced by the monomer ratio of the NIPAAm/acrylamide but not by the cross-linker content [17].

Styrenated chitosan (ST-CS) was prepared via a condensation reaction of a carboxyl or an amino group of the polysaccharides with a vinyl monomer (4-vinylaniline or 4-vinylbenzoic acid) in the presence of a water-soluble condensation agent [18]. Their gels were prepared by polymerization of the vinylated polysaccharides or copolymerization with vinylated gelatin and albumin. In the presence of water-soluble camphorquinone as a photocleavable radical producing agent and under visible light irradiation, aqueous solutions of these vinylated biomolecules were converted to hydrogels. In addition, photocuring characteristics of vinylated polysaccharides were studied, and the photofabrication of tubular constructs derived only from biomacromolecules was demonstrated, which may be useful as a scaffold or a template in tissue-engineered devices.

Methacryloyloxy ethyl carboxyethyl chitosan (EGAMA-CS) was synthesized by Michael addition reaction, and was successively copolymerized with polyethylene glycol dimethacrylate by photopolymerization, because polyethylene glycol segment is one of the most popular applied oligomers for biomaterials due to the solubility in both water and organic solvent, low toxicity, biocompatibility [19]. The hydrogel showed good in vitro biocompatibility, while the result of cell adhesion and morphology suggested that the biomaterials did well in promoting cell attachment and proliferation.

Chitosan can be dissolved in only acidic solution through the interaction between H^+ and NH_2, but it is insoluble under higher pH conditions [20]. It is also of limited solubility in organic solvents. Glycol chitosan (GCS) is soluble over a wide range of pH and, in the few reports to date, has been found to be cytocompatible. The GCS was made photopolymerizable by methacrylation through the reaction with glycidyl methacrylate (MA-GCS). Glycidyl methacrylate has been used to modify chitosan [21] and oligochitosan [22]; however, methacrylated chitosan was not soluble at physiologic pH, while significant water solubility of methacrylated oligochitosan was detected only for tetramers. Furthermore, these materials were not investigated as photopolymerizable pre-polymers. Methacrylated glycol chitosan was prepared at different molar ratios of glycidyl methacrylate to free amine/glycol chitosan residue. The cytotoxicity of the MA-GCS was assessed using an immortalized chondrocyte cell line. Hydrogels prepared from MA-GCS were characterized in terms of their sol content as influenced by degree of methacrylation, UV intensity, time of irradiation, and gel depth [23].

2.2. Dextran

Dextran (DX) is a water-soluble bacterial exo-polysaccharide consisting mainly of α-1,6-linked D-glucopyranose residues, and is an ideal polymer to form hydrogels due to its absent toxicity [24]. In recent years, different dextran hydrogels have been developed. Dextran hydrogels can be formed by the coupling of methacrylic acid to dextran, followed by radical polymerization [25, 26]. However, these hydrogels are not degradable under physiological conditions, and can only be degraded by dextranase [25, 27]. Alternatively, the synthesis of biomaterials can be carried out by radical copolymerization of derivatized dextran with polymerizable groups (Figure 2).

The synthesis of dextran derivatized with polymerizable groups has been described by Edman et al. [25]. In their procedure glycidyl methacrylate was coupled to dextran in a carbonate buffer at pH 11 to provide glycidyl methacrylate-dextran (GMA-DX) (Scheme 1). Despite prolonged reaction time, lower degree of substitution was recorded (less than 10% of the glycidyl methacrylate originally present). The low incorporation of acrylate groups can most likely be ascribed to the aqueous basic reaction conditions. Firstly, the epoxy group can react with water, yielding glyceryl acrylate, which does not react with dextran under these conditions. Secondly, the acrylic ester can be hydrolyzed, before and after reaction with dextran.

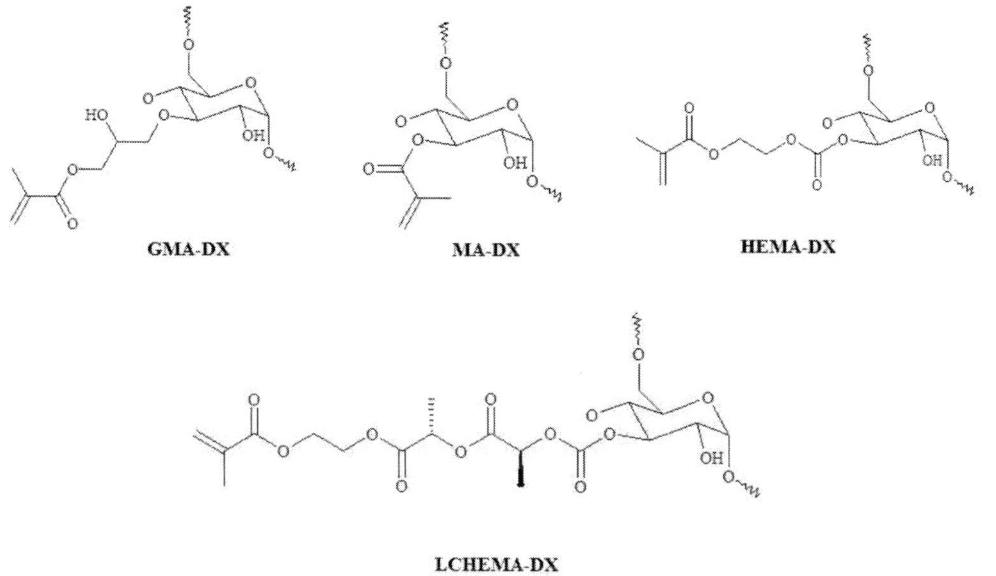

Figure 2.

In order to improve the degree of substitution of the dextran a new synthetic pathway was proposed [26]. Glycidyl methacrylate derivatized dextran was synthesized by coupling of GMA to dextran in the presence of 4-(N,N-dimethylamino)pyridine using dimethyl sulfoxide (DMSO), as an aprotic solvent. Almost quantitative incorporation of GMA (> 90%) was established. Hydrogels were prepared by radical polymerization of aqueous solutions of GMA-DX, using ammonium peroxydisulfate (APS) and TEMED as the initiating system. The reinvestigation of the reaction of GMA with dextran as well as the structure of the reaction product carried out to an unexpected result [28]. The reaction of GMA with dextran proceeds via transesterification and results in the direct attachment of methacryloyl groups at the 2- and 3-hydroxyl group of the glucopyranose ring, in a 1:1 ratio.

On the contrary, Lo and Jiang suggested that DX functionalization with GMA in DMSO involves the polarization of the hydroxyl groups of dextran and subsequently reaction with the less hindered epoxy carbon of glycidyl methacrylate with the aid of catalyst DMAP. These findings were confirmed by ^{1}H-NMR analysis [29].

pH-sensitive hydrogel microspheres composed of methacrylated dextran (MA-DX) and methacrylic acid were prepared by reverse phase suspension copolymerization [30]. Microparticles with a spherical shape, porous surface and a narrow size distribution were obtained. Moreover, a pH dependent swelling, in media that simulate gastrointestinal fluids, was an important characteristic of microspheres. Franssen and Hennink provide a completely aqueous emulsion technique for the preparation of hydrogel microparticles based on polymer–polymer immiscibility. This aqueous polymer immiscibility occurs with many combinations of water-soluble polymers. The polymers stay in solution, but separate in two aqueous phases above a certain concentration. After emulsification, the polymer in the dispersed phase can be crosslinked to form a microparticle with a hydrogel character. This crosslinking can be established by radical polymerization of MA-DX emulsified in an aqueous PEG solution. No organic solvents are used in this technology, which is very attractive for the preparation of protein-loaded microparticles [31].

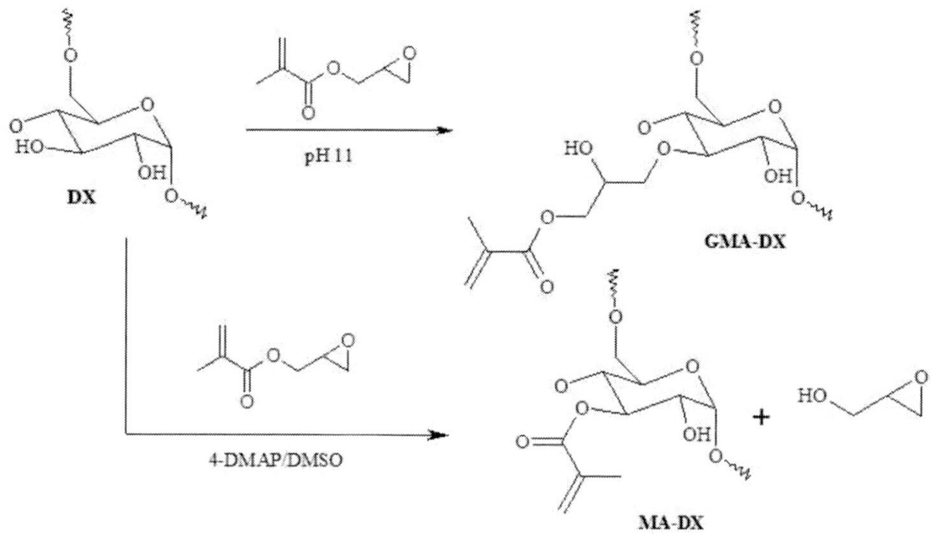

Scheme 1.

Semi-interpenetrating polymer systems (IPNs) based on calcium alginate hydrogel and dextran methacrylate derivative, showing potential applicability in the field of pharmaceutics, were synthesized by UV irradiation [32]. The semi-IPNs, obtained by a dispersion of MA-DX chains into a Ca(II) hydrogel, leads to a hydrogel with rheological properties quite different from those of Ca(II)-alginate, allowing to inject the semi-IPN easily through an hypodermic needle.

Methacrylic groups were introduced onto the dextran polymer backbone by reaction with methacryloyl chloride in a DMF solution in the presence of pyridine as catalyst, to produce MA-DX [33]. Depending on the reaction conditions (time, temperature and derivatization agent concentration), methacrylic derivatives with different functionalization degree were obtained and evaluated by volumetric analysis. Antioxidant dextran hydrogel was successfully prepared introducing ferulic acid moieties onto MA-DX crosslinked with aminoethyl methacrylate. Its antioxidant activity, evaluated through rat liver microsomal membranes, suggested that ferulate material possesses an excellent antioxidant activity. Moreover, preparation of ferulate hydrogel-based dextran was found to be well suited and a sound approach to obtain carrier that preserves Vitamin E during its release.

A next generation of hydrogels with hydrolytically sensitive carbonate ester groups based on dextran derivatized with hydroxyethyl methacrylate (HEMA-DX) was developed [28, 34]. It has been shown that by varying the degree of substitution, the number of methacrylates per 100 glycopyranose residues and the initial water content of the gel, the degradation rate of the hydrogels could be tailored from days to months [35]. Moreover, the feasibility to load injectable HEMA-DX microspheres with protein drugs was demonstrated. The preparation of these protein-loaded microspheres was achieved in an all aqueous environment [36]. HEMA-DX microspheres are formed by a polymerization reaction initiated by potassium peroxodisulfate (KPS) and catalyzed by TEMED. It is therefore likely that the polymerization rate, which is in turn dependent on the polymerization conditions (initiator and catalyst concentration, temperature and pH), may have a significant effect on network properties (cross-link density, pore size) of the microspheres[37]. This can affect both the degradation rate of the microspheres and the release rate of an entrapped protein. Chung et al. showed that HEMA-DX microspheres were mainly elastic up to a deformation of 80%. Increased KPS initiator concentrations caused a faster polymerization rate, and the formed hydrogel was stronger, had a smaller molecular weight between cross-linkings and a smaller pore size [38]. However, although an increase in polymerization temperature also led to a faster reaction rate, the formed hydrogel was weaker, had a larger molecular weight between cross-linkings and a bigger pore size. The pH of the polymerizing solution did not have a significant effect on both the polymerization rate and the hydrogel network properties. The network characteristics are dependent on the polymerization conditions. These altered network properties likely results in different release profiles of entrapped proteins. In order to overcome the problem of the instability of the hyaluronic acid scaffold Pescosolido et al. combine the viscoelastic and bioactive properties of hyaluronic acid with the photo-crosslinkable dextran derivates. For this aim a semi-IPN hydrogel based on hyaluronic acid and a dextran derivate was synthesized by UV irradiation [39]. The mechanical properties and the degradation time of this hydrogel can be tuned by varying the degree of substitution and the concentration of HEMA-DX [40]. Although these gels contain methacrylate esters in their crosslinks, the hydrolysis of these groups is very slow under physiological conditions. An alternative approach to degradable, interpenetrating networks of dextran and polymethacrylate is incorporation of hydrolytically labile spacers between the polymerized methacrylate groups and DX. Lactate-HEMA derivatized dextran (LCHEMA-DX) can be used for the synthesis of hydrogels easily degradable under physiological conditions by chemical hydrolysis of the labile ester groups present in the crosslinks [35].

2.3. Hyaluronic Acid

Hyaluronic acid (HA) is an endogenous polysaccharide which consists of repeating disaccharide units composed of β(1-4) linked N-acetyl-D-glucosamine and β(1-3) linked D-glucuronic acid [41]. Its biocompatibility, biodegradability and immunoneutrality make HA an attractive polymer for biomedical and pharmaceutical applications. Polymerizable methacrylate groups can be introduced on the chain of this polysaccharide by chemical modification of hydroxyl and carboxylic groups (Figure 3).

This crosslinkable HA can be used to form hydrogels for drug delivery and tissue engineering purposes. The synthesis of methacrylated HA (MA-HA) is performed in an aqueous environment with an excess of methacrylic anhydride with respect to the hydroxyl groups of HA [42]. Masters et al. photopolymerized MA-HA with various degrees of methacrylate modification and poly(ethylene glycol) diacrylate to produce scaffolds tested as biologically active carriers for valvular interstitial cells [43]. The photopolymerization of MA-HA and poly-(ethylene glycol) dimethacrylate (PEGDMA) provided networks with a wide range of properties, useful in the tissue engineering field [44]. The authors studies the effects of HA molecular weight, the degree of methacrylation and macromer concentration on the physical properties (e.g., swelling, mechanics, and degradation) of the resulting hydrogels.

In addition, photoencapsulated cell viability and neocartilage tissue formation were investigated as a preliminary test for the potential use of photopolymerizable HA networks as chondrocyte carriers for cartilage regeneration. Finally, methacrylated hyaluronic acid was radical copolymerized, using AIBN as initiator agent, with methacrylated derivative of gelatin to create stable gelatin scaffolds at physiological temperature (37°C) [45].

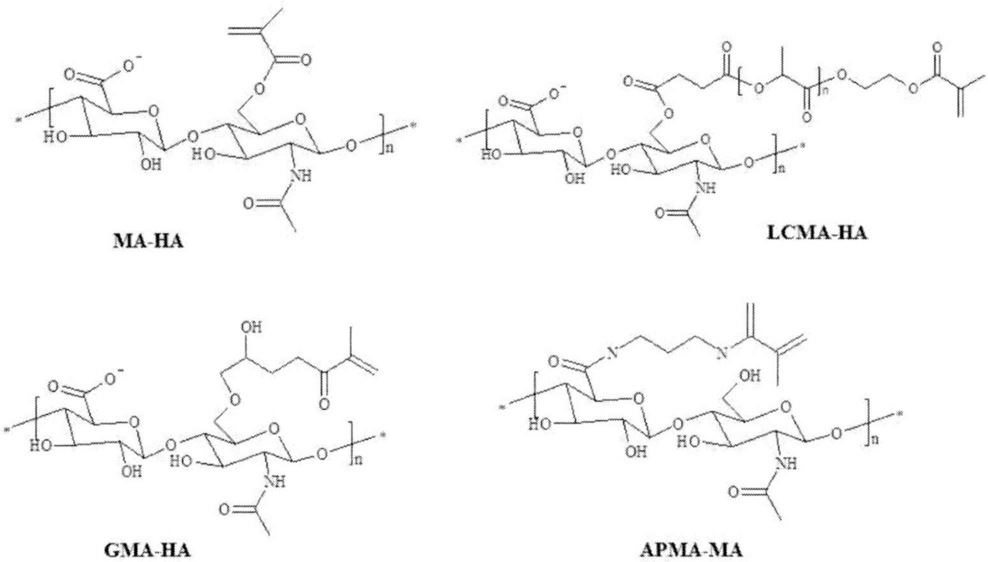

Figure 3.

These hydrogels can be degraded by hyaluronidases, either added exogenously or produced by cells. However, may be inhibitory and their applications are limited if the appropriate enzymes are not present. Sahoo et al. synthesized HA macromers (LCMA-HA) and hydrogels that are both hydrolytically (via ester group hydrolysis) and enzymatically degradable [46]. This approach involved the inclusion of hydrolytically degradable repeat units of lactic acid between the HA and the polymerizing moiety. Since poly(lactic acid) is highly versatile in design, hydrolyzable, and approved by the FDA for several biomedical applications it was an ideal group to incorporate into the hydrogel.

Glycidyl methacryl hyaluronate (GMA-HA) was prepared in the aqueous basic (pH 11) reaction conditions in order to involve the nucleophylic opening of the epoxidic ring. Hyaluronate-hydroxyethyl acrylate blend hydrogels were prepared by radical copolymerization of GMA-HA and hydroxyethyl acrylate (HEA) under various compositions (weight ratios of HEA and GMA-HA: 1–20) in the presence of a photoinduced initiator [47]. The major drawback of this synthesis lies in the aqueous basic reaction conditions, this makes it difficult to control the degree of substitution. Hence, there is need for a method in which the functionalization degree of methacrylated HA can accurately be controlled as was achieved in the past for the modification of polysaccharides allowing the preparation of hydrogels with tailored properties. Methacrylate derivatized HA with precise control over the substitution degree can be synthesized in a suitable aprotic solvent (DMSO) by substitution of the polysaccharide with glycidyl methacrylate, providing MA-HA derivative [48]. The hydrogels were obtained by crosslinking aqueous solutions of MA-HA in the presence of KPS and TEMED as initiator and catalyst, respectively. The radical polymerization resulted into opaque hydrogels, independent of the HA substitution degree used. The opacity of the hydrogels indicates that phase separation (i.e. water-poor domains containing relatively hydrophobic polymerized methacrylates and water-rich hydrated HA domains) has occurred. Characterization of these MA-HA hydrogels showed that the elastic modulus and the dimensional stability of the gels increased with higher substitution degree.

HA-based hydrogels were prepared via photopolymerization of pendent methacrylic esters, previously introduced through functionalization of the carboxylic groups, for tissue engineering applications [49]. HA was modified by using N-(3-aminopropyl) methacrylamide as an acrylating agent and the water soluble N-(3-dimethylpropyl)-N-ethylcarbodiimide hydrochloride as a coupling agent, providing APMA-HA derivative. In the photopolymerization experiments, eosin Y was used as a visible light sensitizer, triethanolamine was employed as an initiator, and PEGDMA and N-vinyl pyrrolidone (NVP) were used as comacromonomer and comonomer, respectively. The hydrogels were degraded by hyaluronidase and their mechanical properties were modulated by HA molecular weight and concentration of PEGDMA.

2.4. Cellulose

Cellulose (CL) is a polysaccharide consisting of a linear chain of several hundred to over ten thousand β(1→4) linked D-glucose units. Cellulose macromer (A-CL) was obtained by heterogeneous synthesis with acryloyl chloride (Figure 4) [50].

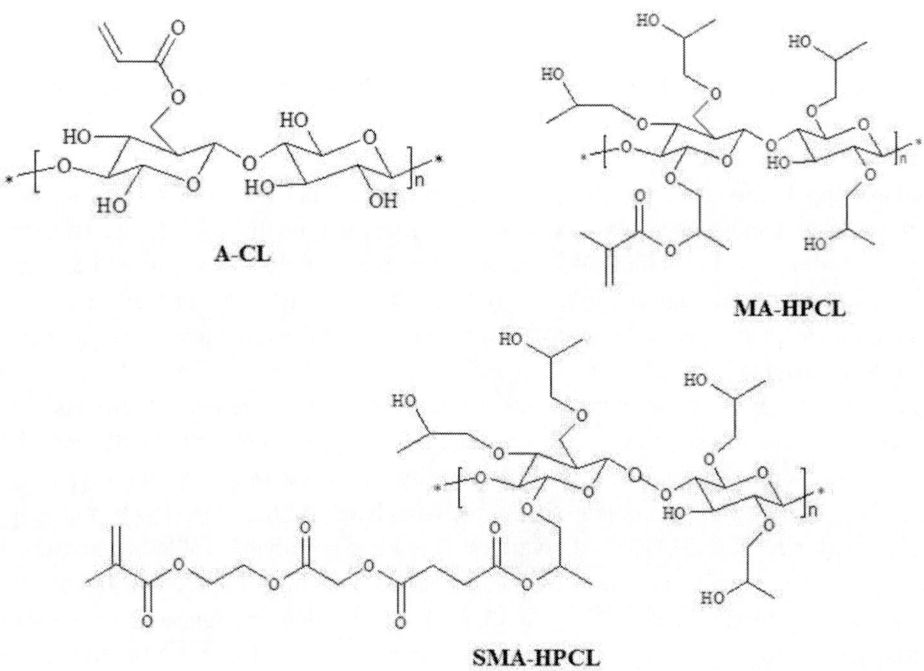

Figure 4.

Preparation of hydrogels was carried out by radical polymerization of A-CL, the crosslinking agent, with N,N-dimethylacrylamide (DMAA), using APS as initiator agent [51]. With the aim of taking advantage of the antioxidant properties of the trans-ferulic acid, an hydrogel containing this residue has been synthesized. The linking of antioxidant groups on the preformed hydrogel rather than on its precursor was effected to avoid the inhibitory action of trans-ferulic acid, a scavenger of radicals species, on the radical polymerization process. The antioxidant activity of cellulose hydrogels in inhibiting the lipid peroxidation in rat-liver microsomal membranes was recorded. The same experiment was performed on a non-derivatized hydrogel and on a commercial trans-ferulic acid. The results revealed that the hydrogel without ferulic groups has no antioxidant activity.

Cai and Hu proposed the synthesis of modified hydroxypropylcellulose (HPCL) by covalently attaching either vinyl groups linked by degradable esters or methacrylate groups, providing methacrylated hydroxypropylcellulose (MA-HPCL) and spacer methacrylated hydroxypropylcellulose (SMA-HPCL) (Figure 4) [52]. The vinyl groups allowed for chemical linking of the HPCL chains into degradable nanoparticles through a free radical polymerization process. The nanoparticle networks of various compositions and/or different biodegradable cross-linker types and amounts containing a specific active compound can be combined together. The resulting combination has provided an overall synergistic effect of drug delivery due to the differences in degradability and diffusion rates depending on the structural characteristics of the networks.

2.5. Starch

Starch (ST) is a material which occupies a prominent position in the group of natural polymers. It is a complex polysaccharide composed of two structural components: amylose and amylopectin [53]. The amylose is a linear starch macromolecule consisting of 250–300 α-(1-4)-linked-D-glucose residues. The amylopectin is a lesser hydrophilic and branched starch macromolecule consisting of nearly 1400 D-glucose residues with α(1-4) and α(1-6) linkages. It constitutes nearly 80% of the total starch and can easy be hydrolyzed. Resistant ST is a special group of polymers, which are not digested in small intestine of healthy persons. Nevertheless, it undergoes a process of fermentation within the colon. The resistance to enzymatic hydrolysis allows the amylopectin to achieve the colon, enabling it to be used as a drug delivery device through the oral cavity [54]. Crosslinking of ST can be an outstanding way to produce hydrogels with exceptional properties. However, there is no any efficient route to form a stable hydrogel when native starch is used, which may be attributed to lacking of functional groups in its structure for crosslinking. Reis et al. prepared hydrogels from ST for potential uses in transporting and in preserving of drugs responsive to acidic environment [55]. A chemical modification of starch was used by reaction with GMA in a solvent mixture of DMSO and water, to obtain both methacrylated starch (MA-ST) and glycidyl methacrylated starch (GMA-ST) derivatives (Figure 5). Polymerizable macromers were coupled onto a polymeric structure of the polysaccharide by further crosslinking reaction using sodium persulfate (NaPS) as an initiating agent. The authors suggested this hydrogel as potential candidate to be used to transport and to preserve acid-responsive drugs, such as corticoids, for the treatment of colon-specific diseases, for example, Crohn's disease. Starch derivative, such as hydroxyethylstarch (HEST), was functionalized in DMSO/dimethyl formamide mixture with hydroxyethylmethacrylate-imidazoyl carbamate to provide a polymerizable polysaccharide, the hydroxymethacrylathydroxyethylstarch, (HEMA-HEST) (Figure 5) [56].

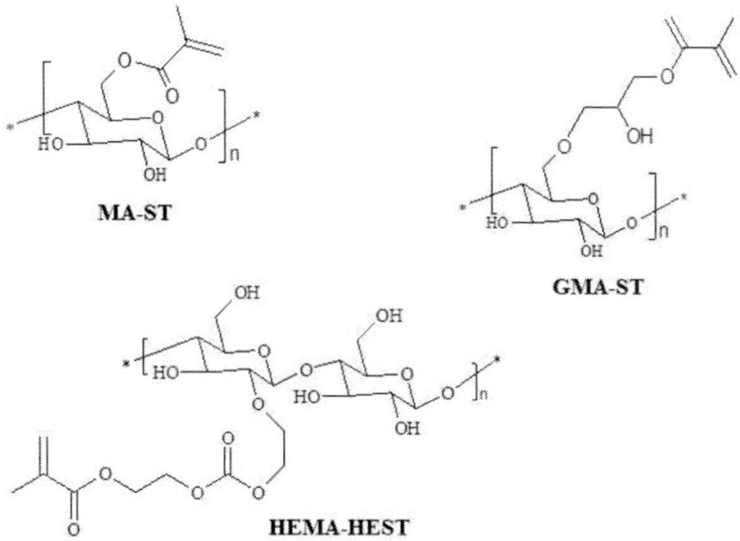

Figure 5.

HEMA-HEST was employed as starting material for the synthesis of hydrogels, used in medicine for a long time as volume therapy, so that a good physiological compatibility of the hydrogels can be assumed. Schwoerer et al. report on the preparation of protein-loaded hydrogel microspheres, by radical polymerization process in all aqueous two-phase system consisting of PEG and photo cross-linkable HEMA-HEST with a monomodal particle size distribution [57]. The release of active ingredients can be controlled by the choice of HEMA-HEST derivatives of different degree of substution.

2.6. Galactomannan, Guar Gum, Inulin and Pectin

Galactomannans (GA) are polysaccharides obtained from exudates or seeds of different vegetable species. They have been employed in research activities as pharmaceutical matrices, especially for drug delivery systems [58].

Reis et al. describe chemical modification of galactomannan, obtained from *Prosopis juliflora*, by reaction with glycidyl methacrylate (MA-GA) and further synthesis of the hydrogel (Figure 6) [59]. The reaction proceeds via transesterification, leading to the addition of methacryloil groups on GA chains. The MA-GA is the main product and glycidol is a by-product. Hydrogels from the MA-GA can be prepared by polymerization just by addition of NaPS and TEMED in a deoxygenated MA-GA aqueous solution. During the polymerization process, the cross-linking occurs on methacryloyl groups, resulting in a three-dimensional hydrogels network. The swelling properties of hydrogels depend slightly on the amount of

TEMED and NaPS in feed solution while the pH does not affect significantly the equilibrium swelling ratio of the hydrogels.

Guar gum (GG) is a non-ionic, water-soluble, biodegradable and biocompatible hetero polysaccharide composed of a β(1-4) D-mannopyranose backbone linked with α(1-6) D-galactopyranose units in a 1:2 M ratio [60]. GG has various commercial applications due to its unique ability to alter rheological properties [61].

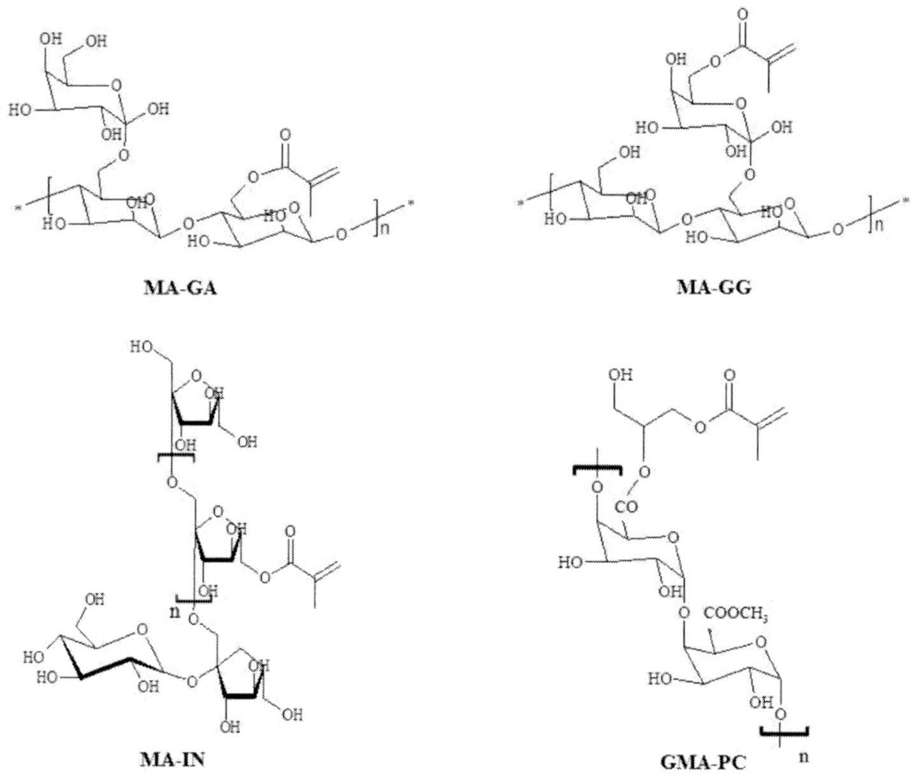

Figure 6.

In the biomedical field GG and modified GG have both been used as a carrier for colon targeted [62] and transdermal drug delivery [63], however, its potential applications for tissue engineering scaffolds have not yet been explored. Tiwari et al. proposed the synthesis of water-soluble and photopolymerizable guar gum–methacrylate (MA-GG) macromonomers by reaction of GG with GMA in DMSO, using DMAP as catalyst (Figure 6) [64]. The goal of the research is to combine the benefits of photo-cross-linked hydrogels (UV-initiated) with the advantageous properties of GG for tissue engineering scaffold applications. GG–MA hydrogels scaffolds are inexpensive and easy to synthesize, process and scale up. The degree of *in vitro* enzymatic biodegradation of the hydrogels decreased linearly with increasing gel content and the degree of methacrylation of the respective macromonomers. Cell viability and cell proliferation were tested with a human endothelial cell line to explore the potential use of GG–MA hydrogels for *in situ* tissue engineering scaffold applications.

Inulin (IN) is a dietary fiber composed of a mixture of oligo- and/ or polysaccharides consist of fructose unit chains (linked by (2-1)-β-D-fructosyl-fructose bonds) of various length, terminated generally by a single glucose unit (linked by an α-D-glucopyranosyl bond) [65]. This natural polymer is widely distributed in some edible plants including asparagus, garlic, chicory, leek, onion and artichoke as storage carbohydrates [66]. Inulin hydrogel formation has been described by Vervoort et al. [67]. Vinyl groups were incorporated in the fructose chains by reaction of inulin with glycidyl methacrylate at room temperature using DMAP, as catalyst. Transesterification occurred during the reaction of inulin with glycidyl methacrylate, since the reaction product was characterized as methacrylated inulin (MA-IN), instead of glyceryl methacrylated inulin (Figure 6). This confirmed the findings which have been published previously by van Dijk-Wolthuis et al. for the derivatization of dextran with glycidyl methacrylate [28]. By varying the molar ratio of glycidyl methacrylate to inulin, the degree of substitution, i.e., the amount of methacryloyl groups per 100 fructose units, of MA-IN could be tuned. Aqueous solutions of MA-IN were subsequently converted into three-dimensional networks by free radical polymerization using APS and TEMED, as radical generating compounds. The resulting hydrogels are characterized by a higher mechanical strength, arising from the increased amount of intermolecular cross-links formed. Increasing concentrations of APS and TEMED also shorten the process of hydrogel formation, but result in hydrogels with lower mechanical strength [68]. Finally, pH-responsive inulin hydrogels were prepared by radical copolymerization of MA-IN with acrylic acid in aqueous solution using APS and TEMED as an initiation system. The results show that the change in swelling in response to pH changes increases with increasing the acrylic acid content [69].

Pectin (PC) is mainly a linear polymer chemically constituted by D-galacturonic acid monomers in α-(1–4) bonds, occasionally interrupted by L-rhaminose α-(1–2) bonds. However, other monomers also may make part of the side chains such as neuter sugars as D-galactose, L-arabinose, D-xylose, L-rhamnose, L-fucose, and traces of 2-*O*-methylfucose [70]. Moreover, pectin has high molecular weight, ranging from 50 to 180 KDa. Depending on the degree of substitution of D-galacturonic acid carboxyl groups by methoxyl groups (–OCH_3), pectin may be classified as high (over 50%) or low (below 50%) methoxylated/esterified. Thus, pectin may be described as a "canonic" structure for presenting a rather heterogeneous and complex chemical structure [62, 70]. However, the greatest challenge found in the use of pectin in the development of drug coatings is to overcome its solubility in aqueous medium, which may contribute to the premature and local undesirable release of the active principle as a polysaccharide in the support system. Maior et al. proposed the chemical modification of pectin by the reaction with GMA [71]. The objective of this modification reaction is the introduction of vinylic groups in the polysaccharide structure to produce glycidyl methacrylated pectin (GMA-PC) (Figure 6). These vinylic groups will later react through free radicals and generate polymeric chains and produce polysaccharide hydrogels [67, 72]. The addition of GMA to pectin in acid aqueous medium (pH 1.2) occurs by opening the epoxide ring of GMA without the formation of glycidol, which results from the transesterification reaction in aprotic solvent at alkaline pH. GMA-PC, with reduced solubility was employed for the development of free films associated to Eudragit® RS 30 D polymethacrylate producing a material with attractive characteristics for the development of drug oral administration systems. When applied as a pharmaceutical coating, this material may prevent the early release of the drug in the proximal GIT, besides ensuring the effective control of the target-site-specific release of drugs due to the prospective specific degradation of the film by enzymes produced by the colonic microflora.

2.7. Albumin

Albumin is an attractive macromolecular specie to prepare pharmaceutical carriers used for the sustained delivery of therapeutic agents since this protein is biodegradable, biocompatible, non-toxic and non-immunogenic. Chemical groups susceptible to radical polymerization were introduced onto bovine serum albumin (BSA) by acylating the BSA with methacrylic anhydride (MA) in water at 0°C and neutral pH [73]. The nucleophilic chemical groups in BSA that could react with MA are the thyolic groups of cysteine, hydroxil groups of serine and tyrosine, and amino groups in the side chain of lysine. The first are involved in disulfide bridges, except cys-34, the latter are the least nucleophilic, and do not react in mild experimental conditions. Then, the sterically accessible amino groups of lysine only react chiefly with acylant agent at controlled pH and temperature to produce water-soluble MA-BSA (Figure 7). If the reaction is carried out without pH and temperature control, denaturation of BSA was observed, and its water solubility is lost. The water-solubility of the derivatized protein was essential to provide spherical polymeric microparticles via reverse phase suspension copolymerisation with DMAA, using APS and TEMED as initiator system. The reverse-phase suspension polymerisation technique represents a simple method for obtaining spherical microparticles whose size can be conveniently varied, according to particular needs, by changing the reaction conditions (speed of stirring, shape and dimension of reactor, shape of stirring-rod).

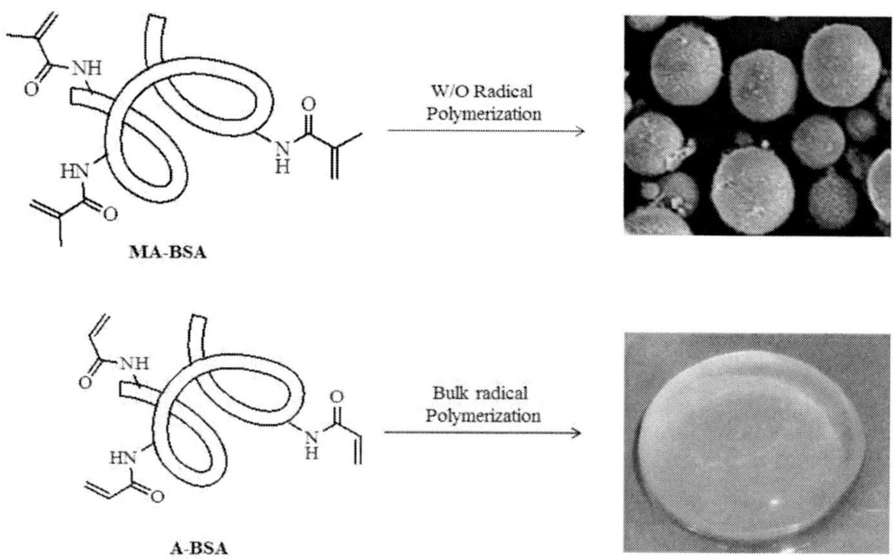

Figure 7.

The polymerization reaction, owing to steric and geometric constraints, involves only the methacrylic functions of MA-BSA which are accessible to the growing chains. The microparticle structure is characterised by a network where the BSA chains are linked by some hydrocarbon bridges. The MA-BSA samples were also polymerized in the presence of various amounts of comonomer (DMAA) in order to study the effects of a different chemical structure and degree of crosslinking on the physical properties of the microparticles. It can be supposed that in the copolymerisation reaction the chains obtained consist of DMAA units randomly interrupted by methacrylic BSA-MA functions which are sterically and geometrically attainable. The applicability of these materials as drug delivery systems has been evaluated by loading drugs with different chemical properties by a soaking procedure. In particular, for this material, the drug release features depend principally on crosslinking degree, ratio among albumin and DMAA, and interactions of loaded drug-beads [74].

Using comonomers with different chemical properties, the polymerization technique allowed to synthesize versatile materials for biomedical and pharmaceutical applications. In order to prepare pH-sensitive microspheres, suitable for oral drug administration, MA-BSA was copolymerized with methacrylic acid sodium salt (NaMA) by reverse phase suspension copolymerization [75]. Additionally, thermo-responsive hydrophilic microspheres were prepared by free radical polymerization of MA-BSA and NIPAAm, as cross-linker and functional monomer, respectively [76]. Thermal analyses of the samples showed negative thermo-responsive behaviour with pronounced water affinity of microspheres at temperature lower than LCST. Furthermore, the influence of the hydrophilic/hydrophobic balance of the monomers in the polymerization feed on the transition temperature of the macromolecular device was investigated [77]. Thermoresponsive microspheres were synthesized by copolymerization of HEMA and NIPAAm, as hydrophilic and thermoresponsive monomers respectively, in presence of a proteic crosslinker, such as MA-BSA. The synthetic approach allows to modify the polymeric network composition producing hydrogels with appropriate and modulable physicochemical properties and a LCST close to the physiological temperature. Finally, dual stimuli-responsive microspheres (pH and temperature) were prepared by free radical polymerization of BSA-MA and NaMA and NIPAAm, as hydrophilic/pH-sensitive and thermo-responsive monomers, respectively [78]. In order to test the preformed materials as site-specific (oral or transdermal) drug carriers, *in vitro* experiments using diclofenac diethyl ammonium-loaded microgels were performed.

Acrylated BSA (A-BSA) (Figure 7), as referred by Tada et al., was prepared via reaction of albumin with N-succinimidylacrylate in aqueous medium at room temperature [79]. BSA with acrylate groups was then copolymerized with acrylamide (AAm) and N,N-methylenebis (acrylamide) (MEBA), using APS/TEMED as initiator system, to produce a biodegradable hydrogel potentially useful as sustained drug release carrier for albumin binding substances.

2.8. Gelatin

Gelatin (GL) is a natural protein with many desirable properties for application as a biomaterial, including scaffolds for tissue engineering. Methacrylated gelatin (MA-GL) was prepared by reaction of the gelatin primary amines groups with an excess of MA in phosphate buffer at 45°C. Thermal polymerization (AIBN as initiator system), in a concentrated oil-in-water emulsion, of the continuous phase gave rise to a polyHIPE, a porous material possessing a highly interconnected, trabecular morphology, useful as scaffolds for tissue engineering applications [80]. The obtained scaffold by radical polymerization possessed a morphology characterized by relatively large voids and interconnects, and as a consequence, it was more suitable for hepatocytes colonization [81].

Koepff et al. reported the derivatization of enzymatic gelatin hydrolisate (HGL) with GMA that reacts with both amino and acidic groups and leads to polymerizable gelatin

(GMA-HGL) (Scheme 2) [82]. Comparing to the native protein, the hydrolysates are characterized not only by enhanced water solubility, but also by a greater number of nucleophilic groups disposable for the reaction with the acylating agent. Unfortunately, solutions of pure GMA-HGL in water form brittle hydrogels when exposed to light in the presence of a photoinitiator. Hence, the use of flexibilizing reactive diluents is necessary to break the intermolecular H-bridges, to lower network density and increase the mechanical stability. Therefore, further modifications of HGL were attempted to increase the monomer tolerance and lower the amount of water or even avoid it entirely. HGL was modified with an PEG-300-monomethacrylate previously derivatized with maleic anhydride and further activated with N-hydroxysuccinimide (PMA-HGL) (Scheme 2) [83].

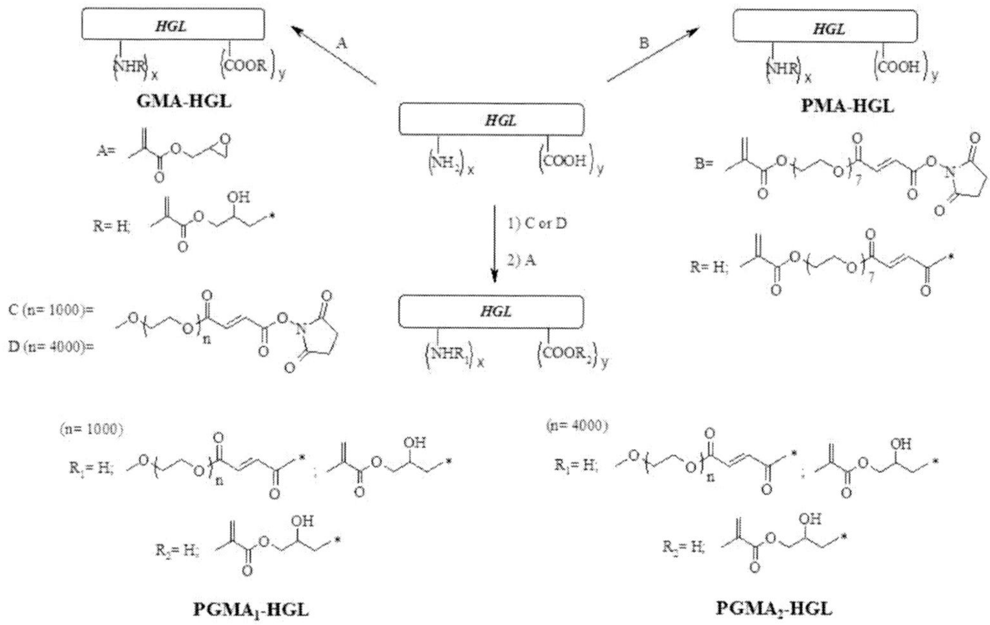

Scheme 2.

Because of the two different reactive sites on HGL-amino and carboxylic groups-introduction of two different moieties is possible with selective reagents. PEG-monomethylethers with molecular weights of 1000 and 4000 were primarily modified with maleic anhydride and subsequently activated with N-hydroxysuccinimide to give the reactive intermediates. HGL could be converted with these products (Scheme 2) under mild conditions in very good yields providing HGL derivatives. The reaction of the remaining carboxylic groups with GMA resulted in polymerizable gelatin derivatives, labelled $PGMA_1$-HGL, $PGMA_2$-HGL. [84]. The authors photocopolymerized the gelatin derivatives with HEMA and poly(ethylene glycol 400) dimethacrylate to provide hydrogels useful as scaffold for tissue engineering field.

Curcio et al. employed methacrylated gelatin hydrolysates (MA-HGL) as pro-hydrophilic macromer in the synthesis of thermo-responsive microspheres by free-radical suspension polymerization using NIPAAm and MEBA, as thermo-responsive functional monomer and crosslinking agent, respectively (Scheme 3) [85]. The potential application of these materials as drug carriers was demonstrated by performing Diclofenac release experiments at temperatures around the LCST. Depending on the temperature of the surrounding environment, the release of the drug across the hydrogels takes place by rapid and reversible modification of volume hydrogels and by diffusion of the therapeutic through the polymeric network.

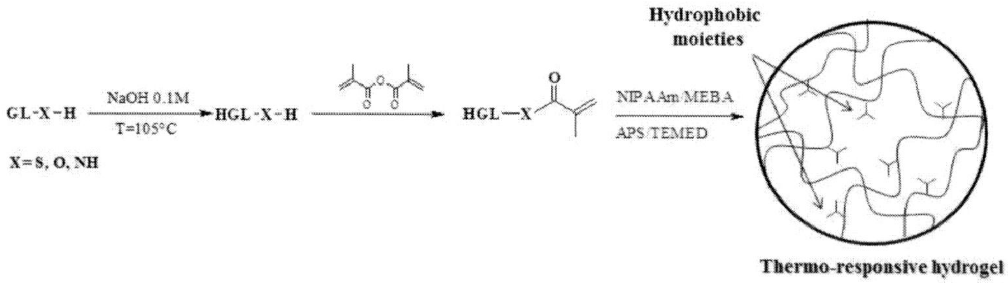

Scheme 3.

3. Grafting of Biomacromolecules

Of late, graft copolymerization is attracting a great interest in the scientific community because it is a convenient method to add new properties to a natural polymer with minimum loss of the initial properties of the substrate. Grafting method allows the formation of functional derivatives by covalent binding of a molecule onto the macromolecular backbone. A graft copolymer can be described as having the general structure, where the main polymer backbone, has branches of another polymeric chain emanating from different points along its length. This technique may be considered as an approach to achieve novel polysaccharide or protein-based materials with improved properties including all the expected usefulness of these biomaterials [86]. In this section, the techniques employed for the graft copolymerization of polysaccharides and proteins and the application of these graft copolymers in pharmaceutical and biomedical fields were reviewed.

3.1. Chitosan

Recently, there has been a growing interest in the chemical modification of CS in order to improve its properties and wide its applications [87, 88]. Among the various methods of modification, graft copolymerization has been the most used. CS has two types of reactive groups that can be grafted. First, the free amine groups on deacetylated units and secondly, the hydroxyl groups on the C_3 and C_6 carbons on acetylated or deacetylated units.

CS-graft-poly (ε-caprolactone) (CS-g-PCL) was synthesized with a facile one-step manner by grafting ε-caprolactone oligomers onto the hydroxyl groups of CS via ring-opening polymerization by using methanesulfonic acid as solvent and catalyst [89]. Then, CS-g-PCL/PCL nanofibrous mats were obtained by electrospinning of CS-g-PCL/PCL mixed solution. The nanofibers were extensively characterized and studies on cell–scaffold interaction were carried out by culturing mouse fibroblast cells (L929) on CS-g-PCL/PCL nanofibrous mats. The obtained results suggested the potential utilization of CS-g-PCL/PCL (2/8) nanofibrous mats for skin tissue engineering.

Several research papers report on grafting of vinyl monomers on the chitosan backbone using free radical initiators (Scheme 4).

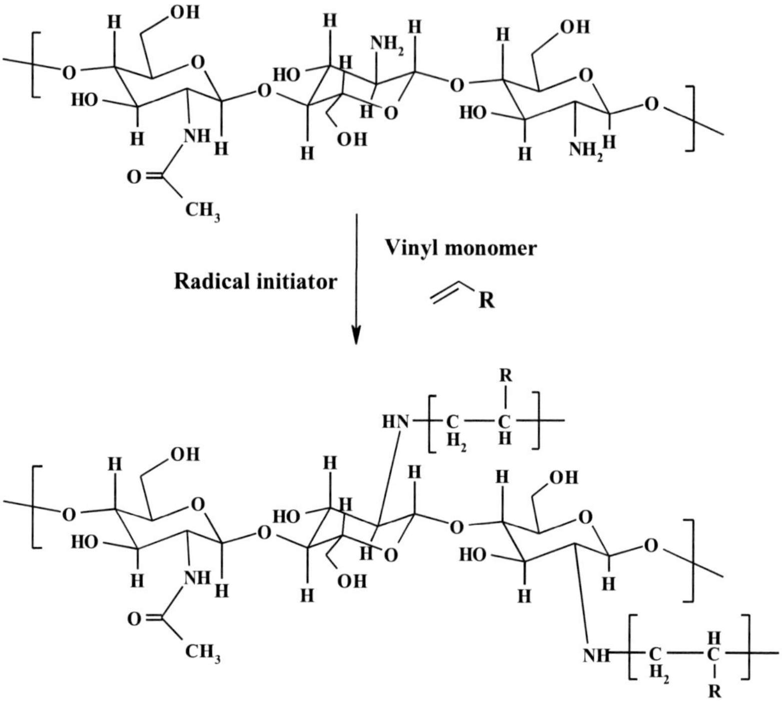

Scheme 4.

Carboxymethyl chitosan-grafted with MAA was prepared by Sun et al. by using APS as initiator in aqueous solution [90]. The effects of APS, MAA, reaction temperature and time on graft copolymerization were analyzed by determining the grafting percentage and efficiency. After grafting reaction, the CS derivatives had much improved water solubility. Similarly, Xie et al. prepared hydroxypropyl chitosan-grafted with MAA by using APS initiator, obtaining a derivative that also presented a good water solubility [91]. In 2004 the same authors published the graft copolymerization of maleic acid sodium (MAS) onto carboxymethyl chitosan and hydroxypropyl chitosan using APS as initiator [92]. The antioxidant activity of these derivatives was evaluated as superoxide anion scavengers by chemiluminescence technology. Compared with CS, the graft chitosan derivatives were found to have an improved scavenging ability against superoxide anion.

In a recent work, Ganji and co-workers used KPS as the initiator to graft monomethoxypoly(ethylene glycol) macromer on chitosan [93]. The resultant hydrogel behaved like a liquid at temperatures below room temperature and like a gel at temperatures close to that of the human body (36°C). The CS-g-PEG copolymer was considered to have the potential to be applied in injectable biomedical systems.

Moreover, Choochottiros and colleagues prepared amphiphilic chitosan nanospheres that possessed a responsive performance [94]. These nanoparticles were formed by grafting of phthalic anhydride, as a hydrophobic group and PEG, as a hydrophilic group. Chitosan nanosphere surfaces, in aqueous solution, are negatively charged. This property promotes a specific behavioral type to these nanospheres, i.e. good solution dispersion from neutral to high pH values and significant precipitation at low pH. The size distributions and morphologies were deeply investigated and the authors also determined the effect of the deacetylation percentage on degree of phthaloylation and PEGylation.

Similar nanoparticles were synthesized by Chuang et al [95]. Here, AA was copolymerized with NIPAAm in a chitosan solution. These nanoparticles were prepared using surfactant-free emulsion polymerization with the encapsulation of doxycycline hyclate. In this work the structure, particles size, morphology, surface charge, responsive properties and *in vitro* drug release behavior of nanoparticles were studied. The release profile of doxycycline hyclate from the drug-loaded nanoparticles at two different pH values (7.0 and 2.0) was distinct, showing faster release at the higher pH.

Vinyl monomers were grafted onto CS also using redox initiator systems. Don et al. employed cerium ammonium nitrate (CAN) to graft vinyl acetate onto CS in order to improve the properties of chitosan material [96]. The reaction was carried out for 2 h at 60°C, obtaining a monomer conversion between 70 and 80%. The experimental results indicated that the grafting efficiency increased with increasing amount of CS and that the polysaccharide acts as a surfactant, providing the stability of the dispersed particles. The data also showed that the incorporation of poly(vinyl acetate) (pVAc) to the chitosan chains increased the toughness and decreased the water absorption of chitosan. CAN was also found to be a suitable initiator for grafting NIPAAm into CS. The copolymer was then crosslinked with glutaraldehyde (GA). The efficiency and percentage of copolymerization increased as the monomer concentration (NIPAAm) increased [97]. Moreover, a redox system, based on potassium diperiodatocuprate (III), was employed to initiate the graft copolymerization of methyl acrylate onto chitosan in alkali aqueous solution [98]. In this work, chitosan acts as reductant and Cu (III) as oxidant in the redox system used to initiate the grafting reaction. High grafting efficiency and percentage were verified.

Aiming to obtain CS-based membranes with dual effects: to accelerate wound healing due to the bioactivities of chitosan and at the same time a drug delivery system, dos Santos and coauthors prepared different chitosan graft copolymers with HEMA and AA using also CAN as initiator [99]. The grafting of different amounts of AA and/or HEMA showed a remarkable influence in the swelling degree, cytotoxicity, thrombogenicity, haemolytic activity and thermal properties [100]. At the same time, El-Sherbinyl and co-workers prepared a chitosan graft poly(ethylene glycol) diacrylate copolymer hydrogel, aiming to improve the hydrophilicity of chitosan, by free radical polymerization, using the same initiator system [101]. Release tests with 5-Fluorouracil at pH values that were similar to those of gastric (pH 2.1) and intestinal (pH 7.4) fluids were performed at 37°C. The hydrogels presented negative pH-responsive behavior and positive temperature responsive behavior. In drug release tests, the hydrogels were able to deliver greater percentages of drug to provide longer release times, with increase in the poly(ethylene glycol) diacrylate content on the copolymer matrix.

Great interest has been made to graft natural polymers using the radiation method. Copolymers of CS and AA produced by γ-radiation have been reported. The systems included CS grafted with AA and with AAm [102]. Systems based on CS, grafted with poly(AA), poly(hydroxy propyl methacrylate), PVA and GL have also been prepared [103]. These materials exhibited smart behavior, responding to pH and to both pH and temperature. The results demonstrated the potential of these hydrogels as drug carriers for oral administration. Singh et al. grafted poly(acrylonitrile) onto chitosan using the microwave irradiation technique under homogeneous conditions [104]. They have obtained 70% grafting yield within 1.5min. The effects of reaction variables as monomer and CS concentration, microwave power, and exposure time on the graft copolymerization were studied. The grafting was found to increase with an increase in the monomer concentration. Grafting was also found to increase up to 80% microwave power and thereafter decreased.

Cai et al, reported the grafting reaction of NIPAAm on CS by gamma-radiation [105]. The resulting copolymer exhibited pH-responsive behavior and temperature responsive behavior, with swelling ratios higher at pH 4 than at pH 7.

3.2. Dextran

Graft polymers of dextran can be synthesized by the reaction of dextran with polypeptides, biologically active proteins (e.g., enzyme, antigen and hormones), starch and carbon chain polymers [106]. Novel CS-N-DX graft copolymers (CS-*g*-DX) with a degree of substitution of CS varying from 16 to 60% were prepared via reductive amination [107]. In particular, DX was grafted with galactosylated chitosan (GaCS) through a reductive amination reaction to enhance stability of DNA/GaCS in water; moreover, *in-vitro* studies were performed for the applicability of GaCS-*g*-DX as a liver-specific DNA carrier [108].

Graft copolymers of dextran and polyacrylamide were synthesized through the grafting of polyacrylamide (pAAm) chains onto a DX backbone with a ceric-ion induced solution polymerization technique [109]. By the variation of the amount of the initiator (CAN), four different grades of graft copolymers were synthesized. The partial alkaline hydrolysis of DX-g-pAAm was carried out in an alkaline medium. Three grades of partially hydrolyzed products were synthesized through the variation of the amount of alkali. Microspheres of acrylamide grafted on dextran (AAm-g-DX) were prepared by emulsion-crosslinking method using glutaraldehyde, as a crosslinker. The grafting efficiency was found to be 94%. The polymers were used as carrier for the sustained release of Acyclovir, an antiviral drug with limited water solubility. Acyclovir encapsulation of up to 79.6% was achieved as measured by UV spectroscopy. By performing *in vitro* release studies, the dependence of drug release rates on both the extent of crosslinking and amount of AAm-g-DX used in preparing microspheres was verified.

In a recent work, DX graft poly(N-methacryloylglycylglycine) copolymer-tyrosine conjugates were synthesized and characterized [110]. Dynamic light scattering results indicated that the graft copolymers are soluble in pH 7.4 (PBS) and 0.9% saline solutions. The graft copolymers were labeled with ^{125}I, and preliminary pharmacokinetics studies were performed to determine the biodistribution of graft copolymer. Biodistribution images confirmed the preferable liver and spleen accumulation within 1 h after injection and rapid clearance from all the organs over time. The graft copolymer with molecular weight of 9.8 kDa was eliminated from the kidney significantly faster than those with higher molecular weight.

Derkaoui et al. have synthesized amphiphilic copolymers based on DX and polybutylmethacrylate with the aim of endothelialization of biomaterials [111]. Grafting of butylmethacrylate onto DX has been carried out using CAN as initiator. Varying the DX percentages, three copolymers were obtained (11, 30 and 37 wt. % dextran), and homogeneous thin films were successfully prepared, obtaining materials with hybrid properties between the starting homopolymers. In contrast to polybutylmethacrylate, where the proliferation of vascular smooth muscle cells (VSMCs) was excellent but that of endothelial cells was very low, the copolymer containing 11% of DX was excellent for endothelial cells but very limited for VSMCs. An *in vitro* wound assay demonstrated that copolymer with 11% DX is even more favorable for endothelial cell migration than tissue-culture polystyrene. Increasing the DX content in the copolymers decreased the proliferation for both vascular cell types. Altogether, these results show that transparent and water-insoluble films made from copolymers of DX and butylmethacrylate copolymers with an appropriate composition could enhance endothelial cell proliferation and migration.

3.3. Starch

Free radical graft copolymerization of low molecular weight hydrophilic vinyl monomers onto the starch substrate in the presence of polymerizable cross-linking agent has been used. Zhang and Zhuo carried out the polymerization of NIPAAm in aqueous gelated corn starch solution using APS and TEMED, as redox pair initiators, and MEBA, as the cross-linker [112]. The materials resulted in temperature-sensitive starch-based hydrogels with improved surface properties. Elvira et al. implemented the free radical polymerization of AAm and AA in a corn starch/ethylene/vinyl alcohol copolymer blend using benzoyl peroxide and 4-dimethylaminobenzyl alcohol, as the initiating system, and MEBA, as the cross-linker [113]. As a consequence of the polymerization, a semi-interpenetrating polymer network with eventual graft-copolymer chains of poly(AA-co-AAm) onto the starch-based blends was formed by transfer reactions of the growing radicals on the side substituents of the pyranosyl cycles.

Bajpai and Saxena prepared the enzymatically degradable and pH-sensitive starch-based hydrogels using KPS-initiated graft copolymerization of AA onto soluble starch in the presence of MEBA, as the cross-linker [114]. The gels exhibited minimum swelling in an acidic pH medium through the formation of a complex hydrogen-bonded structure and underwent enzymatic degradation in a medium of pH 7.4 (i.e., simulating intestinal fluid) along with chain-relaxation-controlled swelling. Therefore, the gels have potential for colon-targeted drug delivery.

To synthesize a biopolymer-based superabsorbent hydrogel, 2-hydroxyethyl acrylate (HEA) and NaMA were grafted on the starch backbone in an aqueous solution by Pourjavadi and coworker [115].

The graft copolymerization reaction was carried out in the presence of APS, as an initiator, and MEBA, as a crosslinker, in a homogeneous medium (Scheme 5). APS is decomposed under heating (80°C) to produce sulfate anion-radicals that abstract hydrogen from the starch backbones and form corresponding macroradicals. These macroradicals initiate grafting of NaMA and HEA onto starch backbones leading to a graft copolymer. pH-responsiveness and swelling-deswelling behavior of the hydrogels make them suitable biomaterials for designing new systems for controlled drug delivery.

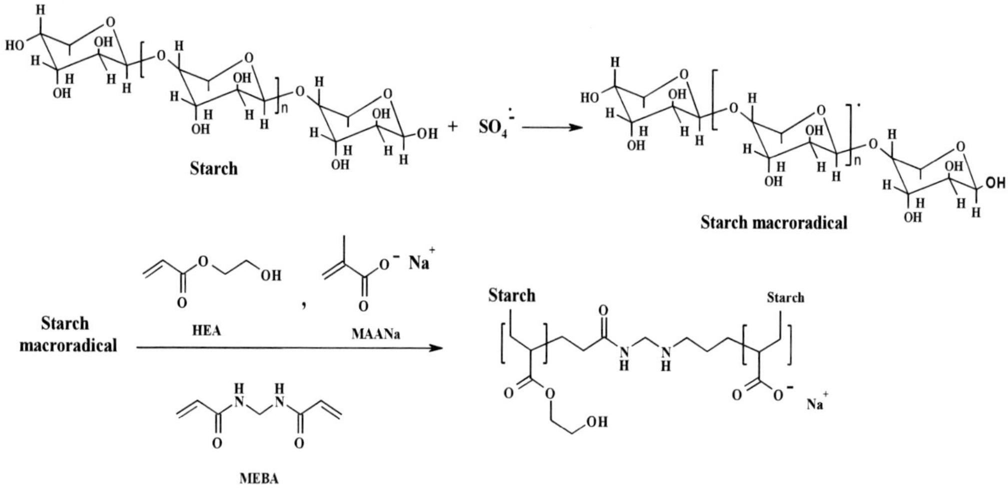

Scheme 5.

3.4. Guar Gum

GG-based formulations developed as controlled release dosage forms have been successful in clinical trials. In order to confer favorable properties to GG, different vinyl monomers such as acrylamide, methyacrylamide, acrylonitrile, methacrylates, 4-vinylpyridine and acrylic acid were grafted onto the polysaccharide backbone to incorporate favorable properties while retaining their desirable properties such as thickening, water/saline retention, and biodegradability. Ceric ion, peroxydiphosphate/metabisulphite, V^{5+}/mandelic acid, Cu^{2+}/mandelic acid and Fe^{2+}/BrO^{3-}, KPS/ascorbic acid are used as initiators of polymerization. A study by Behari et al. revealed that the graft copolymerization of AAm onto xanthan gum could be initiated by the Fe^{2+}/BrO^{3-} redox system in aqueous medium under a nitrogen atmosphere. They observed that grafting takes place efficiently when AAm concentration and temperature were 4.0×10^{-3} M and 35°C, respectively [116].

Mundargi et al. reported graft copolymerization of MAA onto GG by free radical initiation mechanism using two initiators: KPS and CAN (Scheme 6) [117]. KPS is used as an oxidizing initiator and grafting reaction is facilitated due to the generation of free radical site by abstracting hydrogen atom from the -OH group of the polymer. The formed free radicals could then react with the double bond of the vinyl monomer, resulting in a covalent bond between the monomer and GG to propagate the chain. However, termination takes place by a combination of two radicals. When CAN is used as an initiator, ceric ion reacts with GG to form GG–ceric complex. The Ce^{4+} ion in the complex can then be reduced to Ce^{3+} ion with the release of a proton and a subsequent formation of a free radical on the backbone of GG. These free radicals could then react with the monomer to initiate graft copolymerization. Termination of the graft copolymer takes place through the combination of radicals. O-carboxymethyl-O-hydroxypropyl guar gum (CMHPG) is a chemically modified guar gum with high water solubility and functionality [118]. Shi. et al, synthesized pNIPAAm grafted onto CMHPG in aqueous solutions by using KPS and TEMED as initiation system, resulting in new stimuli-responsive grafted polysaccharides[119].

Biswal et co-workers report mutual radiation grafting technique to carry out grafting of AAm onto GG using ^{60}Co γ-radiation, to enhance its flocculating properties for industrial effluents [120]. The grafted product was characterized using analytical probes like elemental, thermal, FT-IR analyses, X-ray diffraction and scanning electron microscope. The grafting extent was observed to decrease with the dose rate and increase with the concentration of AAm.

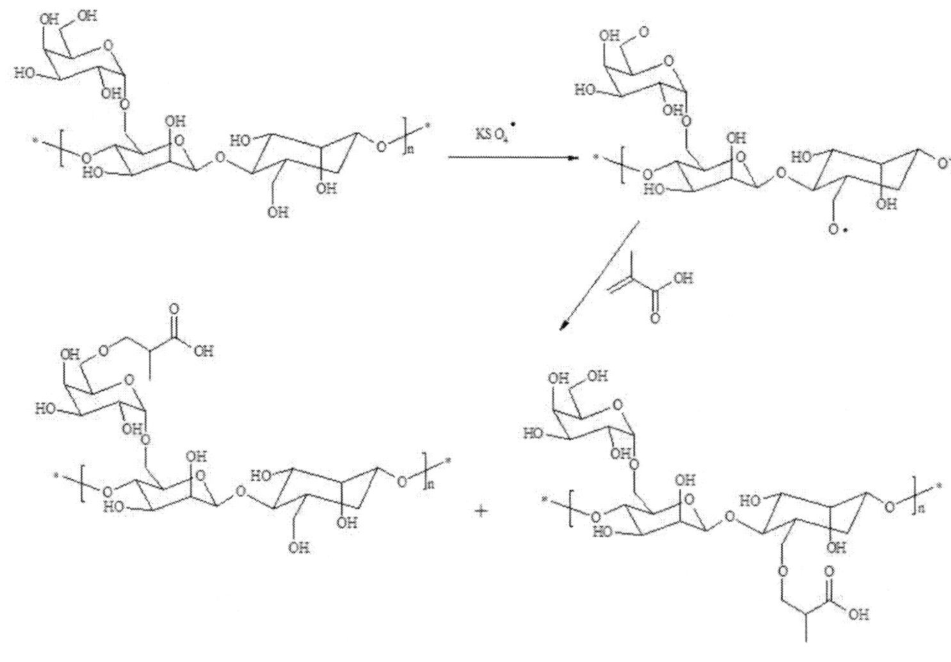

Scheme 6.

3.5. Pectin

Chauhan and coworkers describe preparation by graft copolymerization of hydrogels based on pectin and three different amide monomers: AAm, NIPAAm, and 2-acrylamido-2-methyl-1-propane sulfonic acid [121]. In this work, MEBA, as crosslinker, and APS-ferrous ammonium sulfate, as redox initiator system, were employed. Grafting on polysaccharide takes place usually on the anhydroglucose units, where H is abstracted by the radical generated by the redox initiator system. In the present case SO_4^{2-} is generated by the $S_2O_8^{2-}$ of the initiator. The growing macroradical attaches at these sites to give graft copolymers. Modification of pectin by grafting and crosslinking reactions, offering a wide spectrum of properties (like thermo sensitivity of NIPAAm and ion exchange and self-ionization of $-SO_3H$ group of 2-acrylamido- 2-methyl-1-propane sulfonic acid), is expected to afford a wide range of useful hydrogels. Three common water effluent metal ions, Fe^{2+}, Cu^{2+}, and Cr^{6+} has been explored as model ions to explore the structure-property relationship of these hydrogels for end-uses in metal ion removal and enrichment processes. The hydrogels show sensitivity to the thermal and pH changes made in their swelling environment. Possible use of hydrogels can also be explored as detoxificant agents and also in the loading of some metal ions of medicinal importance. Pectin-*g*-poly-vinylpyrrolidone hydrogels were synthesized by Fares et al. in the presence of MEBA as crosslinker and CAN as initiator. The materials were used as carriers for *in vitro* controllable release of theophylline at two pH values, 5.5 and 7.4 [122].

3.6. Inulin

Various inulin hydrogels have been developed that serve as potential drug delivery carriers. Spizzirri et al. describes the synthesis of a thermo-sensitive antioxidant hydrogels by grafting of catechin, an antioxidant molecule, onto inulin, in the presence of NIPAAm, as thermo-responsive monomer, and N,N-ethylenbisacrylamide, as crosslinker, employing ascorbic acid/hydrogen peroxide as initiator system (Figure 8) [123].

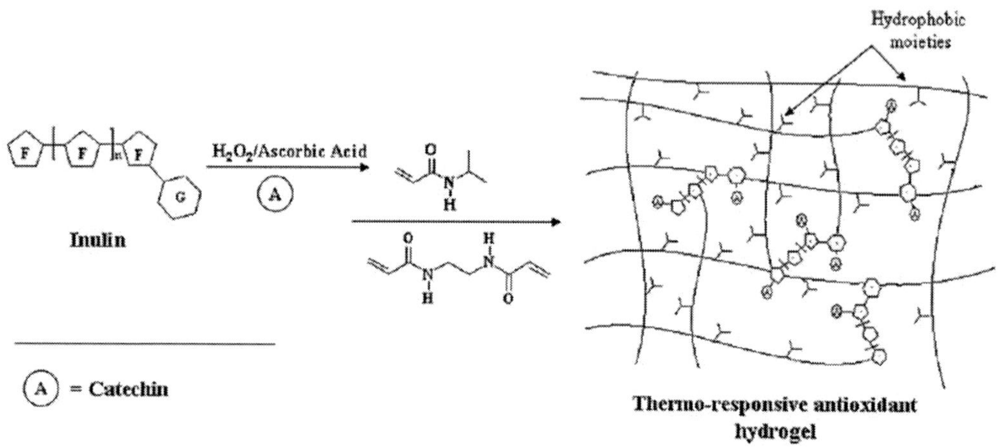

Figure 8.

The hydroxyl radicals, generated by the interaction between redox pair components, attack the sensible residues in the side chains of polysaccharide chain, producing radical species on the inulin structure. These ones react with the antioxidant molecules inducing an antioxidant-inulin covalent bond. The hydrogels transition temperatures were in the range 31.3-33.1°C and their water affinity around these values was evaluated. The results demonstrated that the antioxidant-carbohydrate thermo-responsive conjugates possess peculiar features for specific applications in the food industry.

3.7. Alginate

Alginate (ALG), is a gelatinous substance produced by brown algae and is used in a wide range of food, leather, pharmaceutical, and industrial applications. Because it is one of the few hydrocolloids that are capable of both thickening and gelling water, offers many useful properties, including viscosity control, stabilization of suspensions, emulsions and foams, improved freeze–thaw stability, syneresis and boil out control, film formation, rheology control, and more. Alginate has many useful properties and is very user-friendly and consumer-friendly because it is renewable, biodegradable, vegetable and not animal in origin, and wholly safe by all known tests. The graft copolymerization of sodium AL with pAAm and ethyl acrylate using CAN as an initiator has also been reported [124]. The grafting of vinyl monomers, such as methyl acrylate, acrylamide and acrylonitrile, onto ALG has gained considerable attention and proved of value in preparing new polymeric materials with special properties and enlarging the range of its utilization.

Yin et al. prepared a series of hydrogels by grafting of AAm onto ALG using CAN initiator system. By studying of the swelling kinetics of the hydrogels in different buffer solutions, the overshooting effect was observed in acidic medium, namely the gels firstly swelled to a maximum value, following by a gradual de-swelling until the equilibrium. The phenomenon is interpreted as a cooperative physical cross-linking caused by the hydrogen bond formation between the carboxyl groups of the hydrogels in a hydrophobic environment [125].

Pourjavadi et al. synthesized hydrogels based on sodium AL and polymethacrylamide (pMAAm) through free radical polymerization [126]. The graft copolymerization reaction was performed in a homogeneous medium and in the presence of APS as an initiator and MEBA as crosslinking agent. ALG-*g*-pMAAm, was then partially hydrolyzed by NaOH solution to yield a hydrogel, hydrolyzed alginate-*g*-polymethacrylamide (H-ALG-*g*-pMAM). During alkaline hydrolysis, the carboxamide groups of ALG-*g*-pMAM were converted into hydrophilic carboxylate anions. Either the ALG-*g*-pMAM or the H-ALG-*g*-pMAM was characterized by FT-IR spectroscopy. The effects of the grafting variables and the alkaline hydrolysis conditions were optimized systematically to achieve a hydrogel having the maximum swelling capacity.

More recently, graft copolymers of sodium ALG with itaconic acid has been prepared in aqueous solution using benzoyl peroxide as the initiator. The authors identified as optimum grafting conditions for maximum graft yield a reaction time of 1 h, reaction temperature of 85°C, itaconic acid concentration of 1.38 M, benzoyl peroxide concentration of 1.82×10^{-2} M and percentage of alginate 1.5 g/dl [127].

3.8. Collagen and Gelatin

Collagen (CL) is the primary structural material of vertebrates and plays an important role in the formation of tissues and organs, and is involved in various functional expressions of cells. The structure of CL consists of three polypeptide chains twined around one another and contains significant amounts of glycine, proline, alanine and hydroxyproline. CL has various advantages as a biomaterial and is widely used as carrier systems for delivery of drugs, proteins and genes. CL occurs in many places throughout the body. So far, 29 types of collagen have been identified and described. Over 90% of the collagen in the body, however, is of type I. Type-I collagen is a robust fiber protein and the main component of the extracellular matrix of most tissues. It is commonly immobilized on different substrates for surface modification of biomaterials to improve their biocompatibility. By hydrolysis process of Type I collagen, it is possible to obtain hydrolyzed collagen that is often used in the synthesis of bioconjugates by radical polymerization processes. Hydrolyzed collagen contains

20 amino acids, predominantly glycine, proline and hydroxyproline but, unlike gelatin, does not show any gel strength at all, due to the much lower molecular weight.

Reports in literature have described different methods of initiating monomer grafting onto the CL backbone by employing peroxides, APS or KPS as initiators. The APS can decompose on heating and produce sulfate anion-radicals that abstract hydrogen from one of the existing functional groups in the protein backbone (i.e. COOH, SH, OH, and NH_2) to form corresponding macroinitiator. Pourjavadi et al. have been prepared a novel hydrolyzed collagen-based hydrogel by grafting the binary mixture of AAm and 2-acrylamido-2-methylpropanesulfonic acid (AMPSA) onto the collagen backbone in the presence of a crosslinking agent, MEBA, and using as free-radical initiators, APS [128]. The macroradicals that are generated by APS initiate simultaneous graft copolymerization of AAm and AMPSA onto CL backbone. The reversible swelling/de-swelling behavior in solutions with acidic and basic pH makes the hydrogels a suitable candidate for various applications.

Using as free radical initiator (KPS) Marandi et al. synthesized nanocomposite superabsorbents by simultaneously graft copolymerization of AAm and AA onto hydrolyzed CL backbones in the presence of MEBA and nanoclay sodium montmorillonite, as crosslinker and filler into superabsorbent, respectively [129]. Macroradicals that are formed by KPS, initiate AA/AAm grafting onto the nanoclay and CL backbone, leading to a graft copolymer. The swelling of nanocomposite superabsorbent was measured in solution with pH ranged 1–13 and exhibited a pH-responsive characteristics.

The reaction beetwen the gelatin and a specific functional monomer was proposed to prepare, by radical polymerization, GL-based stimuli responsive hydrogels. The graft copolymerization of GL with various monomers is an effective method to improve its properties. The possibility to insert commercial gelatin in a crosslinked structure bearing thermo-sensitive moieties by reverse phase suspension radical polymerization represents an interesting innovation that significantly improves device performance, opening new applications in biomedical and pharmaceutical fields. Curcio et al. proposed the synthesis of thermo-responsive microspheres based on commercial GL as potential drug delivery systems. In this work, NIPAAm and MEBA were covalently inserted in the GL structure by radical grafting, using APS as initiator [130]. The thermal dissociation of the initiator formed anionic radicals that attack H-atoms in hydroxyl, thiolic or amino groups in the side chain of gelatin, forming a macroradical with several active sites. At those sites, polymer chain of NIPAAm starts and propagates as regular radical polymerization of polyacrylates (Figure 9).

Recently, a novel pH-sensitive hydrogel based on GL and AA, using γ-radiation as super clean source for polymerization and crosslinking, was synthesized [131]. Ionizing radiations are able to generate radicals on the monomer and polymer in addition to the production of $OH^{\cdot}$ and $H^{\cdot}$ radicals, as the primary products of water radiolysis. On exposure to γ-ray, as ionizing radiation, monomers radicals combine and propagate to form linear or branched but soluble polymers. In the same time, GL and the formed pAA are also radiolyzed forming macro radicals that contribute to chain initiation and crosslinking formation. Ketoprofen, an anti-inflammatory and analgesic agent, was used as model drug to evaluate the prepared hydrogel as drug carrier. *In vitro* release studies in different pH similar to the gastrointestinal fluids have been made to show the influence of the environmental pH and the preparation conditions on the release profiles. Because of the use of protein as a natural backbone, it is expected that the resulted pH-sensitive hydrogel show more compatibility with body when they use as drug delivery systems.

Sadeghi et al. report the study of a novel GL-based hydrogel synthesized through graft copolymerization of MAA onto gelatin, using APS, as a free radical initiator, and MEBA, as a crosslinker [132]. The macroradicals produced from thermal decomposition of APS, initiate graft copolymerization of MAA, leading to the graft copolymer. GL-*g*-pMAA hydrogel exhibited a pH-responsiveness character so that a swelling-deswelling pulsatile behavior was recorded at pH 2.0 and 8.0. The on-off switching behavior makes the hydrogel as a good candidate for controlled delivery of bioactive agents.

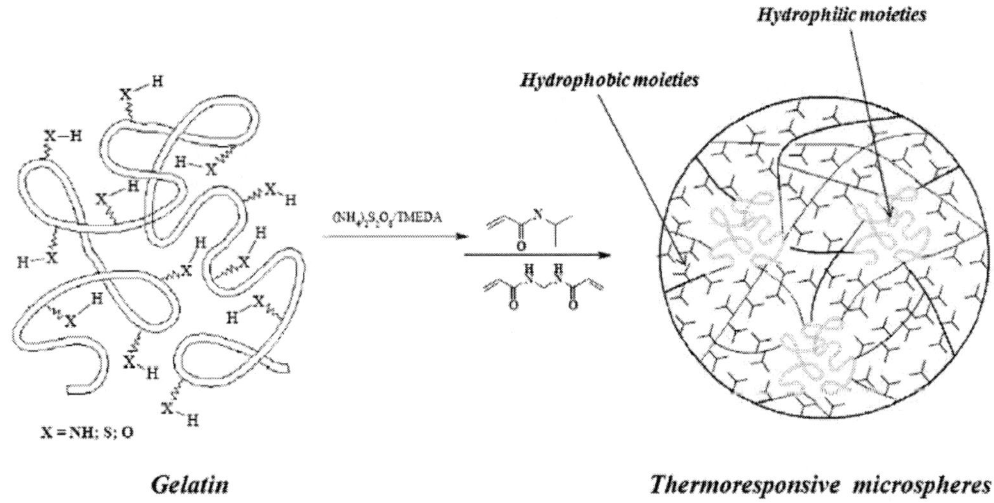

Figure 9.

REFERENCES

[1] Okada, H. & Toguchi, H. (1995). Biodegradable microspheres in drug-delivery. *Critical Reviews in Therapeutic Drug Carrier Systems, 12,* 1-99.

[2] Cleland, J.L.; Johnson, O.L.; Putney, S. & Jones, A.J.S. (1997). Recombinant human growth hormone poly(lactic-co-glycolic acid) microsphere formulation development. *Advanced Drug Delivery Reviews, 28,* 71-84.

[3] Chen, L.; Apte R.N. & Cohen, S. (1997). Characterization of PLGA microspheres for the controlled delivery of IL-1 alpha for tumor immunotherapy. *Journal of Controlled Release, 43,* 261.

[4] Klouda, L. & Mikos, A.G. (2008). Thermoresponsive hydrogels in biomedical applications. *European Journal of Pharmaceutics and Biopharmaceutics, 68,* 34–45.

[5] Ravikumar, M.N.V. (1999). Chitin and chitosan fibres: A review. *Bulletin of Materials Science, 22,* 905-915.

[6] Ueno, H.; Mori, T. & Fujinaga. T. (2001). Topical formulations and wound healing applications of chitosan. *Advanced Drug Delivery Reviews, 52,* 105-115.

[7] Jameela, S.R. & Jayakrishnan, A. (1995). Glutaraldehyde cross-linked chitosan microspheres as a long-acting biodegradable drug-delivery vehicle-studies on the in-

vitro release of mitoxantrone and *in vivo* degradation of microspheres in rat muscle. *Biomaterials*, *16,* 769-775.

[8] Bodnar, M.; Hartmann, J.F. & Borbely, J. (2006). Synthesis and study of cross-linked chitosan-N-poly(ethylene glycol) nanoparticles. *Biomacromolecules*, *7,* 3030–3036.

[9] Kim, M.S.; Choi Y.J.; Noh Y.; Noh I. & Tae G. (2007). Synthesis and characterization of in situ chitosan-based hydrogel via grafting of carboxyethyl acrylate. *Journal of Biomedical Materials Research Part A, 83,* 674-682.

[10] Renbutsu, E.; Hirose, M.; Omura, Y.; Nakatsubo, F.; Okamura, Y.; Okamoto, Y.; Saimoto, H.; Shigemasa, Y. & Minami, S. (2005). Preparation and biocompatibility of novel UV-curable chitosan derivatives. *Biomacromolecules*, *6,* 2385–2388.

[11] Dong, Y.M.; Mao, W.; Wang, H. W.; Zhao, Y.Q.; Li, X.J.; Bi, D.X.; Yang, L.L.; Ge, Q.; & Fang, X.M. (2006). Measurement of critical concentration for mesophase formation of chitosan derivatives in both aqueous and organic solutions. *Polymer International*, *55,* 1444-1449.

[12] Huang, M.F.; Shen, X.F.; Sheng, Y.; & Fang, Y.E. (2005). Study of graft copolymerization of N-maleamic acid-chitosan and butyl acrylate by gamma-ray. *International Journal of Biological Macromolecules*, *36,* 98-102.

[13] Don, T.M. & Chen, H.R. (2005). Synthesis and characterization of AB-crosslinked graft copolymers based on maleilated chitosan and N-isopropylacrylamide. *Carbohydrate Polymers*, *61,* 334-347.

[14] Zhao, L.; Mitomo, H.; Zhai, M.L.; Yoshii, F.; Nagasawa, N. & Kume, T. (2003). Synthesis of antibacterial PVA/CM-chitosan blend hydrogels with electron beam irradiation. *Carbohydrate Research*, *53,* 439-446.

[15] Mesquita, A.C.; Mori, M.N. & Silva, A.L.G. (2004). Polymerization of vinyl acetate in bulk and emulsion by gamma irradiation. *Radiation Physics and Chemistry*, *71,* 253-256.

[16] Fan, J.; Chen, J.; Yang, L.;, Lin, H.; & Cao, F. (2009). Preparation of dual-sensitive graft copolymer hydrogel based on N-maleoyl-chitosan and poly(N-isopropylacrylamide) by electron beam radiation. *Bulletin of Materials Science*, *32,* 521–526.

[17] Yu, Y.; Li, Y.; Zhu, C. & Liu L. (2010). Synthesis and characterization of temperature-sensitive and biodegradable hydrogel based on *N*-isopropylacrylamide. *Central European Journal of Chemistry, 8,* 426–433.

[18] Matsuda, T. & Magoshi, T. (2002). Preparation of vinylated polysaccharides and photofabrication of tubular scaffolds as potential use in tissue engineering. *Biomacromolecules, 3,* 942–850.

[19] Li, Q.; Yang, D.; Ma, G.; Xu, Q.; Chen, X.; Lu, F. & Nie, J. (2009). Synthesis and characterization of chitosan-based hydrogels. *International Journal of Biological Macromolecules*, *44,* 121–127.

[20] Yin, Y.J.; Yao, K.D.; Cheng, G.X. & Ma, J.B. (1999). Properties of polyelectrolyte complex films of chitosan and gelatin. *Polymer International*, *48,* 429-432.

[21] Flores-Ramirez, N.; Elizalde-Pena, E.A.; Vasquez-Garcia, S.R.; Gonzalez-Hernandez, J.; Martinez-Ruvalcaba, A.; Sanchez, I. C.; Luna-Barcenas, G. & Gupta, R.B. (2005). Characterization and degradation of functionalized chitosan with glycidyl methacrylate. *Journal of Biomaterials Science-Polymer Edition*, *16,* 473–488.

[22] Cha, S.Y.; Lee, J.K.; Lim, B.S.; Lee, T.S. & Park, W.H. (2001). Conjugated vinyl derivatives of chitooligosaccharide: Synthesis and characterization. *Journal of Polymer Science Part A-Polymer Chemistry*, *39*, 880–887.

[23] Amsden, B.G.; Sukarto, A.; Knight, D.K. and Shapka S.N. (2007). Methacrylated glycol chitosan as a photopolymerizable biomaterial. *Biomacromolecules, 8,* 3758–3766.

[24] Mehvar, R. (2000). Dextrans for targeted and sustained delivery of therapeutic and imaging agents. *Journal of Controlled Release, 69,* 1–25.

[25] Edman, P.; Ekman, B. & Sjoholm, I. (1980). Immobilization of proteins in microspheres of biodegradable polyacryldextran. *Journal of Pharmaceutical Sciences, 69,* 838–842.

[26] Van Dijk-Wolthuis, W.N.E.; Franssen, O.; Talsma, H.; Van Steenbergen, M.J.; Kettenes-Van den Bosch, J.J. & Hennink, W.E. (1995). Synthesis, characterization and polymerization of glycidyl methacrylate derivated dextran. *Macromolecules, 28,* 6317–6322.

[27] Franssen, O.; Van Ooijen, R.D.; De Boer, D.; Maes, R.A.A. & Hennink, W.E. (1999). Enzymatic degradation of cross-linked dextrans. *Macromolecules, 32,* 2896–2902.

[28] Van Dijk-Wolthuis, W.N.E.; Van Steenbergen, M.J.; Hoogeboom, C.; Tsang, S.K.Y. & Hennink, W.E. (1997). Reaction of dextranwith glycidyl methacrylate: an unexpected transesterification. *Macromolecules, 30,* 3411–3413.

[29] Lo, C.-W. & Jiang, H. (2010). Photopatterning and degradation study of dextran-glycidyl methacrylate hydrogels. *Polymer Engineering and Science, 50,* 232-239.

[30] Castiglione, M.; Puoci, F.; Iemma, F.; Privitera, L.; Spizzirri, U.G.; Curcio, M. & Picci, N. (2007). pH-sensitive microspheres obtained by derivatized dextran. *Chemistry Today, 25,* 4, 48-50.

[31] Franssen, O. & Hennink, W.E. (1998). A novel preparation method for polymeric microparticles without the use of organic solvents. *International Journal of Pharmaceutics*, *168,* 1–7.

[32] Matricardi, P.; Pontoriero, M.; Coviello, T.; Casadei, M.A. & Alhaique, F. (2008). In situ cross-linkable novel alginate-dextran methacrylate IPN hydrogels for biomedical applications: mechanical characterization and drug delivery properties. *Biomacromolecules, 9,* 2014–2020.

[33] Cassano, R.; Trombino, S.; Muzzalupo, R.; Tavano, L. & Picci, N. (2009). A novel dextran hydrogel linking trans-ferulic acid for the stabilization and transdermal delivery of vitamin E *European Journal of Pharmaceutics and Biopharmaceutics, 72,* 232–238.

[34] Van Dijk-Wolthuis, W.N.E.; Tsang, S.K.Y.; Kettenes-van den Bosch, J.J. & Hennink, W.E., (1997). A new class of polymerizable dextrans with hydrolysable groups: hydroxyethyl methacrylated dextran with and without oligolactate spacer. *Polymer, 38,* 6235–6242.

[35] Van Dijk-Wolthuis, W.N.E.; Van Steenbergen, M.J.; Underberg, W.J.M. & Hennink, W.E. (1997c). Degradation kinetics of methacrylated dextrans in aqueous solution. *Journal of Pharmaceutical Sciences, 86,* 413–417.

[36] De Groot, C.; Cadee, J.; Koten, J.W. & Hennink, W.E., Den Otter, W. (2002). Therapeutic efficacy of IL-2-loaded hydrogels in a mouse tumor model. *International Journal of Cancerogenesis, 98,* 134–140.

[37] Stenekes, R.J.H. & Hennink, W.E. (2000). Polymerization kinetics of dextran-bound methacrylate in an aqueous two-phases system. *Polymer, 41,* 5563–5569.

[38] Chung, J.T.; Vlugt-Wensink K.D.F.; Henninkb, W.E. & Zhanga, Z. (2005). Effect of polymerization conditions on the network properties of dex-HEMA microspheres and macro-hydrogels. *International Journal of Pharmaceutics, 288,* 51–61.

[39] Pescosolido, L.; Schuurman, W.; Malda, J.; Matricardi, P.; Alhaique, F.; Coviello, T.; van Weeren, P. R.; Dhert, W.J.A.; Hennink, W.E. & Vermonden, T. (2011). Hyaluronic Acid and Dextran-Based Semi-IPN Hydrogels as Biomaterials for Bioprinting. *Biomacromolecules, 12,* 1831-1838.

[40] Pescosolido, L.; Vermonden, T.; Malda, J.; Censi, R.; Dhert, W. J. A.; Alhaique, F.; Hennink, W. E.; Matricardi, P. (2011). In situ forming IPN hydrogels of calcium alginate and dextran-HEMA for biomedical applications. *Acta Biomaterialia, 494,* 7, 1627-1633.

[41] Lapcik Jr, L.; Lapcik, L.; De Smedt, S.; Demeester, J. & Chabrecek P. (1998). Hyaluronan: Preparation, structure, properties, and applications. *Chemical Reviews, 98,* 2663-2684.

[42] Smeds, K.A. & Grinstaff, M.W. (2001). Photocrosslinkable polysaccharides for in situ hydrogel formation. *Journal of Biomedical Materials Research, 54,* 115-121.

[43] Masters, K.S.; Shah, D.N.; Leinwand, L.A. & Anseth K.S. (2005). Crosslinked hyaluronan scaffolds as a biologically active carrier for valvular interstitial cells. *Biomaterials, 26,* 2517-2525.

[44] Burdick, J.A.; Chung, C.; Jia, X.; Randolph, M.A. & Langer, R. (2005). Controlled degradation and mechanical behavior of photopolymerized hyaluronic acid networks. *Biomacromolecules, 6,* 386-391.

[45] Barbetta, A.; Massimi, M.; Di Rosario, B.; Nardecchia, S.; De Colli, M.; Devirgiliis, L. & Dentini, M. Emulsion templated scaffolds that include gelatin and glycosaminoglycans. (2008). *Biomacromolecules, 9,* 2844-2856.

[46] Sahoo, S.; Chung, C.; Khetan, S. & Burdick, J.A. (2008). Hydrolytically degradable hyaluronic acid hydrogels with controlled temporal structures. *Biomacromolecules, 9,* 1088-1092.

[47] Jin, Y.; Yamanaka , J.; Sato, S.; Miyata, I., Yomota, C. & Yonese, M. (2001). Recyclable characteristics of hyaluronate–polyhydroxyethyl acrylate blend hydrogel for controlled releases. *Journal of Controlled Release, 73,* 173–181.

[48] Oudshoorn, M.H.M.; Rissmann, R.; Bouwstra, J.A. & Hennink, W.E. (2007). Synthesis of methacrylated hyaluronic acid with tailored degree of substitution. *Polymer, 48,* 1915-1920.

[49] Park, Y.D.; Tirelli, N. & Hubbell, J.A. (2003). Photopolymerized hyaluronic acid-based hydrogels and interpenetrating networks. *Biomaterials, 24,* 893-900.

[50] Bojanic, V.; Jovanovic, S.; Tabakovic, R. & Tabakovic, I. (1996). Synthesis and electrochemistry of grafted copolymers of cellulose with 4-vinylpyridine, 1-vinylimidazole, 1-vinyl-2-pyrrolidinone, and 9-vinylcarbazole. *Journal of Applied Polymer Science, 60,* 1719–1725.

[51] Trombino, S.; Cassano, R.; Bloise, E.; Muzzalupo, R.; Tavano, L. & Picci, N. (2009). Synthesis and antioxidant activity evaluation of a novel cellulose hydrogel containing trans-ferulic acid. *Carbohydrate Polymers, 75,* 184–188.

[52] Cai, T. & Hu, Z. (2003). Synthesis and study of and controlled release from nanoparticles and their networks based on functionalized hydroxypropylcellulose. *Macromolecules, 36,* 6559-6564.

[53] Davis, H.; Skrzypek, W. & Khan, A. (1994). Iodine binding by amylopectin and stability of the amylopectin iodine complex. *Journal of Polymer Science Part A-Polymer Chemistry, 32,* 2267-2274.

[54] Peppas, N.A.; Bures, P.; Leobandung, W. & Ichikawa, H. (2000). Hydrogels in pharmaceutical formulations. *European Journal of Pharmaceutics and Biopharmaceutics, 50,* 27-46.

[55] Reis, A.V.; Guilherme, M.R.; Moia, T.A.; Mattoso, L.H.C.; Muniz, E.C. & Tambourgi E.B. (2008). Synthesis and characterization of a starch-modified hydrogel as potential carrier for drug delivery system. *Journal of Polymer Science: Part A: Polymer Chemistry, 46,* 2567–2574.

[56] Heim, E.; Harling, S.; Pöhling, K.; Ludwig, F.; Menzel, H. & Schilling, M. (2007). Fluxgate magnetorelaxometry of supermagnetic nanoparticles for hydrogel characterisation. *Journal of Magnetism and Magnetic Materials, 311,* 150–154.

[57] Schwoerer, A.D.A.; Harling, S.; Scheibe, K.; Menzel, H. & Daniels, R. (2009). Influence of degree of substitution of HES–HEMA on the release of incorporated drug models from corresponding hydrogels. *European Journal of Pharmaceutics and Biopharmaceutics, 73,* 351–356.

[58] Sinha, V.R. & Kumria, R. (2001). Polysaccharides in colon-specific drug delivery. *International Journal of Pharmaceutics, 224,* 19–38.

[59] Reis, A.V.; Cavalcanti, O.A.; Rubira, A. F. & Muniz, E.C. (2003). Synthesis and characterization of hydrogels formed from a glycidyl methacrylate derivative of galactomannan. *International Journal of Pharmaceutics, 267,* 13–25.

[60] Singh, V.; Tiwari, A.; Tripathi, D.N. & Sanghi, R. (2004). Microwave assisted synthesis of guar-g-polyacrylamide. *Carbohydrate Polymers, 58,* 1–6.

[61] Singh, V.; Tiwari, A.; Singh, S.P.; Kumari, P. & Tiwari, S. (2008). Ceric ammonium sulfate/sodium disulfite initiated grafting of acrylamide on to Cassia reticulata seed gum. *Journal of Applied Polymers Science, 110,* 1477–84.

[62] Chourasia, M.K. & Jain, S.K. (2003). Pharmaceutical approaches to colon targeted drug delivery systems. *Journal of Pharmaceutical Sciences, 6,* 33–66.

[63] Reddy, D.N. & Udupa, N. (1993). Formulation and evaluation of oral and transdermal preparations of flurbiprofen and piroxicam incorporated with different carriers. *Drug Development and Industrial Pharmacy, 19,* 843–852.

[64] Tiwari, A.; Grailer, J.J.; Pilla, S.; Steeber, D.A. & Gong, S. (2009). Biodegradable hydrogels based on novel photopolymerizable guar gum–methacrylate macromonomers for in situ fabrication of tissue engineering scaffolds. *Acta Biomaterialia, 5,* 3441–3452.

[65] Roberfroid, M.B. & Delzenne, N.M. (1998). Dietary fructans. *Annual Review of Nutrition, 18,* 117-143.

[66] Kaur, J. & Gupta, A.K. (2002). Applications of inulin and oligofructose in health and nutrition. *Journal of Biosciences, 27,* 703–714.

[67] Vervoort, L.; Van den Mooter, G.; Augustijns, P.; Busson, R.; Toppet, S. & Kinget, R. (1997). Inulin hydrogels as carriers for colonic drug targeting. I. Synthesis and

characterization of methacrylated inulin, and hydrogel formation. *Pharmaceutical Research, 14*, 1730-1737.

[68] Vervoort, L.; Vinckier, I.; Moldenaers, P.; Van Den Mooter, G.; Augustijns, P. & Kinget R.(1998). Inulin hydrogels as carriers for colonic drug targeting. rheological characterization of the hydrogel formation and the hydrogel network. *Journal of Pharmaceutical Sciences, 88,* 209-214.

[69] Chiu, H.C.; Hsu, Y.H. & Lin, P.J. (2002). Synthesis of pH-sensitive inulin hydrogels and characterization of their swelling properties. *Jounal of Biomedical and Material Research, 61,* 146-152.

[70] Perez, S.; Rodrıguez-Carvajal, M.A. & Doco, T. (2003). A complex plant cell wall polysaccharide: rhamnogalacturonan. II. A structure in quest of a function. *Biochimie, 85,* 109–121.

[71] Maior, J.F.A.S.; Reis, A.V.; Muniz, E.C. & Cavalcanti, O.A. Reaction of pectin and glycidyl methacrylate and ulterior formation of free films by reticulation. *International Journal of Pharmaceutics, 355,* 184–194.

[72] Reis, A.V.; Guilherme, M.R.; Cavalcanti, O.A.; Rubira, A.F. & Muniz, E.C. (2006). Synthesis and characterization of pH-responsive hydrogels based on chemically modified Arabic gum polysaccharide. *Polymer, 47,* 1–7.

[73] Iemma, F.; Spizzirri, U.G.; Muzzalupo, R.; Puoci, F.; Trombino, S. & Picci, N. (2004). Spherical hydrophilic microparticles obtained by the radical copolymerisation of functionalised bovine serum albumin. *Colloid of Polymer Sciences, 283,* 250–256.

[74] Iemma, F.; Spizzirri, U.G.; Puoci, F.; Muzzalupo, R.; Trombino, S. & Picci, N. (2005). Radical crosslinked albumin microspheres as potential drug delivery systems: preparation and in vitro studies. *Drug Delivery, 12,* 179–184.

[75] Iemma, F.; Spizzirri, U.G.; Puoci, F.; Muzzalupo, R.; Trombino, S.; Cassano, R.; Leta, S. & Picci, N. (2006). pH-Sensitive hydrogels based on bovine serum albumin for oral drug delivery. *International Journal of Pharmaceutics, 312,* 151–157.

[76] Spizzirri, U.G.; Iemma, F.; Puoci, F.; Cirillo, G.; Curcio, M.; Parise, O.I. & Picci, N. (2009). Synthesis and release profile analysis of thermo-sensitive albumin hydrogels. *Colloid of Polymer Sciences, 287,* 779–787.

[77] Spizzirri, U.G.; Cirillo, G.; Iemma, F.; Puoci, F.; Curcio, M.; Parise, O.I. & Picci, N. (2011). Thermo-responsive albumin hydrogels with LCST near the physiological temperature. *Journal of Applied Polymer Science, 121,* 342–351.

[78] Cirillo, G.; Iemma, F.; Spizzirri, U.G.; Puoci, F.; Curcio, M.; Parisi, O.I. & Picci, N. (2011). Synthesis of stimuli-responsive microgels for in vitro release of diclofenac diethyl ammonium. *Journal of Biomaterial Science Polymer Edition, 22,* 823–844.

[79] Tada, D.; Tanabe, T.; Tachibana, A. & Yamauchi, K.(2005). Drug release from hydrogel containing albumin as crosslinker. *Journal of Bioscience and Bioengineering, 100,* 551–555.

[80] Barbetta, A.; Dentini, M.; Zannoni, E.M. & De Stefano M.E. (2005). Tailoring the porosity and morphology of gelatin-methacrylate polyHYPE scaffolds for tissue engineering applications. *Langmuir*, *21,* 12333-12341.

[81] Barbetta, A.; Massimi, M.; Conti Devirgiliis, L. & Dentini, M. (2006). Enzymatic cross-linking versus radical polymerization in the preparation of gelatin polyHYPEs

and their performance as scaffolds in the culture of hepatocytes. *Biomacromolecules, 7,* 3059-3068.

[82] Koepff, P.; Braumer, K. & Babel, W. (1998). U.S. Pat. Appl US 5,733,994; 3.

[83] Pochlauer, P. & Hendel, W. (1998). One-pot formation of succinimidyl esters by the system chlorophosphate/hydroxysuccinimide/base. *Tetrahedron, 54,* 3489–3494.

[84] Schuster, M.; Turecek, C.; Weigel, G.; Saf, R.; Stampfl, J.; Varga, F.; Liska, R. (2009). Gelatin-based photopolymers for bone replacement materials. *Journal of Polymer Science: Part A: Polymer Chemistry, 47,* 7078–7089.

[85] Curcio, M.; Puoci, F.; Spizzirri, U.G.; Iemma, F.; Cirillo, G.; Parisi, O.I. & Picci, N. (2010). Negative Thermo-responsive Microspheres Based on Hydrolyzed Gelatin as Drug Delivery Device. *AAPS Pharmaceutical Science Technologies, 11,* 652-662.

[86] Maiti, S.; Ranjit, S. & Sa, B.(2010). Polysaccharide-based graft copolymers in controlled drug delivery. *International Journal of Pharmaceutical and Technological Research, 2,* 1350-1358.

[87] Alves N.M. & Mano J.F. (2008). Chitosan derivatives obtained by chemical modifications for biomedical and environmental applications. *International Journal of Biological Macromolecules, 43,* 401–414.

[88] Van Vlierberghe, S.; Dubruel, P. & Schacht E. (2011). Biopolymer-Based Hydrogels As Scaffolds for Tissue Engineering Applications: A Review. *Biomacromolecules, 12,* 1387-1408

[89] Chena, H.; Huanga, J.; Yu, J.; Liuc, S. & Gud, P. (2011). Electrospun chitosan-graft-poly (ε-caprolactone)/poly (ε -caprolactone) cationic nanofibrous mats as potential scaffolds for skin tissue engineering. *International Journal of Biological Macromolecules, 48,* 13-19.

[90] Sun T.; Xu P.; Liu Q.; Xue J. & Xie W. (2003). Graft copolymerization of methacrylic acid onto carboxymethyl chitosan. *European Polymer Journal, 39,* 189-192.

[91] Xie W.M.; Xu P.X.; Wang W. & Lu Q. (2002). Preparation and antibacterial activity of a water-soluble chitosan derivative. *Carbohydrate Polymers, 50,* 35–40.

[92] Sun T.; Xie W. & Xu P. (2004). Superoxide anion scavenging activity of graft chitosan derivatives. *Carbohydrate Polymers, 58,* 379-382.

[93] Ganji, F. & Abdekhodaie, M.J. (2008). Synthesis and characterization of a new thermosensitive chitosan-PEG diblock copolymer. *Carbohydrate Polymers, 74,* 435-441.

[94] Choochottiros, C.; Yoksan, R.; Chirachanchai, S. (2009) Amphiphilic chitosan nanospheres: Factors to control nanosphere formation and its consequent pH responsive performance. *Polymer, 50,* 1877-1886

[95] Chuang, C.Y.; Don, T.M. & Chiu, W.Y. (2009). Synthesis of Chitosan-Based Thermo- and pH-Responsive Porous Nanoparticles by Temperature-Dependent Self-Assembly Method and Their Application in Drug Release. *Journal of Polymer Science Part A-Polymer Chemistry, 47,* 5126-5136.

[96] Don T.M.; King C.F. & Chiu W.Y. (2002). Synthesis and properties of chitosan-modified poly(vinyl acetate). *Journal of Applied Polymer Science, 86,* 3057–3063.

[97] Carreira, A.S.; Gonçalves, F.A.M.M.; Mendonça, P.V.; Gil, M.H. & Coelho, J.F.J. (2010). Temperature and pH responsive polymers based on chitosan: Applications and

new graft copolymerization strategies based on living radical polymerization. *Carbohydrate Polymers, 80,* 618–630.

[98] Liu, Y.; Liu, Z.; Zhang, Y. & Deng, K. (2003). Graft copolymerizaztion of methyl acrylate onto chitosan initiated by potassium diperiodatocuprate (III). *Journal of Applied Polymer Science, 89,* 2283–2289.

[99] dos Santos, K.; Coelho, J.F.J.; Ferreira, P.; Pinto, I.; Lorenzetti, S. G. & Ferreira, E. I. (2006). Synthesis and characterization of membranes obtained by graft copolymerization of 2-hydroxyethyl methacrylate and acrylic acid onto chitosan. *International Journal of Pharmaceutics, 310*, 37–45.

[100] Ferreira, P.; Coelho, J. F.J.; dos Santos, K.; Ferreira, E.I. & Gil, M.H. (2006). Thermal characterization of chitosan-grafted membranes to be used as wound dressings. *Journal of Carbohydrate Chemistry, 25,* 233–251.

[101] El-Sherbinyl, I.M.; Abdel-Bary, E.M. & Harding, D.R.K. (2006). Swelling characteristics and in vitro drug release study with pH- and thermally sensitive hydrogels based on modified chitosan. *Journal of Applied Polymer Science, 102*, 977–985.

[102] Taleb, M.F.A. (2008). Radiation synthesis of polyampholytic and reversible pH-responsive hydrogel and its application as drug delivery system. *Polymer Bulletin, 61*, 341–351.

[103] Sokker, H.H.; Ghaffar, A.M.A.; Gad, Y.H. & Aly, A.S. (2009). Synthesis and characterization of hydrogels based on grafted chitosan for the controlled drug release. *Carbohydrate Polymers, 75*, 222–229.

[104] Singh, V.; Tripathi, D.N.; Tiwari, A. & Sanghi, R. (2005). Microwave promoted synthesis of chitosan-graft-poly(acrylonitrile). *Journal of Applied Polymer Science*, *95,* 820–825.

[105] Cai, H.; Zhang, Z.P.; Sun, P.C.; He, B.L. & Zhu, X.X. (2005). Synthesis and characterization of thermo- and pH-sensitive hydrogels based on Chitosan grafted N-isopropylacrylamide via gamma-radiation. *Radiation Physics and Chemistry, 74*, 26–30.

[106] Shrivastava, P.K. & Shrivastava, S.K. (2009). Dextran polysaccharides: successful macromolecular carrier for drug delivery. *International Journal of Phamaceutical Science, 1,* 353-368.

[107] Janciauskaite, U.; Rakutyte, V.; Miskinis, J. &; Makuska, R. (2008). Synthesis and properties of chitosan-N-dextran graft copolymers. *Reactive and Functional Polymers,* 68, 787-796.

[108] Park, Y.K.; Park, Y. H.; Shin, B.A; Choi, E.S.; Park,Y.R.; Akaike, T. & Cho, C.S. (2000). Galactosylated chitosan.graft.dextran as hepatocyte-targeting DNA carrier, *Journal of Controlled Release, 69*, 97-108.

[109] Krishnamoorthi, S.; Mal, D. & Singh, R.P. (2007). Characterization of graft copolymer based on polyacrylamide and dextran. Carbohydrate Polymers, 69, 371–377.

[110] Wang, D.; Shi, J.; Tan, J.; Jin, X.; Li, Q.; Kang, H.; Liu, R.; Jia, B. & Huang Y. (2011). Synthesis, characterization, and *in vivo* biodistribution of ^{125}I-Labeled Dex-g-PMAGGCONHTyr. *Biomacromolecules, 12,* 1851–1859.

[111] Derkaoui, S.M.; Purnama, A.L.A.; Gueguen, V.; Barbaud, C.; Avramoglou, T. & Letourneur, D. (2010). Films of dextran-graft-polybutylmethacrylate to enhance endothelialization of materials. *Acta Biomaterialia, 6,* 3506–3513.

[112] Zhang, X. & Zhuo, R. (2000). Synthesis of temperature-sensitive poly(N-isopropylacrylamide) hydrogel with improved surface property. *Journal of Colloid and Interface Science*, *223,* 311–313.

[113] Elvira, C.; Mano, J.F.; Roman, J.S. & Reis, R.L. (2002). Starch-based biodegradable hydrogels with potential biomedical applications as drug delivery systems. *Biomaterials*, *23*, 1955–1966.

[114] Bajpai, S.K. & Saxena, S. (2004). Enzymatically degradable and pH-sensitive hydrogels for colon-targeted oral drug delivery. I. Synthesis and Characterization. *Journal of Applied Polymer Science, 92*, 3630–3643.

[115] Pourjavadi, A. ; Samadi, H. & Ghasemzadeh, H.(2008). Fast-swelling superabsorbent hydrogels from poly(2-hydroxy ethyl acrylate-co-sodium acrylate) grafted on starch. *Starch/Stärke, 60*, 79–86.

[116] Behari, K.; Pandey, P.K.; Kumar, R. & Taunk, K. (2001). Graft copolymerization of acrylamide onto xanthan gum. *Carbohydrate Polymer, 46,*185-189.

[117] Mundargi, R.C.; Agnihotri, S.; Patil, S.A. & Aminabhavi T.M. (2006). Graft copolymerization of methacrylic acid onto guar gum, using potassium persulfate as an initiator. *Journal of Applied Polymer Science, 101, 618–623.*

[118] Zhang, L. M.; Zhou, J. F. & Hui, P. S. (2005). A comparative study on viscosity behavior of water-soluble chemically modified guar gum derivatives with different functional lateral groups. *Journal of the Science of Food and Agriculture*, *85*, 2638–2644.

[119] Shi, H. & Zhang, L. (2007). New grafted polysaccharides based on O-carboxymethyl-O-hydroxypropyl guar gum and N-isopropylacrylamide: Synthesis and phase transition behavior in aqueous media. *Carbohydrate polymer, 67*, 337-342.

[120] Biswal, J.; Kumar, V.; Bhardwaj, T.K.; Goel, N.K.; Dubey, K.A.; Chaudhari, C.V. & Sabbarwal, S. (2007). Radiation-induced grafting of acrylamide onto guar gum in aqueous medium: Synthesis and characterization of grafted polymer guar-g-acrylamide. *Radiation Physics and Chemistry*, *76*, 1624-1630.

[121] Chauhan, G.S.; Kumari, A. & Sharma, R. (2007). Pectin and acrylamide based hydrogels for environment management technologies: Synthesis, characterization, and metal ions sorption. *Journal of Applied Polymer Science*, *106*, 2158-2168.

[122] Fares, M.M.; Assaf, S.M. & Abul-Hajja, Y.M. (2010). Pectin grafted poly(N-vinylpyrrolidone): Optimization and *in vitro* controllable theophylline drug release. *Journal of Applied Polymer Science,117,* 1945-1954.

[123] Spizzirri, U.G.; Altimari, I.; Puoci, F.; Parisi, O.I.; Iemma, F. & Picci, N. (2010). Innovative antioxidant thermo-responsive hydrogels by radical grafting of catechin on inulin chain, *Carbohydrate Polymers*, *115*, 784-789.

[124] Shah, S.B.; Patel, C.P. & Trivedi, H.C. (2003). Ceric-induced grafting of ethyl-acrylate onto sodium alginate. *Die. Angewandte. Makromolekulare Chemie*, *214,* 75-89.

[125] Yin, Y.; Ji, X.; Dong, H.; Hui, Ying, Y. & Zheng, H. (2008). Study of the swelling dynamics with overshooting effect of hydrogels based on sodium alginate-g-acrylic acid. *Carbohydrate Polymers, 71*, 682-689.

[126] Pourjavadi, A.; Amini-Fazl, M.S. & Hosseinzadeh, H. (2005). Partially hydrolyzed crosslinked alginate-*graft*-polymethacrylamide as a novel biopolymer-based superabsorbent hydrogel having pH-responsive properties, *Macromolecular Research*, *13*, 45-53.

[127] Işıklan, N.; Kurşun, A. & İnal, M. (2010). Graft copolymerization of itaconic acid onto sodium alginate using benzoyl peroxide. *Carbohydrate Polymers, 79,* 665-672.
[128] Pourjavadi, A.; Salimi, H. & Kurdtabar, M. (2007). Hydrolyzed collagen-based hydrogel with salt and pH-responsiveness properties. *Journal of Applied Polymer Science, 106,* 2371-2379.
[129] Marandi, G.B.; Mahdavinia, G.R. & Ghafary, S. (2010). Collagen-g-poly(sodium acrylate-co-acrylamide)/sodium montmorillonite superabsorbent nanocomposites: synthesis and swelling behavior. *Journal of Polymer Research*, DOI: 10.1007/s10965-010-9554-6.
[130] Curcio, M.; Spizzirri, U.G.; Iemma, F.; Puoci, F.; Cirillo, G.; Parisi, O.I. & Picci N. (2010). Grafted thermo-responsive gelatin microspheres as delivery systems in triggered drug release. *European Journal of Pharmaceutics and Biopharmaceutics, 76,* 48-55.
[131] Amany, I.R. (2010). Gelatin based pH-sensitive hydrogels for colon-specific oral drug delivery: Synthesis, characterization, and *in vitro* release study. *Journal of Applied Polymer Science, 118*, 2642-2649.
[132] Sadeghi, M. & Heidari, B. (2011). Crosslinked graft copolymer of methacrylic acid and gelatin as a novel hydrogel with pH-responsiveness properties, *Materials, 4*, 543-552.

In: Radical Polymerization: New Developments ISBN 978-1-62100-406-6
Editors: I. O. Paulauskas, L. A. Urbonas, pp. 97-148

Chapter 3

NITROXIDE-MEDIATED PHOTO CONTROLLED/LIVING RADICAL POLYMERIZATION

Eri Yoshida
Department of Environmental and Life Sciences,
Toyohashi University of Technology; 1-1 Hibarigaoka,
Tempaku-cho, Toyohashi, Aichi 441-8580, Japan.

1. INTRODUCTION

Controlled/living radical polymerizations have made great progress in the past two decades based on their advantages over ionic polymerizations as a simple procedure without severe conditions and widely applicable monomers [1]. Examples of the controlled radical polymerizations include the iniferter polymerization [2], the atom transfer radical polymerization (ATRP) using transition metal complexes, such as Cu [3], Ni [4], Co [5], Fe [6], Ru [7], Rh [8], Pd [9], and Re [10], the reversible addition fragmentation chain transfer polymerization (RAFT) [11, 12], the iodide-transfer polymerization [13-15], and the nitroxide-mediated polymerization (NMP) [16, 17]. The primary significance of these controlled/living radical polymerizations lies in the fact that the polymerizations can produce precisely designed architectures. A vast number of precisely designed architectures have been prepared by these polymerizations. It is no exaggeration to say that the controlled/living radical polymerization techniques have now been necessary for the molecular design and creation of new polymer materials.

The NMP has disadvantages over the ATRP and RAFT regarding the limited number of monomers and the severe conditions of a high polymerization temperature over 120°C [18-22], although the scope of the applicable monomers has been extended by improving the structure of the nitroxide [23-25] using the alkoxyamine adducts [26, 27] and utilizing additives [28, 29]. However, the NMP also has advantages in using nonmetallic catalysts, thus creating a great variety of architectures through designing the nitroxide catalysts, and producing uncolored or less-colored polymers. While the nitroxides with the 2,2,5-trimethyl-4-phenyl-3-azahexane-3-oxyl skeleton provide highly controlled molecular weight distributions [26], 2,2,6,6-tetramethylpiperidine-1-oxyl (TEMPO) also has been often used as the mediator for the NMP, because TEMPO is easily available with a low cost and can be converted into a variety of derivatives that have functional groups [30, 31] and are supported on polymers [32-36]. The recent findings that the TEMPO-mediated NMP is induced not only by heat, but also by photo irradiation involves the significance of the TEMPO mediator [37-49].

The photopolymerization has significant advantages over the thermal polymerization in applications in local places and environmentally clean processes. Hence, considerable attention has been paid to the controlled photoradical polymerization to create macromolecules with well-defined structures. For the purpose of establishing the controlled photoradical polymerization to well-control the molecular weight, new photoinitiators have been prepared; the dithiocarbamate derivatives [50, 51], trithiocarbonate [52, 53], dithiodiethanol [54], xanthate [55], and a benzophenone derivative [56]. The potential of photopolymerization has also been explored for the thermal ATRP and RAFT polymerizations using photosensitive initiators [57-60] and a catalyst containing dithiocarbamate [61]. More recently, it was found that the TEMPO-mediated photo controlled/living radical polymerization provided comparatively narrow molecular weight distribution for methyl methacrylate (MMA) [37, 38, 42, 44]. This paper describes the TEMPO-mediated photo controlled/living radical polymerization and the molecular design through this polymerization.

2. Molecular Weight Control of Polymers [37, 38]

It has been reported that the thermal NMP cannot be applied to methacrylate monomers, with the exception of the fact that it inserts a small amount of the monomers into styrene and acrylate polymers through the copolymerization [19]. This is because the disproportionation termination predominates over the coupling termination during the methacrylate polymerization at high temperature over 120°C [62]. The photopolymerization eliminates such side reaction caused by the high temperature during the polymerization.

The photoradical polymerization of MMA was performed by azobis(4-methoxy-2,4-dimethylvaleronitrile) (AMDV) as the initiator and 4-methoxy-TEMPO (MTEMPO) as the mediator in the presence of bis(alkylphenyl)iodonium hexafluorophosphate (BAI) as the photo-acid generator. BAI in 50 wt.% propylene carbonate solution was supplied from Wako Pure Chemical Industries Ltd. and was used without further purification. The bulk polymerization was carried out at room temperature by irradiation with a high-pressure mercury lamp. The 10-h half-life temperature of AMDV is 30°C, while that of 2,2'-azobisisobutyronitrile (AIBN) is 65°C [63]. AMDV is expected to produce a polymer with a molecular weight distribution (MWD) narrower than AIBN by its rapid decomposition during the polymerization at room temperature. The orange-colored monomer solution turned colorless after the polymerization. These results are shown in Table 2-1. The polymerization in the absence of MTEMPO produced the PMMA with a broad MWD. BAI had a slight effect by decreasing the MWD, although the MWD was still broad. It was found that the MWD dramatically decreased in the presence of MTEMPO. Figure 2.1 shows the GPC profiles of the PMMA obtained in the presence and absence of MTEMPO. The PMMA in the presence of MTEMPO provided a unimodal GPC with a comparatively narrow MWD, whereas the

PMMA prepared in its absence showed a bimodal GPC. BAI also affected the MWD in the presence of MTEMPO. As a result of increasing of the molar ratio of BAI to MTEMPO (BAI/MTEMPO), the MWD of the resulting polymer was broadened (Figure 2-2). The increase in BAI also accelerated the polymerization. BAI should facilitate the scission between MTEMPO and the growing polymer radical. On the other hand, an increase in the molar ratio of MTEMPO to AMDV (MTEMPO/AMDV) decreased the polymerization rate, indicating that an increase in MTEMPO shifted the equilibrium between MTEMPO and the growing polymer radical toward their recombination. It was confirmed that the polymerization proceeded in accordance with a living mechanism based on the first order time-conversion plots and the plots of the PMMA molecular weight vs. the conversion. Figure 2.3 shows the first order time-conversion plots for the polymerization at MTEMPO/AMDV of 1.1 and BAI/MTEMPO of 0.5. The $ln([M]_0/[M])$ almost linearly increased over time. The number of polymer chains was constant throughout the course of the polymerization. The plots of the molecular weight of the resulting PMMA vs. the monomer conversion also linearly increased (Figure 2.4). The MWD of the PMMA retained ca. 1.6 throughout the polymerization. As can be seen in Figure 2.5, the GPC curve was shifted to the higher side of the molecular weight with an increase in the conversion, also supporting the living mechanism.

Table 2-1. The photoradical polymerization of MMA by AMDV and MTEMPO in the presence of BAI[a]

Run	MTEMPO/AMDV	BAI/MTEMPO	Time (h)	Conv. (%)	Mn[b]	Mw/Mn[b]
1	–	–	1	85	33,000	6.94
2	–	–[c]	1	87	32,300	5.86
3	1.1	–	31	40	11,600	1.47
4	1.1	0.5	3	68	16,200	1.66
5	1.1	0.75	2.5	88	19,800	2.88
6	1.1	1.5	2	95	20,600	3.16
7	1.4	0.5	9	68	13,600	1.41
8	2.0	0.5	12	68	10,600	1.55

[a] $[AMDV]_0 = 0.0454$ M.
[b] Estimated by GPC based on PMMA standards.
[c] $[BAI]_0 = 0.0249$ M.

The relationship between the molecular weight of the resulting PMMA and the initial concentration of the initiator is summarized in Table 2-2. As a result of increasing the initial concentration of AMDV ($[AMDV]_0$), the molecular weight of the PMMA decreased. The MWDs were retained in the range of 1.5–1.7. Based on a plot of the molecular weight of the PMMA vs. the reciprocal of $[AMDV]_0$, a linear correlation was obtained (Figure 2.6). It was deduced that the photoradical polymerization mediated by MTEMPO took place by a living mechanism.

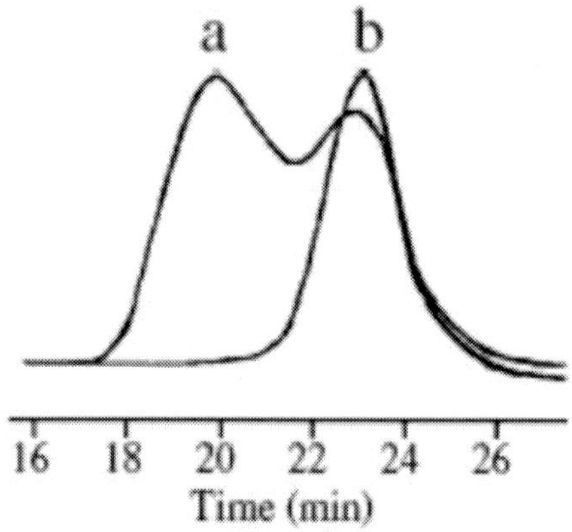

Figure 2.1. GPC profiles of the PMMA obtained by the photoradical polymerization in the absence (a, Run 1 in Table 1) and presence (b, Run 4) of MTEMPO.

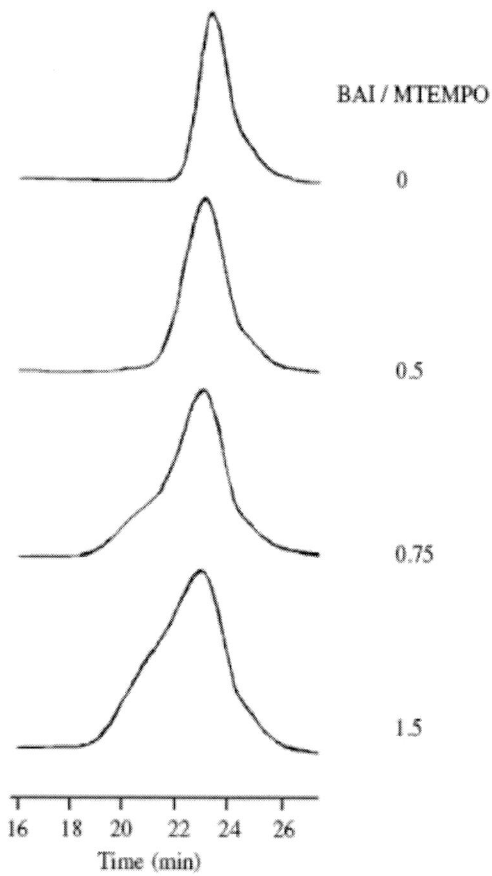

Figure 2.2. GPC profiles of the PMMA obtained by the polymerization in the presence of BAI. MTEMPO/AMDV=1.1.

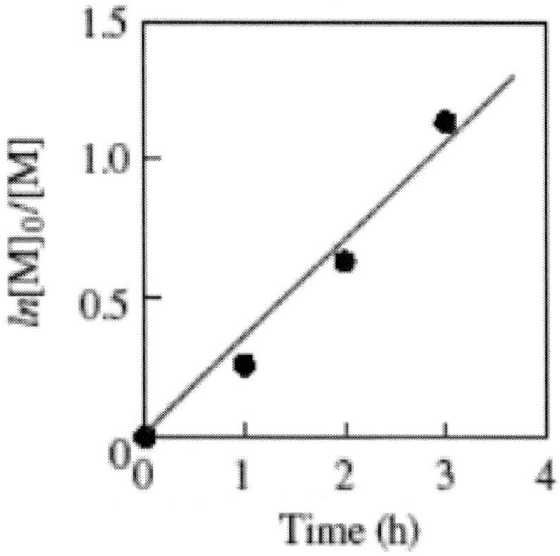

Figure 2.3. The first order timeconversion plots for the polymerization of MMA. MTEMPO/AMDV=1.1, BAI/MTEMPO=0.5.

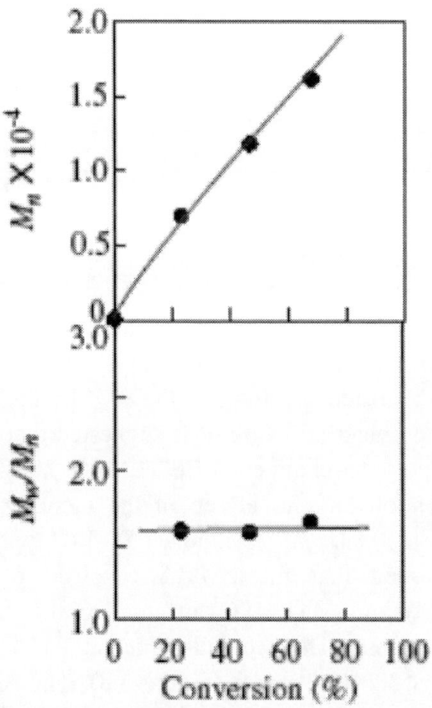

Figure 2.4. The plots of the molecular weight of the PMMA vs. the conversion. MTEMPO/AMDV=1.1, BAI/MTEMPO=0.5.

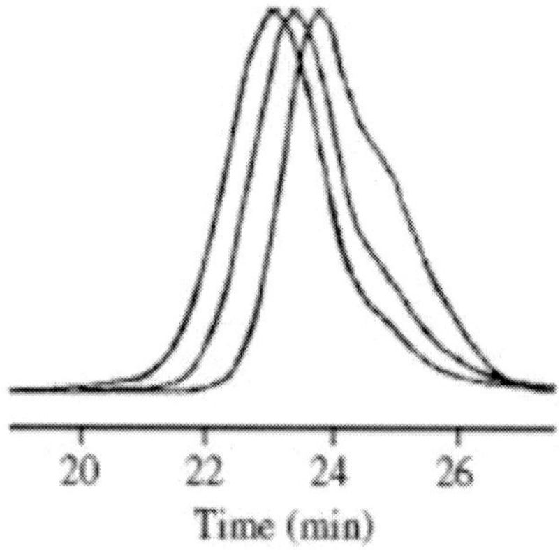

Figure 2.5. The variation in the GPC curves vs. the conversion: 23% (1 h), 47% (2 h), and 68% (3 h) from the right.

Table 2.2. The Relationship between the initial concentration of AMDV and molecular weight of the resulting PMMA[a]

$[AMDV]_0$ ($X\ 10^2$ M)	Time (h)	Conversion (%)	Mn[b]	Mw/Mn[b]
2.59	19	76	23,800	1.62
4.54	3	68	16,200	1.66
6.48	5	72	10,600	1.72
12.3	3	79	7,900	1.53

[a] MTEMPO/AMDV = 1.1, BAI/MTEMPO = 0.5.
[b] Estimated by GPC based on PMMA standards.

The ^{1}H NMR analysis clarified the structure of the resulting PMMA. Figure 2.7 shows the ^{1}H NMR spectra of PMMA (*Mn*=7,030, *Mw/Mn*=1.60), BAI, and AMDV. No observed signals originating from BAI in the spectrum of the PMMA indicates that no fragments of BAI were contained in the polymer structure. It is suggested that the BAI fragments did not combine with the growing polymer chain end or MTEMPO, but just had the interaction with MTEMPO. The polymerization had no effect on the tacticity of PMMA based on the observation of the signals at 0.84 ppm (syndiotactic), 1.01 ppm (atactic), and 1.25 ppm (isotactic) [46]. It was found that the PMMA involved the 1-cyano-1,3-dimethyl-3-methoxybutyl group (CDM) and MTEMPO attached to the chain ends. Signals of the methoxy protons originating from CDM were discerned at 3.13–3.22 ppm, while those from MTEMPO were observed at 3.29–3.39 ppm (Figure 2.8). The molar ratio of MTEMPO to CDM in the PMMA was estimated to be 0.952 based on their respective methoxy protons, indicating that the PMMA had the CDM and MTEMPO at the 1:1 molar ratio. This good agreement in the molar ratio of MTEMPO/CDM indicates that the growing radical generated by the initiation with the CDM radical was completely captured by MTEMPO (Scheme 2-1). The molecular weight of the PMMA was estimated to be *Mn*=6,500 based on the methoxy protons of CDM and the MMA units. The theoretical molecular weight based on the initial concentration of AMDV was *Mn*=2,370. This significant difference in the molecular weight between the theoretical and experimental values implies that the initiator efficiency was not unity. Accordingly, the initiator efficiency (IE) was determined by a UV analysis. Figure 2.9 shows the UV spectra of MTEMPO before and after the reaction with AMDV in methyl isobutylate. The reaction was performed by irradiation at room temperature for 1 h. An excess of MTEMPO to AMDV was used for the reaction to effectively capture the CDM radicals. The reaction was carried out in the absence of BAI in order to prevent the CDM radicals captured by MTEMPO from regenerating by BAI. In addition, no change in the absorbance of MTEMPO by the irradiation in the absence of AMDV was confirmed. The IE was determined to be 0.378 based on the absorbance at 470 nm. The theoretical molecular weight when this IE was taken into account was calculated to be *Mn*=6,270, being in a close agreement with the molecular weight estimated by ^{1}H NMR (*Mn*=6,500).

The polymerization was revealed to have a photoswitching ability based on the investigation of the polymerization in the dark. Figure 2.10 shows the variation in the monomer conversion and molecular weight of PMMA when the irradiation was interrupted for a time during the polymerization. There were negligible changes in the conversion and molecular weight during the dark reaction. However, the conversion and molecular weight increased again by the further irradiation. The progress of the polymerization can be controlled by on–off of the irradiation.

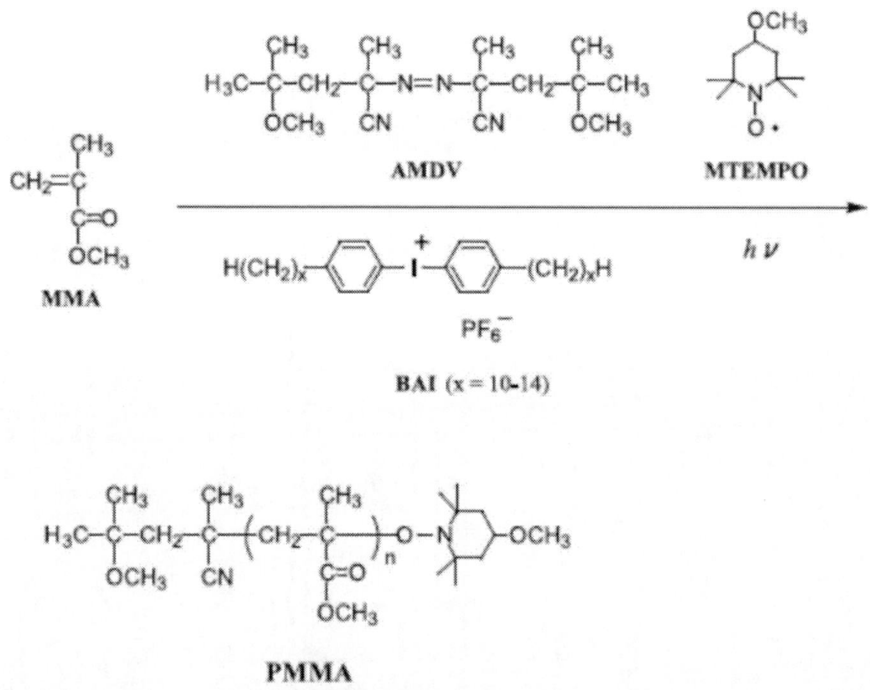

Scheme 2-1. The photoradical polymerization of MMA by AMDV and MTEMPO in the presence of BAI.

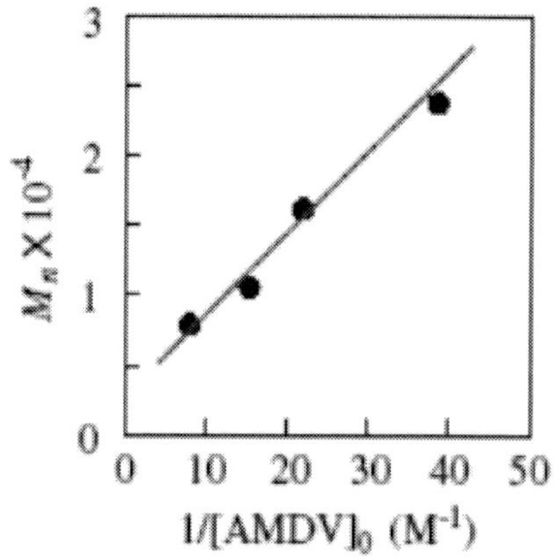

Figure 2.6. The plots of the molecular weight of the PMMA vs. the reciprocal of $[AMDV]_0$. MTEMPO/AMDV=1.1, BAI/MTEMPO=0.5.

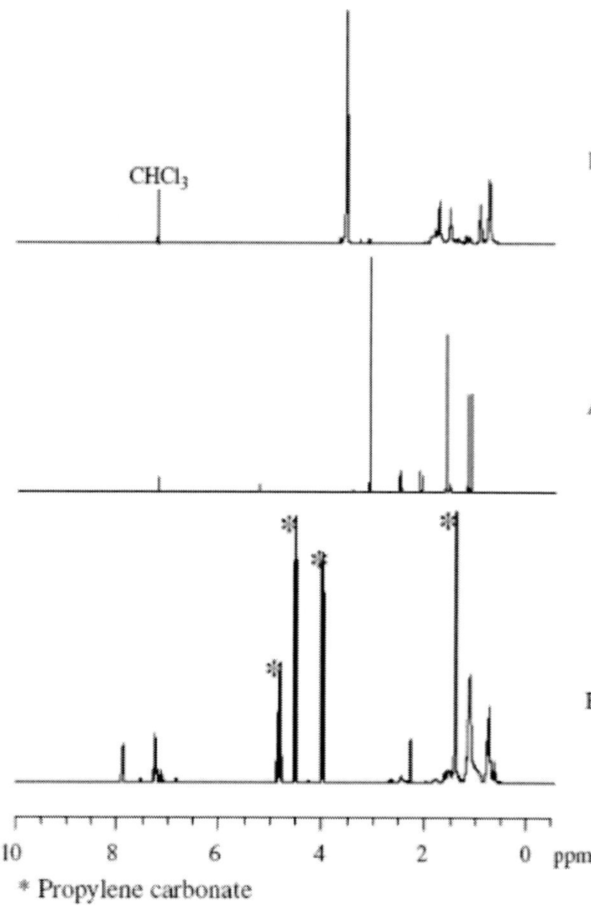

Figure 2.7. ^{1}H NMR spectra of the PMMA, BAI, and AMDV. PMMA (Mn=7,030, Mw/Mn=1.60). Solvent: $CDCl_3$.

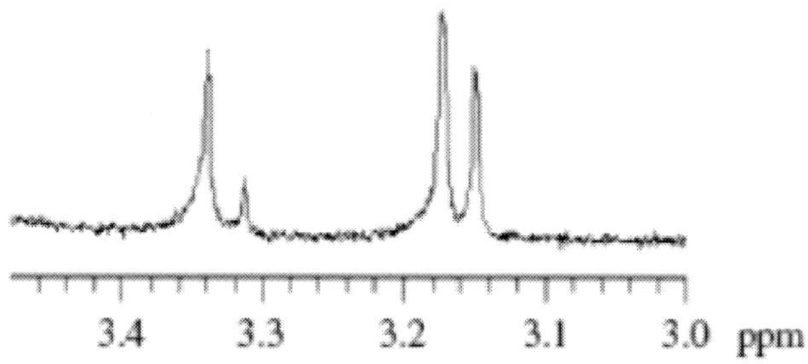

Figure 2.8. ^{1}H NMR spectra of the methoxy protons originating from CDM and MTEMPO attached to the PMMA chain ends. Solvent: $CDCl_3$.

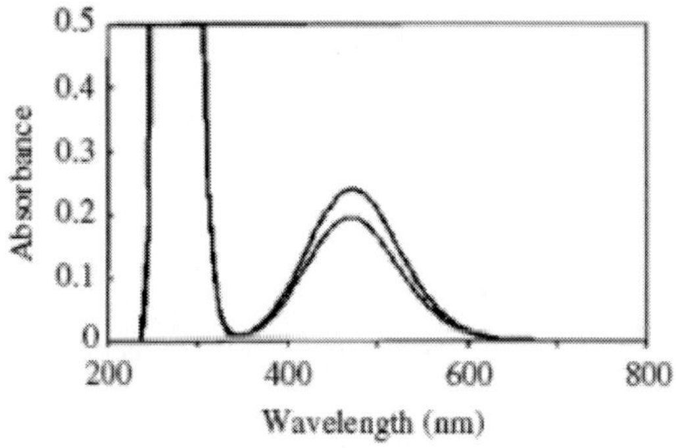

Figure 2.9. UV spectra of MTEMPO before (upper) and after (lower) the reaction with AMDV. Solvent: methyl isobutylate.

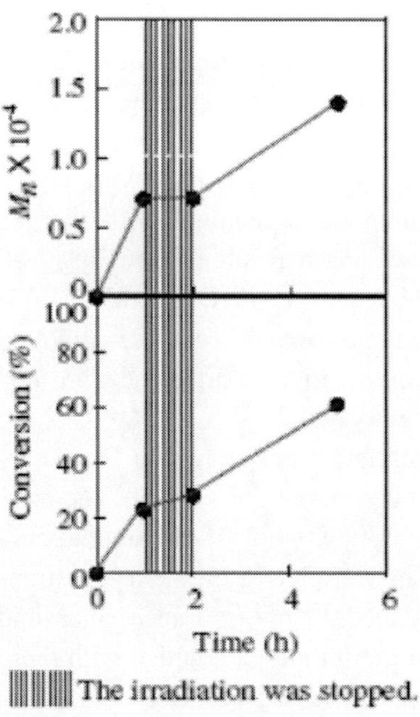

Figure 2.10. The variation in the molecular weight of PMMA and conversion when the light was turned off in the middle of the polymerization. MTEMPO/AMDV=1.1, BAI/MTEMPO=0.5.

3. Effect of Initiators [43]

The nitroxide-mediated photo controlled/living radical polymerization was determined using the AMDV initiator and the MTEMPO mediator in the presence of the BAI photo-acid generator. While the polymerization provided comparatively narrow MWD, there was a significant difference in the molecular weight between the theoretical and experimental values. This difference was based on the fact that the initiator efficiency was not unity. Furthermore, the decomposition of the azoinitiator partly produces ketenimine [64, 65] that has the potential to cause broadening of the MWD of a polymer by the polymerization, because the keteninime has a different initiation rate constant from the parent azoinitiator [66-68].

In order to clarify which characteristic of the azoinitiator dominates the MWD of the PMMA, the MTEMPO-mediated photopolymerization of MMA was performed using 8 different kinds of azoinitiators in the presence of BAI. The polymerization was carried out at the BAI/MTEMPO molar ratio of 0.52 at room temperature for 3 h by irradiation. The azoinitiators used for the polymerization are shown in Scheme 3-1. The results are summarized in Table 3-1. When 2,2'-azobisisobutyronitrile (AIBN), 2,2'-azobis(2-methylbutyronitrile) (V-59), 2,2'-azobis(2,4-dimethylvaleronitrile) (V-65), and 1,1'-azobis(cyclohexane-1-carbonitrile) (V-40) were used as the initiator, the MMA conversions were very high at a molar ratio of unity for the MTEMPO/the initiator. However, the molecular weights of the resulting PMMA were not controlled at this ratio, and the broad MWDs were observed in the GPC (Figure 3.1). Among these azoinitiators, only V-65 provided a slightly narrower MWD. The concentration of the growing polymer chain ([P]) was estimated on the basis of the conversion and the molecular weight of the resulting PMMA. Based on the [P], the IE was also determined. It was found that V-65 had a higher [P] and IE than AIBN, V-59, and V-40. Racemic-AMDV (*r*-AMDV), meso-AMDV (*m*-AMDV), dimethyl 2,2'-azobis(2-methylpropionate) (V-601), and 2,2'-azobis(*N*-butyl-2-methylpropionamide) (VAm-110) provided much narrower MWDs with moderate conversions, although VAm-110 produced only a 25% conversion. In particular, *r*-AMDV and *m*-AMDV produced the PMMA with the narrowest MWD. *r*-AMDV had a higher [P] and IE than *m*-AMDV, being in good agreement with the fact that *r*-AMDV has a higher reactivity than *m*-AMDV [69]. As a result of doubling MTEMPO to the initiator, all the initiators produced the PMMA with a MWD below 1.7. As can be seen in Figure 3.2, the resulting PMMAs displayed sharp GPC curves, although the polymers had different molecular weights. There were negligible differences in the [P] and IE values between the MTEMPO/initiator ratios of unity and 2.

The UV spectra of the initiators are shown in Figure 3.3. The UV measurements were performed using methyl isobutyrate as the solvent. The azoinitiators containing the nitriles had a λ_{max} in the 340-350-nm range, while V-601 and VAm-110 showed it at a much higher wavelength. These absorptions were based on the $n \rightarrow \pi^*$ transition. *r*-AMDV, *m*-AMDV, V-601, and VAm-110 also displayed the $n \rightarrow \sigma^*$ transition around 250 nm. The characteristics of the UV absorption spectra of the initiators, coupled with their 10-h-half-life temperatures are listed in Table 3-2. The initiators having higher ε values tended to more strictly control the molecular weight and provide a higher IE. The reason that *r*-AMDV, *m*-AMDV, V-601, and VAm-110 provided much narrower MWDs than the other initiators may include the fact that these initiators exhibit the n $\rightarrow\sigma^*$ transition in addition to the high ε. It was found that the half-lives of the initiators had little effect on the molecular weight control. The [P] and IE were also independent of the half-life temperature.

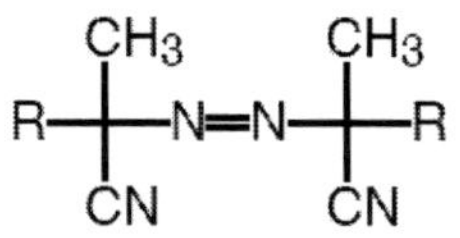

AIBN: R = CH_3

V-59: R = CH_2CH_3

V-65: R = $CH_2CH(CH_3)_2$

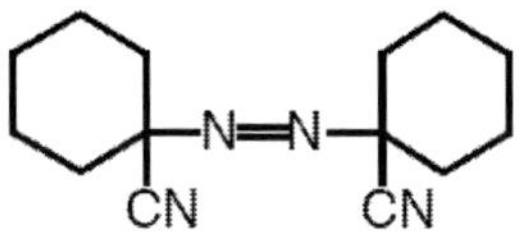

V-40

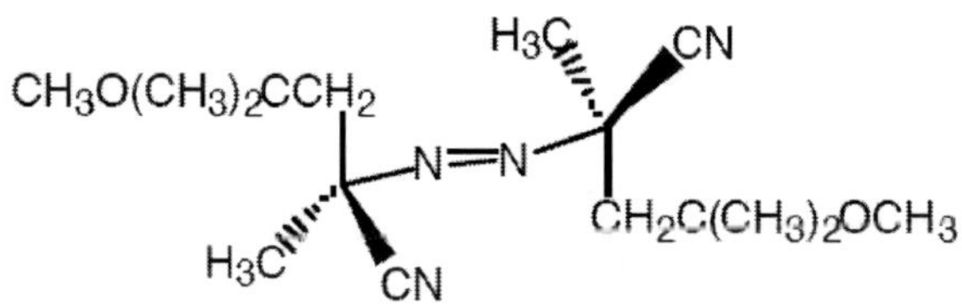

***r*-AMDV**

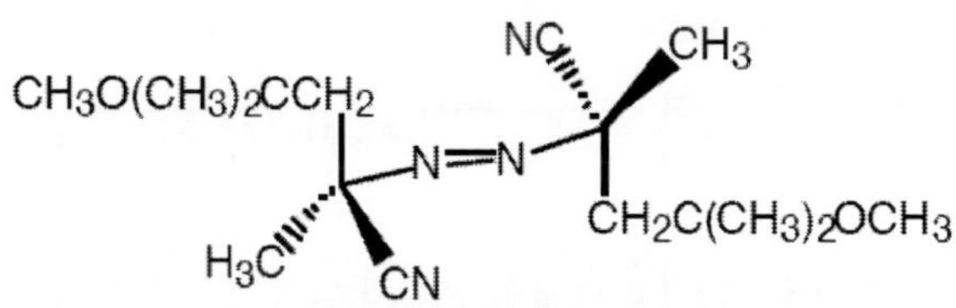

***m*-AMDV**

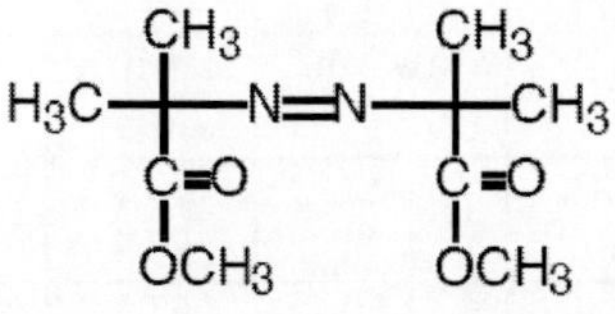

V-601

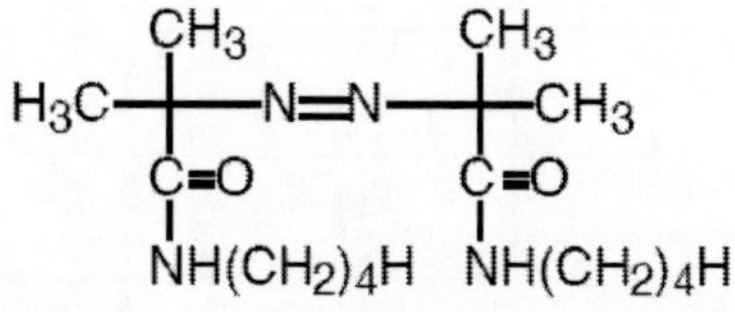

VAm-110

Scheme 3.1. Azoinitiators used for the MTEMPO-mediated photopolymerization.

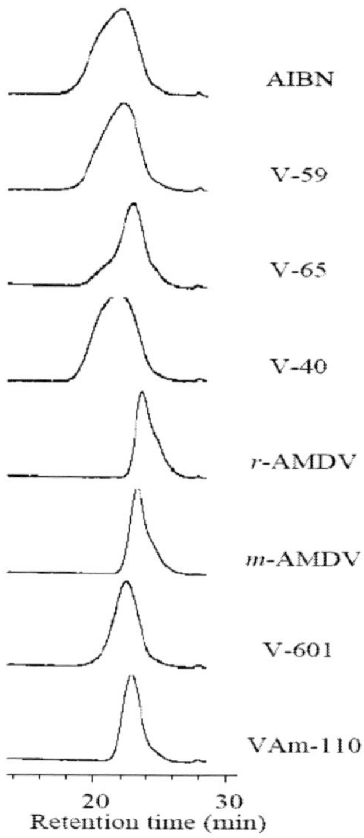

Figure 3.1. GPC profiles of the PMMAs obtained by the MTEMPO-mediated photopolymerization in the presence of BAI. MTEMPO/initiator = 1.

Table 3.1. The MTEMPO-mediated photoradical polymerization of MMA by the azoinitiators in the presence of BAI[a]

Initiators	MTEMPO/ Initiator	Conversion (%)	Mn[b]	Mw/Mn[b]	[P] (mM)	IE
AIBN	1	95	39,700	3.37	22.4	0.131
	2	36	15,300	1.68	22.0	0.129
V-59	1	97	31,600	3.37	28.8	0.173
	2	46	15,700	1.66	27.4	0.165
V-65	1	96	18,000	2.33	50.0	0.296
	2	62	9,410	1.53	61.7	0.365
V-40	1	98	38,200	3.24	24.0	0.140
	2	67	26,300	1.65	23.9	0.139
Initiators	MTEMPO/ Initiator	Conversion (%)	Mn[b]	Mw/Mn[b]	[P] (mM)	IE
r-AMDV	1	60	7,640	1.52	73.5	0.436
	2	37	3,530	1.28	98.2	0.582
m-AMDV	1	68	10,100	1.50	63.3	0.375
	2	40	4,580	1.34	81.8	0.485
V-601	1	84	26,300	1.99	29.9	0.181
	2	35	10,700	1.68	30.6	0.185
VAm-110	1	25	16,600	1.53	14.1	0.0847

[a] $[Initiator]_0$ = 84.3 mM, BAI/MTEMPO = 0.52. Irradiated for 3 h.
[b] Estimated by GPC based on the PMMA standard.

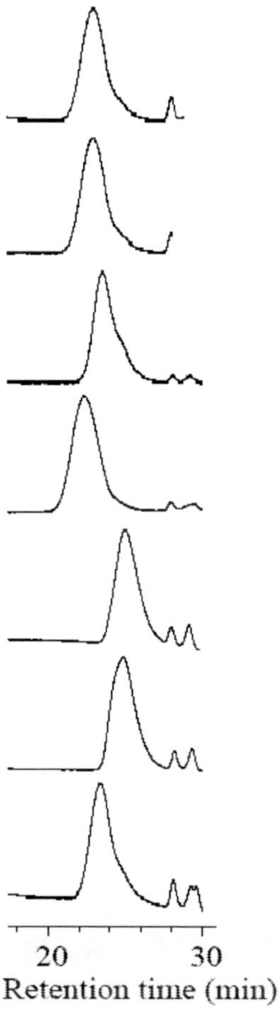

Figure 3.2. GPC profiles of the PMMAs obtained by the MTEMPO-mediated photopolymerization in the presence of BAI. MTEMPO/initiator = 2.

Table 3.2. The UV absorption and 10-h half-life temperature of the azoinitiators

Initiators	λ_{max} (nm)	Absorbance	ε	$T_{1/2}$ (10 h)[a]
AIBN	345	0.176	12.3	65
V-59	348	0.226	15.8	67
V-65	348	0.291	20.4	51
V-40	350	0.236	16.5	88
r-AMDV	348 253	0.404 0.050	28.3 3.50	30
m-AMDV	341 253	0.245 0.078	17.2 5.47	30
V-601	363 253	0.273 0.164	19.1 11.5	66
VAm-110	376 258	0.455 2.392	31.9 167.7	110[b]

[a] Ref. 63.
[b] Ref. 70.

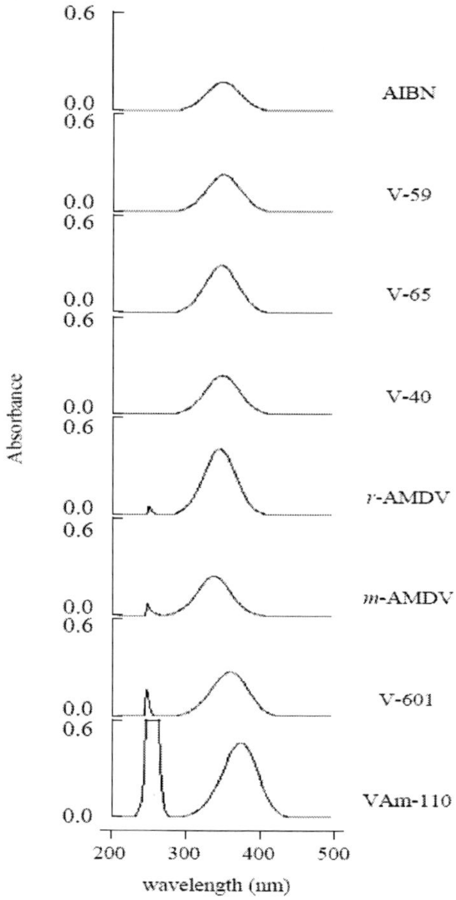

Figure 3.3. UV spectra of the initiators. Solvent: methyl isobutyrate.

4. Mechanisms [40]

The MTEMPO-mediated photo controlled/living radical polymerization efficiently proceeded in the presence of the BAI photo-acid generator in spite of the fact that the resulting polymer contained no BAI fragments in its structure. In order to clarify the mechanism of the polymerization, the polymerization was explored using 1-(cyano-1-methylethoxy)-4-methoxy-2,2,6,6-tetramethylpiperidine (CMTMP), the alkoxyamine adduct of MTEMPO and the 1-cyano-1-methylethyl radical generated from AIBN.

CMTMP was prepared by the reaction of MTEMPO and AIBN in benzene. The ^{1}H NMR spectrum of CMTMP is shown in Figure 4.1. The white crystal of CMTMP was stable at room temperature and had a melting point at 53.7°C. The UV spectra of CMTMP and AIBN, coupled with the illumination spectrum of a high-pressure mercury lamp are shown in Figure 4.2. AIBN exhibited an absorption at 345 nm asλ_{max} and overlapped with the illumination of the lamp. On the other hand, CMTMP had a slight absorption at 257 nm and almost no overlap with it, implying a lower photo activity than AIBN.

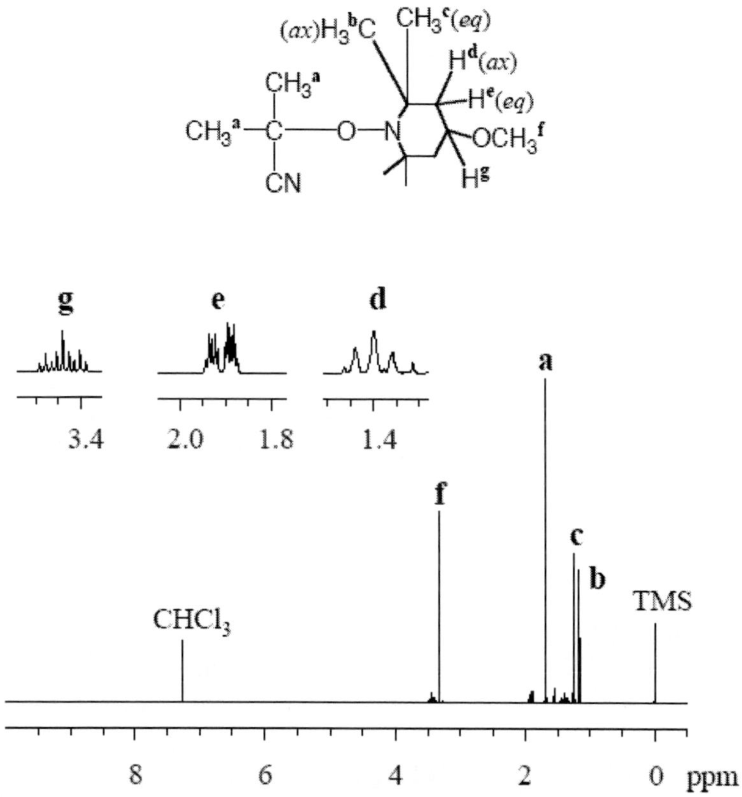

Figure 4.1. The ^{1}H NMR spectrum of CMTMP. Solvent: $CDCl_3$.

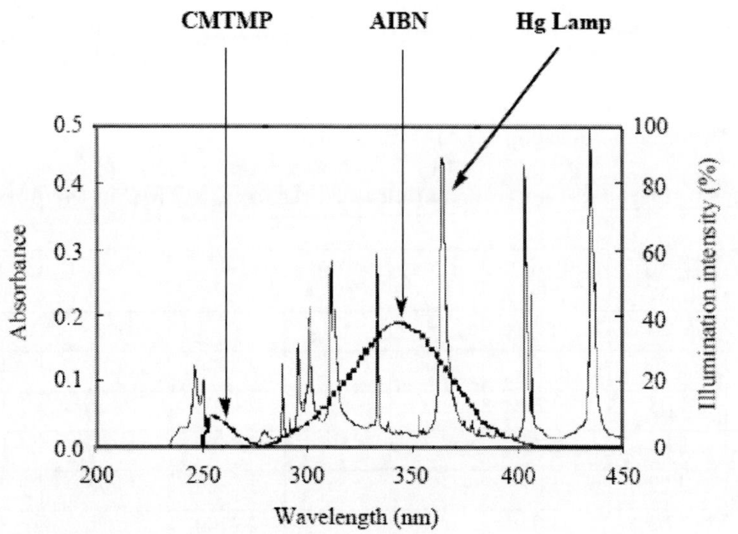

Figure 4.2. The UV spectra of CMTMP and AIBN, coupled with the illumination spectrum of a high-pressure mercury lamp. Solvent: methyl isobutyrate.

The photopolymerization of MMA was performed by CMTMP as the initiator at room temperature. The results are shown in Table 4-1. The polymerization only slightly occurred using CMTMP itself, however, it smoothly proceeded in the presence of BAI. The monomer conversion increased with the result of the increasing molar ratio of BAI to CMTMP (BAI/CMTMP), indicating that the polymerization was accelerated by BAI. On the other hand, the molecular weight of the resulting polymer decreased with the increasing BAI/CMTMP ratio, suggesting that the BAI/CMTMP ratio affected the initiator efficiency. The initiator efficiency (I.E.) of CMTMP was determined on the basis of the concentration of the growing polymer chain ([P]) calculated by the monomer conversion and the molecular weight of the resulting polymer. The I.E. increased with the increase in the BAI/CMTMP ratio, although it was not quantitative even at 1.0 of BAI/CMTMP. This nonquantitative I.E. should be caused by the cage effect and the deactivation of CMTMP by the disproportionation into the hydroxylamine and methyl methacrylonitrile (Scheme 4-1). In fact, the formation of a significant amount of the hydroxylamine lowered the yield of CMTMP in the reaction of MTEMPO and AIBN. The molecular weight distributions of the resulting polymers were somewhat broadened as compared to those of the polymers obtained by the photopolymerization by AIBN and 4-methoxy-TEMPO in the presence of BAI [37, 71]. This broad molecular weight distribution should be due to the slow initiation by CMTMP having a low photo activity.

The polymerization was carried out for the different initial concentrations of CMTMP ($[CMTMP]_0$) at a constant BAI/CMTMP ratio, resulting in the fact that the molecular weight of the polymer decreased with an increase in $[CMTMP]_0$. The plots of the molecular weight vs. the reciprocal of $[CMTMP]_0$ provided a linear correlation, implying the living nature of the polymerization (Figure 4.3).

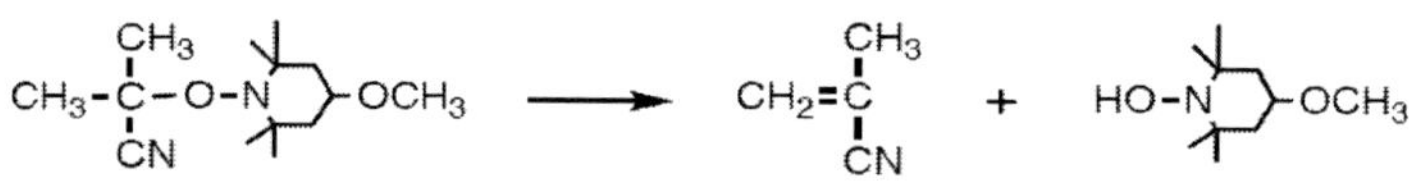

CMTMP

Scheme 4.1. The disporportionation of CMTMP.

Table 4.1. Photo-radical polymerization of MMA by CMTMP in the presence of BAI

$[CMTMP]_0$ (mM)	$[BAI]_0$ (mM)	BAI/ CMTMP	Time (h)	Conv. (%)	Mn[a]	Mw/Mn[a]	[P] (mM)	I.E.
47.2	0	0	24	2	–	–	–	–
47.2	5.24	0.11	5	32	27,300	1.68	11.0	0.23
47.2	11.8	0.25	5	51	29,300	1.66	13.7	0.29
47.2	24.9	0.53	5	50	22,600	1.69	17.4	0.37
47.2	47.2	1.0	5	55	16,800	1.78	30.6	0.65
47.2	47.2	1.0	10	94	25,700	2.30	34.2	0.73
31.5	31.5	1.0	11	82	33,400	1.88	23.0	0.73
94.4	94.4	1.0	3.5	92	17,300	2.67	49.8	0.53

[a] Estimated by GPC based on the PMMA.

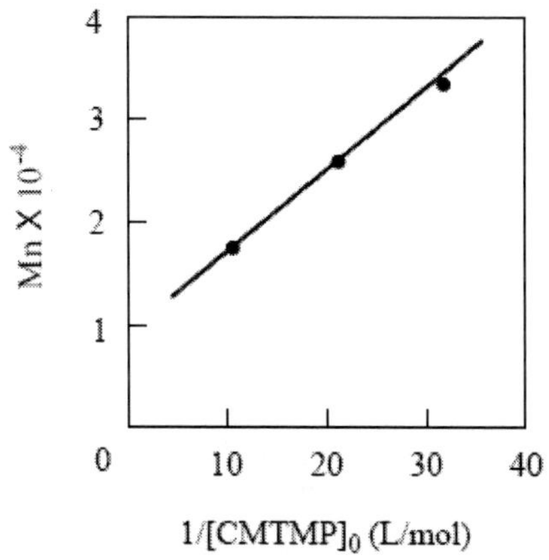

Figure 4.3. The plots of the molecular weight of the resulting polymer vs. the reciprocal of $[CMTMP]_0$. BAI/CMTMP = 1.0.

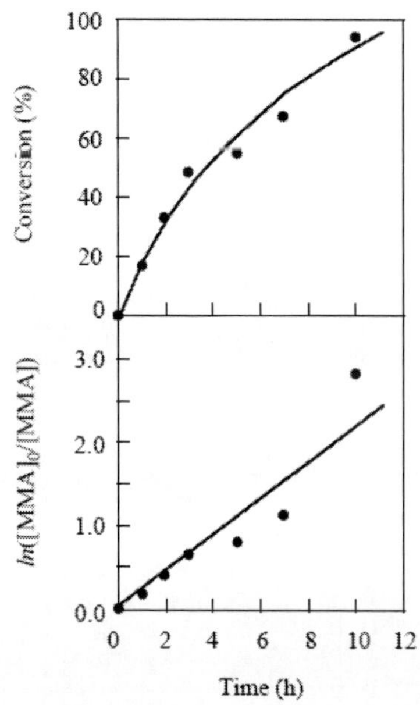

Figure 4.4. The first order time-conversion plots for the polymerization of MMA.$[CMTMP]_0 = 47.2$ mM, BAI/CMTMP = 1.0.

The correlations of the time-conversion and conversion-molecular weight were explored in order to determine the living mechanism of the polymerization. Figure 4.4 shows the first order time conversion plots for the polymerization. The $ln([MMA]_0/[MMA])$ almost linearly increased over time, although the plots somewhat varied over the 50% conversion. The number of polymer chains was almost constant throughout the course of the polymerization.

The GPC analysis of the resulting polymers also supported the living mechanism of the polymerization. Figure 4.5 shows the GPC profiles of the polymers produced for each conversion. The curves were shifted to the higher molecular weight side with an increase in the conversion. The plots of the molecular weight of the polymer vs. the conversion are shown in Figure 4.6. The molecular weight linearly increased with an increase in the conversion, however, the line did not pass through the origin and was almost parallel to the theoretical line. This phenomenon can be accounted for by the fact that a small amount of a polymer with a lower molecular weight was produced during the very early stage before the system reached the steady state of the living polymerization.

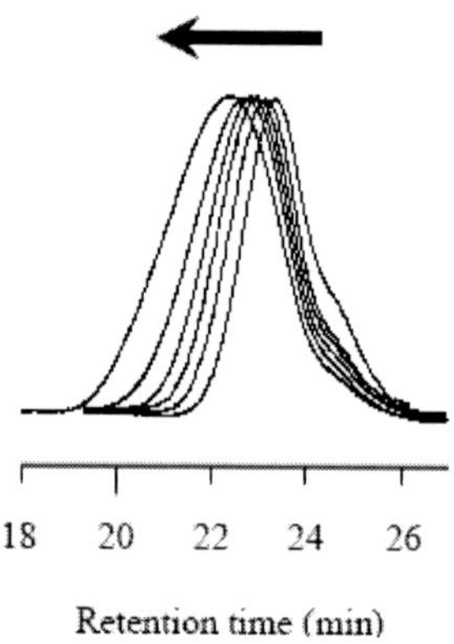

Figure 4.5. The GPC profiles of the resulting polymer obtained at each conversion: 17% (1 h), 33% (2 h), 48% (3 h), 55% (5 h), 67% (7 h), and 94% (10 h) from the right.

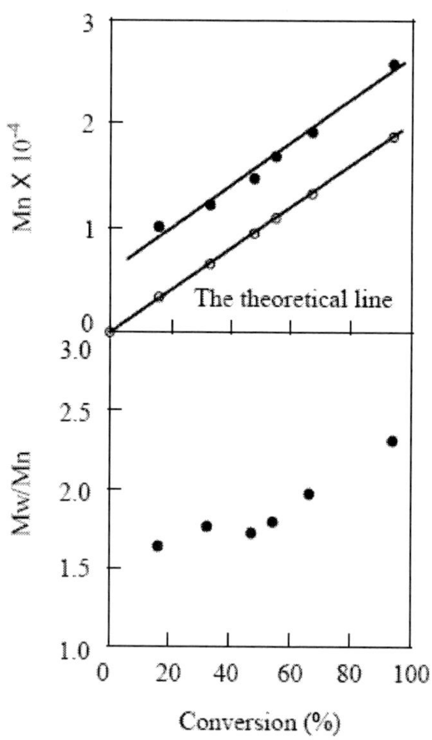

Figure 4.6. The plots of the molecular weight vs. the conversion. $[CMTMP]_0 = 47.2$ mM, BAI/CMTMP = 1.0.

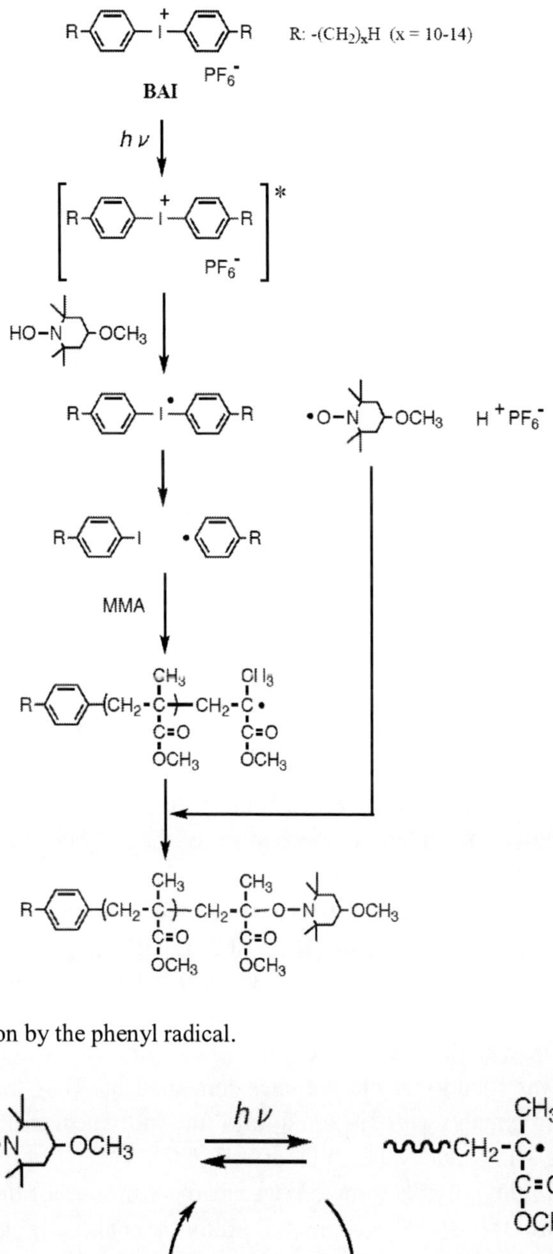

Scheme 4.2. The initiation by the phenyl radical.

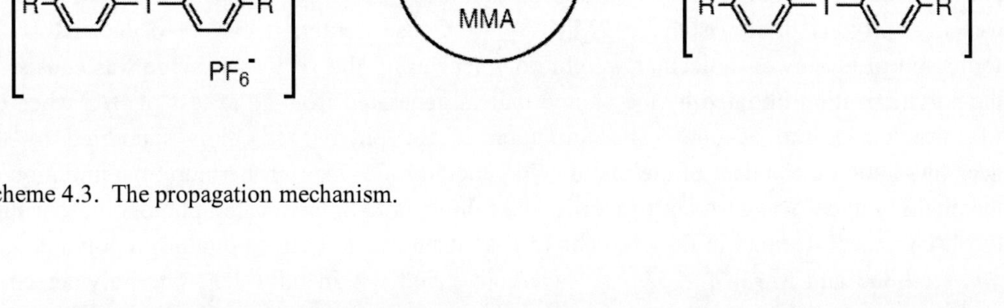

Scheme 4.3. The propagation mechanism.

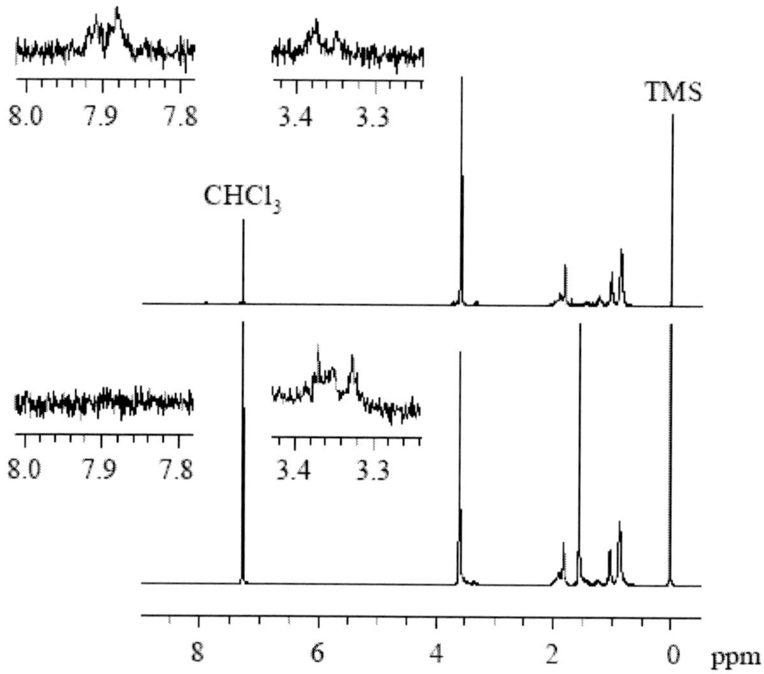

Figure 4.7. The ^{1}H NMR spectra of the polymers obtained by the polymerizations using CMTMP (upper, *Mn* = 10,195 and *Mw/Mn* = 1.63) and AIBN (lower, *Mn* = 9,140 and *Mw/Mn* = 1.57). Solvent: $CDCl_3$.

The ^{1}H NMR analysis revealed the mechanism of the polymer formation during the very early stage of the polymerization. Figure 4.7 shows the ^{1}H NMR spectrum of the polymer obtained by the polymerization for 1 h. The molecular weight and the molecular weight distribution were *Mn* = 10,195 and *Mw/Mn* = 1.63 by GPC based on the PMMA standards. The molecular weight estimated by ^{1}H NMR was *Mn* = 10,166 based on the signal intensity for the methyl protons of the MMA unit at 3.60 ppm and the methoxy protons of CMTMP at 3.33 ppm. There was a negligible difference in the molecular weight estimated by the ^{1}H NMR and GPC. It was found that the polymer contained the BAI fragment in its structure, whereas no BAI fragments were inserted into the polymer structure obtained by the polymerization using AIBN and MTEMPO instead of CMTMP (Figure 4.7). The no insertion of BAI fragments into the polymer structure was also confirmed for the polymerization using AMDV and MTEMPO [38]. The aromatic proton signals originating from BAI were discerned at 7.21-7.40 and 7.85-7.92 ppm. The signals of the alkyl groups attached to the phenyl were also observed at 1.15-1.35 ppm, although the alkyl signals overlapped with the α-methyl protons of the isotactic PMMA [72]. Consequently, it can be deduced that the formation of the lower molecular weight polymer during the very early stage was caused by the polymerization initiated by the phenyl radical generated from an excess of BAI when the I.E. was taken into account. The formation of the phenyl radical is supported by the decomposition mechanism of the photo-acid generator [73-75]. Furthermore, the initiation by the phenyl radical was also confirmed by the following experiment: the photopolymerization of MMA was performed in the presence of BAI without CMTMP to produce a polymer with *Mn* = 62,800 and *Mw/Mn* = 3.77 at 84% conversion within only 2 h. The polymer chain formed by the phenyl radical should participate in the living polymerization controlled by MTEMPO by obtaining the counter radical from the hydroxylamine (Scheme 4-2). This is because the GPC showed that the curve on the right side from the peak top was also shifted to the higher molecular weight side with the increasing conversion. The propagation should proceed by repeating the dissociation and recombination of the C-ON bond, as well as the thermal polymerization, since no BAI fragments were contained in the polymer structure obtained by the polymerization using AIBN and MTEMPO. However, during the photopolymerization, electron transfer is expected to occur between MTEMPO and BAI in the excited state during the dissociation and recombination when it is taken into consideration that almost no polymerization proceeded in the absence of BAI and that the excited diphenyliodonium salt receives an electron [73-75]. Consequently, the mechanism of the polymerization was thus proposed (Scheme 4-3).

5. Photo-Acid Generators

5.1. Triarylsulfonium Salts [42, 44]

Triarylsulfonium salts are known as efficient photoinitiators for cationic polymerization as well as diaryliodonium salts. While the triarylsulfonium salts show a lower photosensitivity than the diaryliodonium salts [73, 76], the sulfonium salts have still attracted more attention because of their extraordinary thermal stability [77] and easier preparation [78]. In order to precisely control of the molecular weight of polymers, the MTEMPO-mediated photopolymerization was performed using the triarylsulfonium salt instead of the diaryliodonium salt.

The polymerization of MMA was performed at room temperature using (4-*tert*-butylphenyl)diphenylsulfonium triflate (^{t}BuS) as the photo-acid generator by the *r*-AMDV initiator and the MTEMPO mediator. The results are shown in Table 5-1-1. While the polymerization slowly occurred in the absence of ^{t}BuS, the polymerization smoothly proceeded in its presence. The ^{t}BuS/MTEMPO molar ratio increased [P], resulting in a decrease in the molecular weight of the resulting polymer. There was a tendency that the polymerization was accelerated as the ^{t}BuS/MTEMPO ratio increased. The polymerization rate was also dependent on MTEMPO because the large excess of MTEMPO to *r*-AMDV retarded the polymerization even in the presence of ^{t}BuS. The MWD was significantly broadened at a MTEMPO/*r*-AMDV less than 1, suggesting less control of the polymerization. Furthermore, no polymerization occurred in the absence of *r*-AMDV, although ^{t}BuS solely initiated the polymerization to produce a polymer with a very broad MWD. It is implied that MTEMPO controlled the polymerization.

The livingness of the polymerization was explored at the MTEMPO/*r*-AMDV of 1.06 and ^{t}BuS/MTEMPO of 0.487. The first order time-conversion plots for the polymerization are shown in Figure 5.1.1. The $ln([MMA]_0/[MMA]_t)$ almost linearly increased with time, suggesting that the number of polymer chains was constant throughout the course of the polymerization. Figure 5.1.2 shows the plots of the molecular weight, MWD, and [P] vs. the conversion. The molecular weight linearly increased with the increasing conversion and was in a good agreement with the theoretical values. The MWD remained constant at around 1.45, although it slightly increased with the increase in the conversion. The [P] also remained at ca. 45 mM, indicating the constant number of polymer chains. The GPC profiles of the resulting polymer for each conversion are shown in Figure 5.1.3. The curves were shifted to the higher molecular weight side with the increasing conversion. Based on the GPC analysis, coupled with the linear correlations of the first order time-conversion and conversion-molecular weight plots, it can be deduced that the polymerization proceeded in accordance with a living mechanism. It was deduced that ^{t}BuS more effectively controlled the molecular weight than BAI.

Table 5.1.1. Photo-radical polymerization of MMA by MTEMPO and ^{t}BuS[a]

$[MTEMPO]_0$ (mM)	$[^tBuS]_0$ (mM)	MTEMPO/ AMDV	^{t}BuS/ MTEMPO	Time (h)	Conv. (%)	Mn[b]	Mw/Mn[b]	[P] (mM)
48.3	–	1.06	–	31	40	11,600	1.47	32.3
48.3	12.8	1.06	0.265	9	63	13,200	1.46	44.8
48.3	23.5	1.06	0.487	9	63	11,100	1.46	53.0
48.3	34.1	1.06	0.706	9	64	10,200	1.44	59.0
48.3	47.0	1.06	0.973	6	77	15,700	1.58	46.0
48.3	70.4	1.06	1.46	6	71	11,100	1.45	60.1
80.5	66.2	1.77	0.822	23	74	9,350	1.51	74.1
91.3	91.8	2.01	1.01	23	82	7,790	1.70	98.5
91.3	47.0	2.01	0.515	14	26	4,670	1.36	52.1
$[MTEMPO]_0$ (mM)	$[^tBuS]_0$ (mM)	MTEMPO/ AMDV	^{t}BuS/ MTEMPO	Time (h)	Conv. (%)	Mn[b]	Mw/Mn[b]	[P] (mM)
21.5	10.7	0.473	0.498	5	71	18,300	1.76	36.4
48.3	25.6	–[c]	0.530	6	0	–	–	–
–	94.4	–[c]	–	3	73	417,000	19.6	1.64

[a] $[AMDV]_0$ = 45.4 mM.
[b] Estimated by GPC based on the PMMA standard.
[c] Without AMDV.

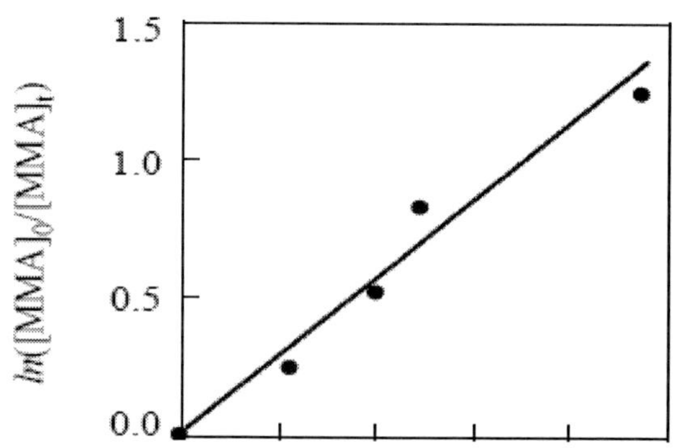

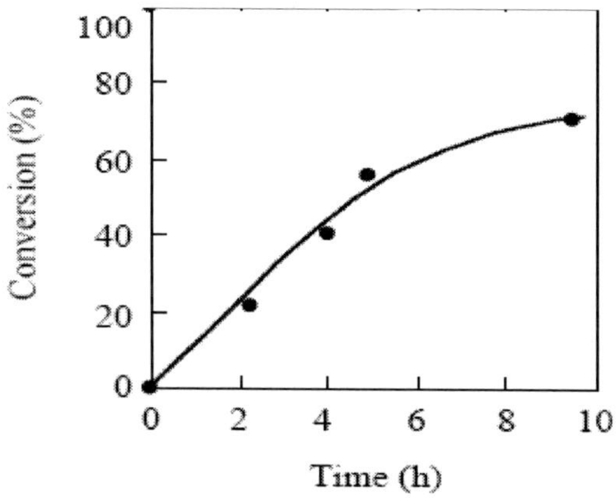

Figure 5.1.1. The first order time-conversion plots for the polymerization of MMA. MTEMPO/*r*-AMDV = 1.06, *ᵗ*BuS/MTEMPO = 0.487.

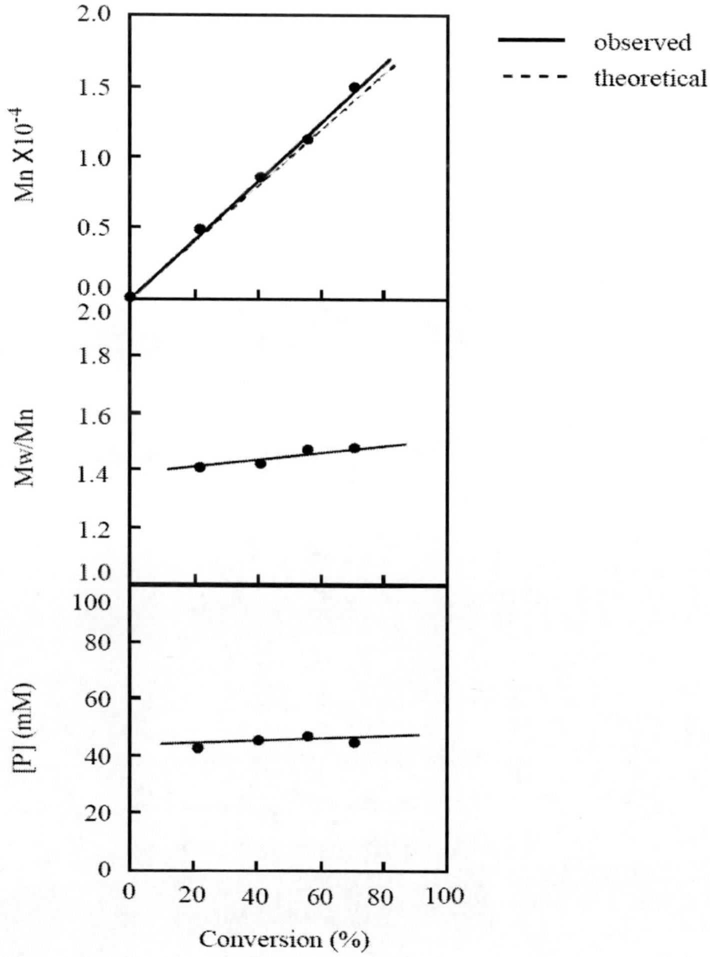

Figure 5.1.2. The plots of the molecular weight, molecular weight distribution, and [P] vs. the conversion. MTEMPO/*r*-AMDV = 1.06, *ᵗ*BuS/MTEMPO = 0.487.

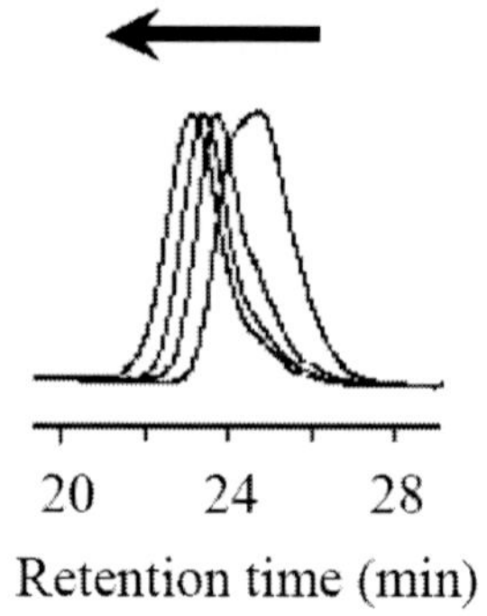

Figure 5.1.3. GPC profiles of the PMMA obtained at each conversion: 22% (2.25 h), 41% (4.00 h), 56% (4.92 h), and 71% (9.50 h) from the right.

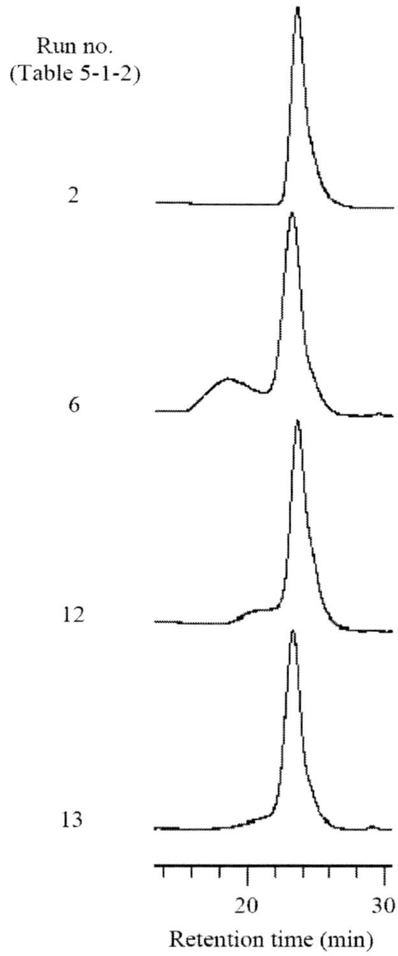

Figure 5.1.4. The GPC profiles of the polymers obtained by the MTEMPO-mediated photopolymerization initiated by *r*-AMDV.

Table 5.1.2. The MTEMPO-mediated photopolymerization of MMA in the presence of the sulfonium salts[a]

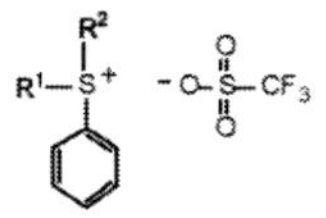

Run no.	R^1	R^2	Solubility[b]	Conversion (%)	Mn[c]	Mw/Mn[c]	[P] (mM)	IE
1	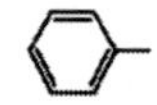	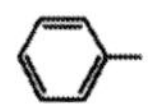	I	63	12,200	1.42	48.2	0.531
2	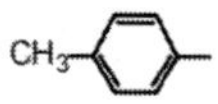	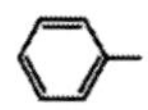	S	52	10,200	1.39	47.8	0.527
3	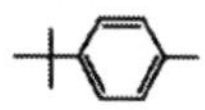	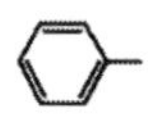	S	58	11,000	1.43	49.5	0.545
4	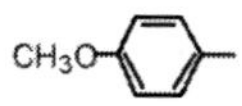	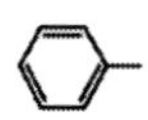	I	62	11,000	1.50	52.9	0.582
5	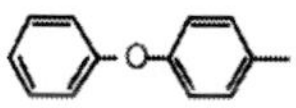	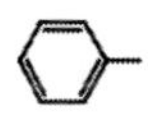	S	46	9,810	1.45	43.9	0.483
6	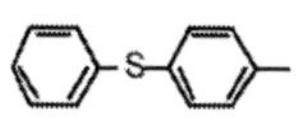	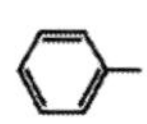	S	80	427,000[d] 14,000	3.64 1.70	40.1	0.442
7	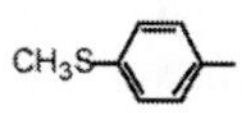	CH_3–	S	66	12,200	1.45	51.0	0.562
8	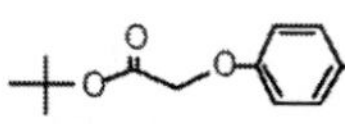	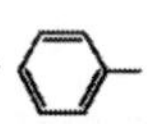	S	58	12,600	1.44	43.0	0.474
9	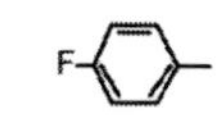	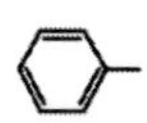	S	64	11,400	1.45	52.3	0.577
10	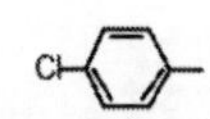	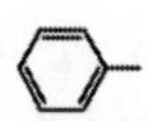	I	53	11,000	1.44	45.0	0.496
11	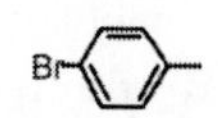	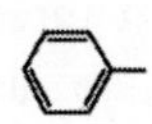	S	70	12,100	1.49	54.0	0.594
12	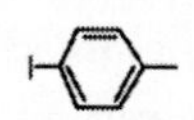	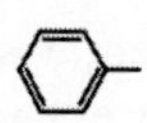	I	62	10,100	3.14	57.5	0.634
13	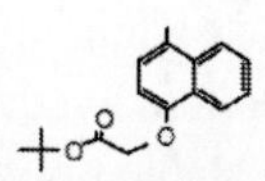	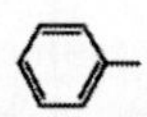	I	73	13,200	2.40	51.9	0.571

[a] MTEMPO/*r*-AMDV = 1.1, Sulfonium salt/MTEMPO = 0.5. Irradiated for 6 h.
[b] Solubility of the sulfonium salts. S: Soluble, I: Insoluble.
[c] Estimated by GPC based on the PMMA standard.
[d] Bimodal GPC. The area ratio: *Mn*(427,000/14,000) = 0.26/0.74.

In order to explore the influence of the structure of the photo-acid generator on the polymerization, the polymerization was performed using 13 different sulfonium salts. These results are shown in Table 5-1-2. The solubility of the sulfonium salt in MMA had no influence on the MWD, because the heterogeneous solution including the insoluble sulfonium salt became homogeneous within 1 h of the irradiation. There was a negligible difference in [P] and IE, independent of the substituents. The IEs were around 0.5. The substituents produced a significant difference in the MWD. The sulfonium salts with the alkyl, methoxy, phenoxy, methylthio, and *tert*-butoxycarbonylmethoxy groups had no effect on the molecular weight control because the MWDs of the polymers obtained by these sulfonium salts were close to that by the triphenylsulfonium triflate (Run 1). Halogens with the exception of the iodide also had a negligible effect on it. These polymers had similar GPC profiles (Figure 5.1.4). On the other hand, the the iodide, phenylthio, and naphthyl groups on the sulfonium salts caused broad MWDs with the bimodal GPCs including a peak on the higher molecular weight side. These functional groups should have competitively participated in the electron transfer between the sulfonium salt and MTEMPO during the propagation, resulting in the formation of polymers with uncontrolled high molecular weights.

5.2. An Iron-Arene Complex [47]

Iron-arene complexes are known to act as photoinitiators for cationic polymerizations as well as diaryliodonium salts and triarylsulfonium salts. While the photolysis of the iodonium salts and sulfonium salts yields the strong acids that remain in the product and cause the undesired ether cleavage, the iron complexes produce no such acids during the photolysis [74]. A significant variety of the iron-arene complexes were prepared; e.g., (η^6-benzene)(η^5-cyclopentadienyl)Fe^{II} salts, (η^6-naphthalene)(η^5-cyclopentadienyl)Fe^{II} salts, (η^6-pyrene)(η^5-cyclopentadienyl)Fe^{II} salts, and their derivatives having different counter anions, such as BF_4^-, PF_6^-, AsF_6^-, and SbF_6^- [79, 80], and various supporting substituents, such as alkyl, alkoxy, hydroxy, mercapto, carbonyl, and halogens [81]. Among these iron-arene complexes, the (η^6-benzene)(η^5-cyclopentadienyl)Fe^{II} hexafluorophosphate (BzCpFe^{II}) has been the most widely investigated for determining the initiation mechanism of the epoxide polymerization and the dependence of the epoxide structure on the polymerization rate [74, 82, 83].

For the MTEMPO-mediated photo controlled/living radical polymerization, the diaryliodonium salts and triarylsulfonium salts did not directly engage in controlling the growing polymer radicals, but enhanced the polymerization rate by interacting with MTEMPO accompanied by the growing polymer radical. It was considered that the interaction was attributed to the electron transfer between these onium salt and MTEMPO in their excited states. This electron transfer mechanism was based on the fact that MTEMPO forms the redox systems in which MTEMPO is converted into the oxoaminium cation by the one-electron oxidation, while it is converted into the aminoxy anion by the one-electron reduction (Scheme 5-2-1) [84]. The iron-arene complexes are more stable toward the electron transfer due to the redox formation of the iron than the iodonium salts and sulfonium salts [79].

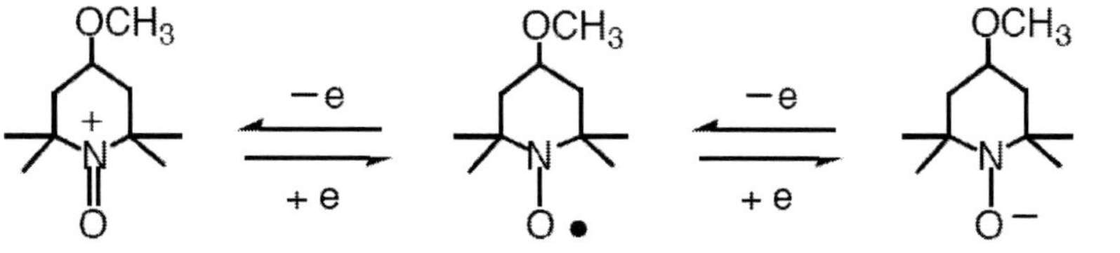

Scheme 5.2.1. The redox systems of MTEMPO.

The MTEMPO-mediated photopolymerization of MMA was performed using BzCpFe^{II} as the accelerator. BzCpFe^{II} was insoluble in MMA so that the polymerization was carried out in acetonitrile. The yellow solution turned brown after a 24-h polymerization. The results are shown in Table 5-2-1. While the conversion was below 50% at the MTEMPO/*r*-AMDV molar ratio of 2.13, it reached 87% as a result of doubling the amount of the initiator. However, the increase in the amount of the initiator also caused an increase in the molecular weight distribution of the resulting PMMA. On the other hand, an increase in the amount of BzCpFe^{II} decreased the conversion, i.e., retarded the polymerization rate. BzCpFe^{II} showed the opposite tendency against the onium salts of BAI and ^{t}BuS that enhanced the polymerization rate [37, 38, 42]. The increase in the amount of BzCpFe^{II} also decreased the molecular weight distribution due to the reduced propagation rate.

The bulk polymerization of MMA was also explored using BzCpFe^{II} based on the fact that the solubilities of the triarylsulfonium salts in the monomer were independent of the molecular weight control (Table 5-2-1). The orange solution including the yellow powder of BzCpFe^{II} turned into a brown mass. The conversion reached 73% at 3 h, however, it did not increase over this time. It is likely that the conversion reached its plateau at 3 h due to the high viscosity of the system. The bulk polymerization provided narrower MWDs than the solution polymerization. This tendency was also observed for the MMA polymerization using ^{t}BuS [42]. It was expected that MTEMPO more effectively trapped the propagating radical in the bulk polymerization, avoiding a normal termination between the propagating radicals at the very early stage, with the result that the polymers with the narrower MWD were produced. There was a tendency similar to the solution polymerization with respect to the rate retardation by BzCpFe^{II}. It is suggested that BzCpFe^{II} is different from BAI and ^{t}BuS regarding the molecular weight control mechanism. Actually, BzCpFe^{II} itself had a poor ability to initiate the polymerization and produced a polymer at only 6% conversion for the 3-h irradiation, whereas both BAI and ^{t}BuS effectively initiated the polymerization [40, 42]. However, the polymerization with BzCpFe^{II} and MTEMPO in the absence of *r*-AMDV produced a polymer with a 20% conversion for the same irradiation period. In addition, BzCpFe^{II} alone had no ability to control the molecular weight because the polymerization without MTEMPO provided a very broad MWD. It is clear that the interaction between BzCpFe^{II} and MTEMPO is involved in the molecular weight control. The interaction should be attributed to the electron transfer between them. When the oxidation state of the iron complex from Fe^{II} into Fe^{III} is taken into account, it is considered that the reduction redox system between MTEMPO and the aminoxy anion was involved in the electron transfer with BzCpFe^{II} (Scheme 5-2-2). For the interaction between MTEMPO and the onium salts of BAI and ^{t}BuS, the oxidation redox system between MTEMPO and the oxoaminium cation was involved in the electron transfer. Therefore, the difference between $BzCpFe^{II}$ and these onium salts in the molecular weight control is based on which redox system of MTEMPO is involved in the electron transfer with them.

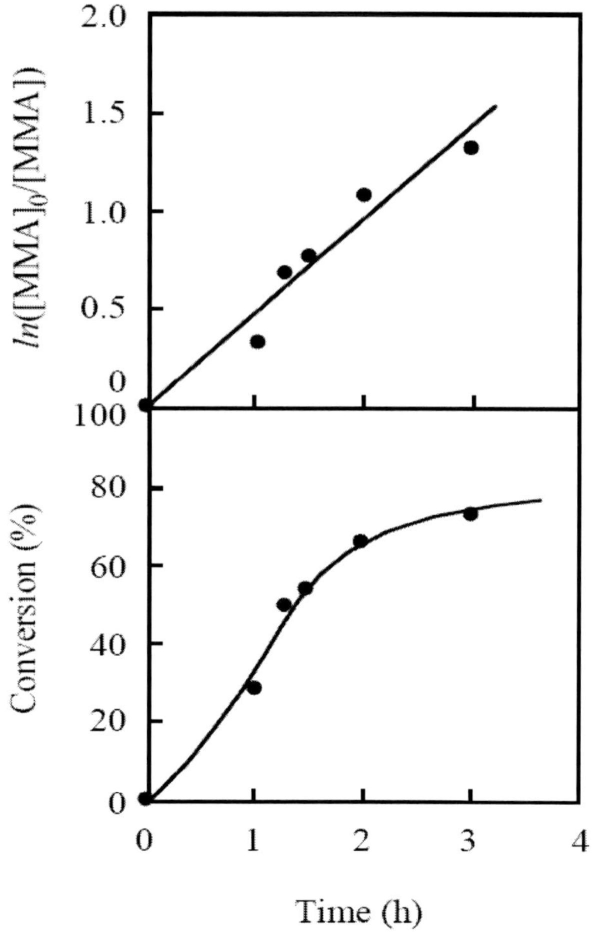

Figure 5.2.1. The time-conversion and its first order plots for the MMA polymerization. $[r\text{-AMDV}]_0$ = 45.4 X 10^{-3} M, MTEMPO/*r*-AMDV = 1.07, $BzCpFe^{II}$/MTEMPO = 1.00.

The living nature of the polymerization was investigated for the bulk polymerization at the MTEMPO/*r*-AMDV of 1.07 and $BzCpFe^{II}$/MTEMPO of 1.00. The time-conversion and its first-order plots for the polymerization are shown in Figure 5.2.1. The $ln([MMA]_0/[MMA])$ plots almost linearly increased with time, although the conversion reached its plateau around 3 h. The plots of the molecular weight and MWD vs. the conversion are shown in Figure 5.2.2. The molecular weight linearly increased with the conversion, indicating that the polymerization proceeded in accordance with a living mechanism. The MWDs were Mw/Mn = 1.4-1.5. Figure 5.2.3 shows the GPC profiles for the PMMA obtained for each conversion. The curves were shifted to the higher molecular weight side as the conversion increased, thus supporting the living mechanism of the polymerization.

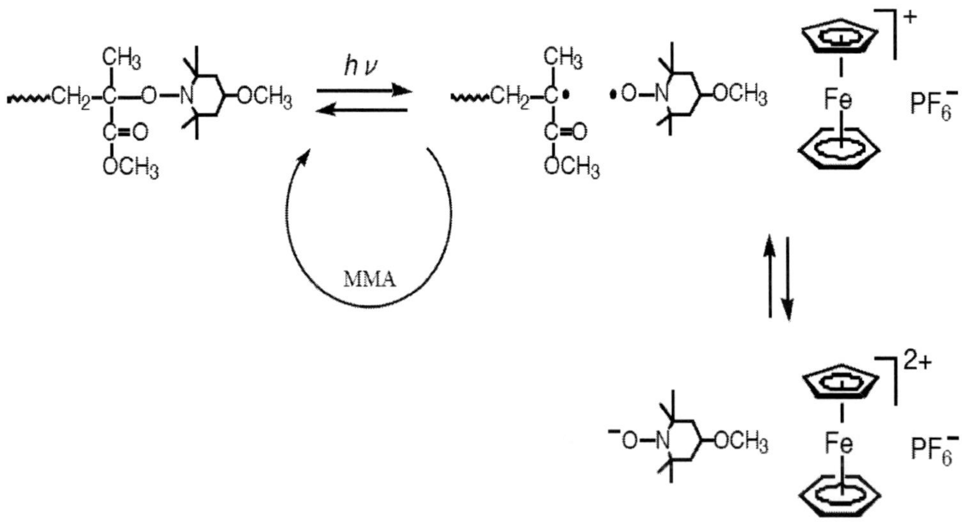

Scheme 5.2.2. The electron transfer between MTEMPO and $BzCpFe^{II}$.

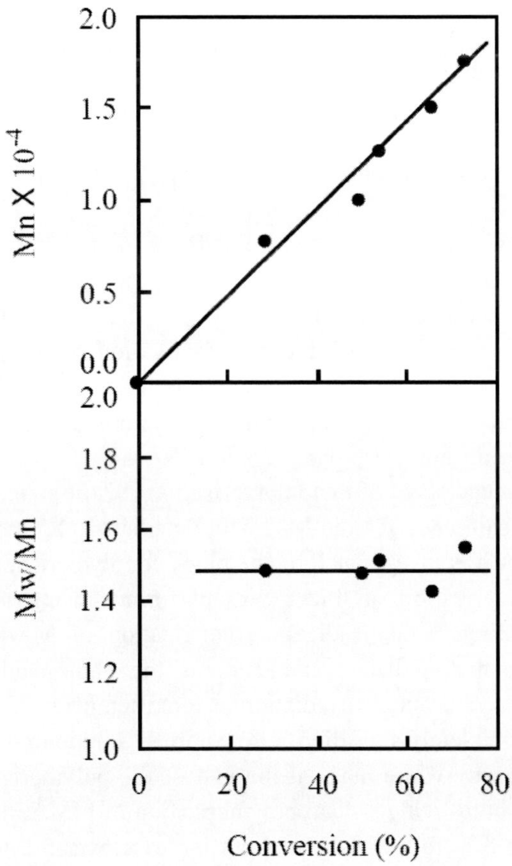

Figure 5.2.2. The plots of the molecular weight and MWD vs. the conversion for the MMA polymerization. $[r\text{-AMDV}]_0 = 45.4 \times 10^{-3}$ M, MTEMPO/*r*-AMDV = 1.07, $BzCpFe^{II}$/MTEMPO = 1.00.

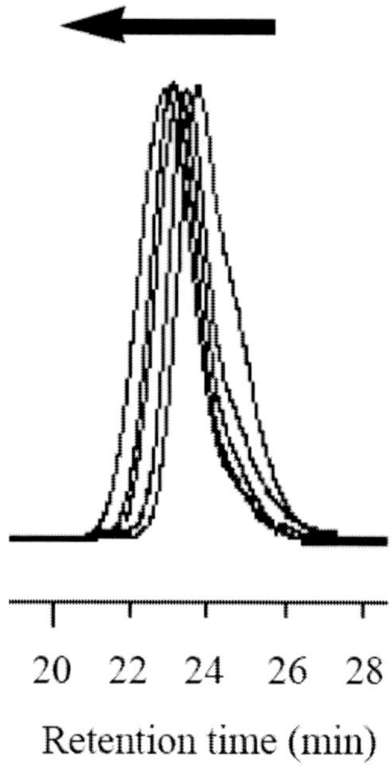

Figure 5.2.3. The GPC profiles of the PMMA obtained at each conversion: 28% (1.0 h), 50% (1.3 h), 54% (1.5 h), 66% (2.0 h), and 73% (3.0 h) from the right.

6. Stability of the Growing Polymer Chain End [45]

For the molecular design using the controlled/living radical polymerization, the stability of the growing polymer chain ends is the key to the quantitative preparation of the architectures. This is because the strict control of the architecture structures and its quantitative preparation are often hindered by deactivation of the growing polymer chain ends, even if the polymerizations that highly controls molecular weights are used [85, 86]. The thermal NMP is often deactivated through the β–elimination from the growing polymer radical by the nitroxide during the polymerization [87, 88]. In particular, this deactivation occurs more often at the end stage of the polymerization [87, 89]. Acordingly, the stability of the growing polymer chain ends for the MTEMPO-mediated photo controlled/living radical polymerization of MMA was investigated by the block copolymerization with ^{i}PMA.

The block copolymerization of ^{i}PMA was performed using the PMMA prepolymer prepared through the photo-living radical polymerization of MMA by *r*-AMDV as the initiator, MTEMPO as the mediator, and ^{t}BuS as the photo-acid generator. The block copolymerization was carried out by irradiation at room temperature and continued until the reaction mixture was completely solidified. The colorless solution containing the prepolymer and ^{i}PMA became a cloudy white mass at the end of the polymerization. For the previous results concerning the photo-living radical polymerization of MMA, the monomer conversion linearly increased up to 5 h, followed by a slight increase over 5 h to 10 h [42]. Figure 6.1 shows the GPC profiles of the prepolymer prepared by the MMA polymerization for 6.5 h and the block copolymer obtained by the ^{i}PMA polymerization using the prepolymer. The block copolymerization took 22 h until the stirrer bar stopped rotating. The GPC demonstrated that the resulting block copolymer contained the prepolymer due to the deactivation of the growing chain end. The bimodal curve consisted of the block copolymer with Mn = 116,000 and Mw/Mn = 1.70 and the prepolymer with Mn = 12,300 and Mw/Mn = 1.31 by GPC based on the PMMA standards. The area ratio of the block copolymer (BL) and the prepolymer (PR) in the GPC was BL/PR = 0.755/0.245. Based on this area ratio and the respective molecular weights, the ratio of the number of the growing polymer chains for the BL and PR was BL/PR = 0.246/0.754. It was found that ca. 75% of the prepolymer was deactivated.

The MMA polymerization was shortened to 5 h in order to prevent the deactivation of the growing chain end. The block copolymerization was completed in 10 h and produced a block copolymer with a unimodal GPC (Figure 6.2). The GPC curve of the block copolymer was shifted to the higher molecular weight side and contained no prepolymer. The molecular weight and its distribution of the block copolymer were estimated to be Mn = 43,200 and Mw/Mn = 2.14, while those of the prepolymer were Mn = 10,100 and Mw/Mn = 1.63. It was found that the deactivation of the growing chain ends was caused by a decrease in the monomer concentration and occurred during the late stage of the polymerization. The deactivation for the MMA polymerization should be caused by the β–elimination mechanism as well as that for the styrene polymerization [87, 88], because the disproportionation termination more often occurs for the MMA polymerization than the styrene polymerization due to the steric repulsion at the growing chain end [62].

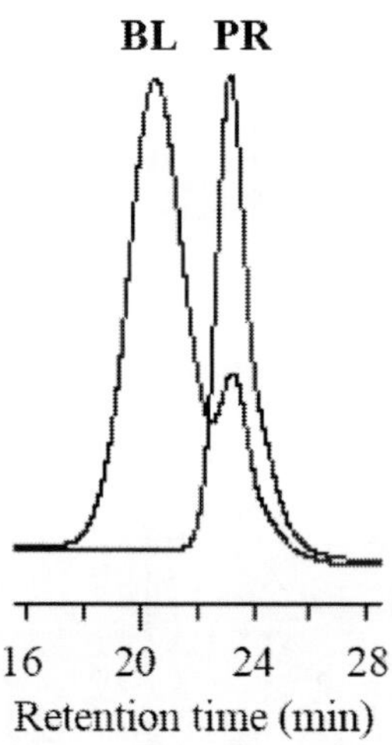

Figure 6.1. The GPC profiles of the resulting block copolymer (BL) and prepolymer (PR) prepared by the MMA polymerization for 6.5 h.

The absolute molecular weight of the PMMA-*b*-P^{i}PMA diblock copolymer obtained by the MMA polymerization for 5 h was determined by ^{1}H NMR. Figure 6.3 shows the ^{1}H NMR spectra of the block copolymer and prepolymer. The proton signals of the methyl and methine originating from the PMMA and P^{i}PMA blocks were observed at 3.63 and 4.87 ppm in the spectrum of the block copolymer. Based on their signal intensities and the molecular weight of the prepolymer, the absolute molecular weight of the P^{i}PMA block was determined to be

Mn(P^{i}PMA block) = 25,900. The total molecular weight of the block copolymer was *Mn* = 36,000.

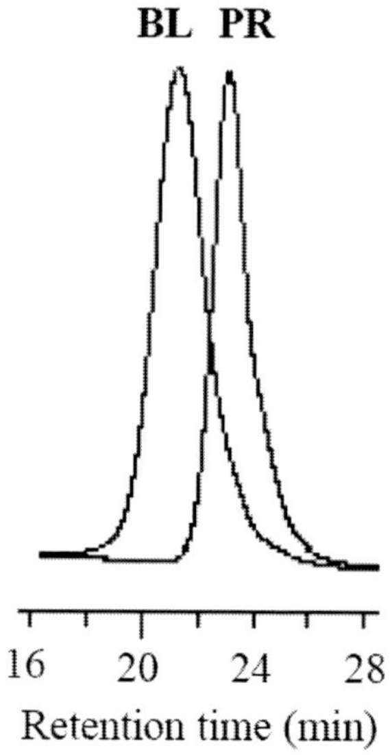

Figure 6.2. The GPC profiles of the resulting block copolymer and prepolymer prepared by the MMA polymerization for 5 h.

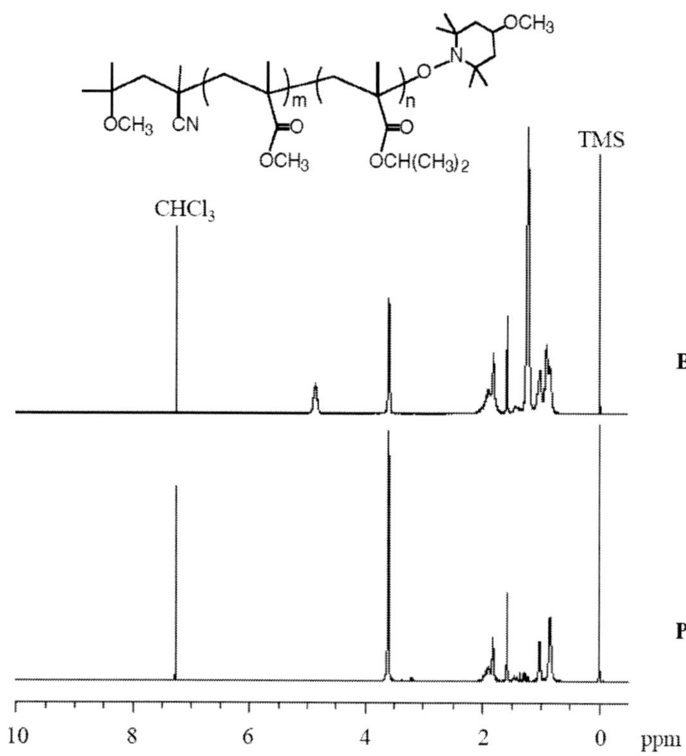

Figure 6.3. The ^{1}H NMR spectra of the block copolymer and the prepolymer obtained by the MMA polymerization for 5 h. Solvent: $CDCl_3$.

7. Solution Polymerization [46]

The deactivation of the growing polymer chain ends occurred during the MTEMPO-mediated photopolymerization, as well as the thermal NMP [45]. The growing polymer chain ends should be more stabilized in a solution than in the bulk, because their deactivation occurs at the end stage of the polymerization where most monomers are consumed [89], although the solution polymerization retards the propagation rate. In order to obtain more stable growing polymer chain ends, the solution polymerization was explored.

The photopolymerization of MMA was performed in acetonitrile at room temperature using the *r*-AMDV initiator and the MTEMPO mediator, and the ^{t}BuS photo-acid generator. The results are shown in Table 7-1. As a result of increasing the amount of MTEMPO, the rate of the polymerization decreased. However, the polymerization rate was enhanced by increasing the amount of ^{t}BuS. The solution polymerization allowed the conversion to reach 97%, whereas it was difficult for the bulk polymerization to increase the conversion over 85% [42].

The livingness of the solution polymerization was explored at the MTEMPO/*r*-AMDV of 2.13 and ^{t}BuS/MTEMPO of 1.02. The time-conversion and its first-order plots for the polymerization are shown in Figure 7.1. The $ln([MMA]_0/[MMA])$ plots showed different lines before and after 4 h, indicating that the radical concentrations were different for these two lines. The polymerization should be under the non-steady-state below 4 h and reached the steady-state over this time. The non-steady-state was not observed in the bulk polymerization [42]. It is suggested that MTEMPO could not effectively trap the propagating radical due to its low concentration in the solution polymerization, causing occurrence of a normal termination between the propagating radicals under the non-steady-state. However, the proportion of the polymers produced by the normal termination is low based on its conversion (< 18%). Figure 7.2 shows the plots of the molecular weight vs. the conversion for the solution polymerization. It was observed that oligomers with several thousand molecular weights were formed under the non-steady-state below the 18% conversion. The molecular weight linearly increased with the conversion under the steady-state, indicating that the polymerization proceeded by the living mechanism. The MWD was around 1.8, somewhat higher than that for the bulk polymerization. This broader MWD should be due to the presence of the oligomers produced under the non-steady-state. The GPC curves obtained under the steady-state were shifted to the higher molecular weight side as the conversion increased, thus supporting the living mechanism of the polymerization.

Table 7.1. The photopolymerization of MMA in acetonitrile

MTEMPO/ *r*-AMDV	^{t}BuS/ MTEMPO	Conversion (%)	Mn[a]	Mw/Mn[a]
1.06	0.53	80	7,930	1.72
2.13	0.53	67	10,100	1.67
2.13	1.02	97	11,700	1.78

Irradiation time: 24 h. $[MMA]_0 = 9.35$ M, $[MTEMPO]_0 = 0.0483$ M.

[a] Estimated by GPC based on PMMA standards.

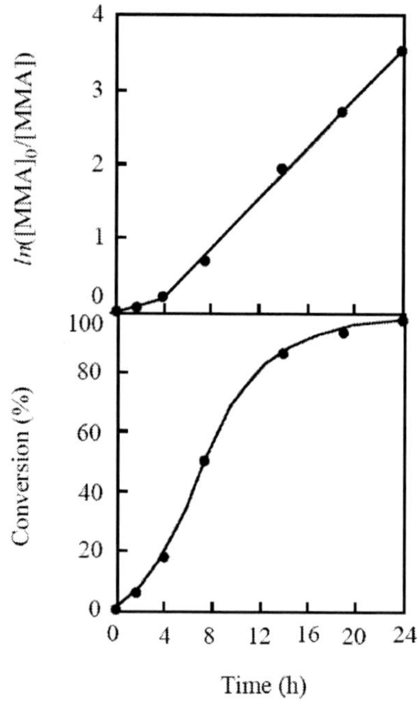

Figure 7.1. The time-conversion and its first order plots for the polymerization of MMA in acetonitrile. $[MMA]_0$ = 9.35 M, MTEMPO/*r*-AMDV = 2.13, ^{t}BuS/MTEMPO = 1.02.

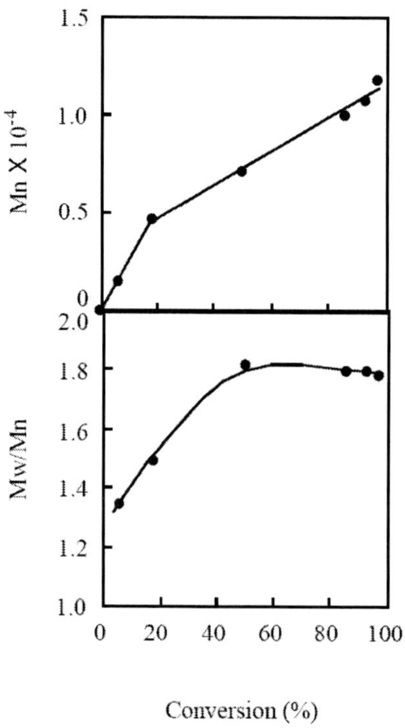

Figure 7.2. The plots of the molecular weight and MWD vs. the conversion for the polymerization of MMA in acetonitrile. $[MMA]_0$ = 9.35 M, MTEMPO/*r*-AMDV = 2.13, ^{t}BuS/MTEMPO = 1.02.

The stability of the growing polymer chain ends for the solution polymerization was also investigated through the block copolymerization. The PMMA prepolymer was prepared by a 14-h polymerization. The block copolymerization was performed for 19 h using ^{i}PMA as the second monomer. The solution containing the PMMA prepolymer and ^{i}PMA in acetonitrile was converted into a white suspension after 19 h. Figure 7.3 shows the GPC profiles of the prepolymer and the block copolymer. The GPC curve of the block copolymer was shifted to the higher molecular weight side, indicating that the prepolymer efficiently initiated the polymerization of ^{i}PMA. The growing polymer chain ends were stabilized even at a high conversion over 85% for the solution polymerization, whereas their stability was retained only for a conversion below 60% for the bulk polymerization. The molecular weights and molecular weight distributions were estimated to be Mn = 8,650 and Mw/Mn = 1.75 for the prepolymer and Mn = 30,400 and Mw/Mn = 2.61 for the block copolymer by the GPC based on the PMMA standards. The molecular weight of the P^{i}PMA blocks was determined to be Mn = 30,300 by ^{1}H NMR based on the signal intensity of the methyl at 3.63 ppm for the PMMA blocks and the methine at 4.87 ppm for the P^{i}PMA blocks [45] and on the molecular weight by GPC for the prepolymer. The total molecular weight of the block copolymer was Mn = 39,000 by ^{1}H NMR.

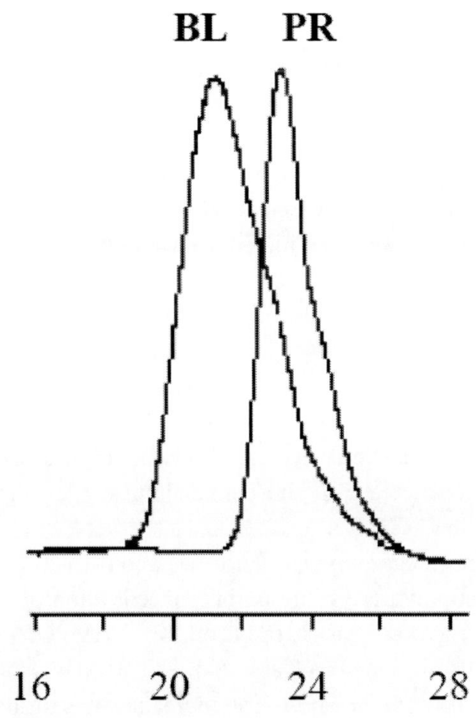

Figure 7.3. The GPC profiles of the resulting block copolymer (BL) and prepolymer (PR).

8. Molecular Design

8.1. Block Copolymerization Using a TEMPO Macromediator [39]

The MTEMPO-mediated photo controlled/living radical polymerization has the potential to create a variety of architectures through designing TEMPO derivatives. The synthesis of a diblock copolymer using a macromediator of TEMPO is described.

Two types of TEMPO-terminated poly(tetrahydrofuran) (PTHF-TEMPO) with different molecular weights were used for the polymerization: L-TEMPO with Mn = 1,720 and Mw/Mn = 1.91 and H-TEMPO with Mn = 12,700 and Mw/Mn = 1.25 [90]. The photopolymerization of MMA was performed by AMDV, BAI, and PTHF-TEMPO as the macromediator. Figure 8.1.1 shows the ^{1}H NMR spectra of the resulting block copolymer with Mn = 11,300 and Mw/Mn = 1.83 and the L-TEMPO of which the terminal TEMPO was reduced by phenylhydrazine. In the spectrum of the copolymer, a signal of the methylene attached to the TEMPO were discerned at 3.36 ppm, confirming that the PMMA and PTHF blocks were connected through the TEMPO (Scheme 8.1.1). This methylene signal was observed on the basis of the restriction of the piperidine ring conformation by the combination of the terminal TEMPO with PMMA. Signals of the methoxy protons (MeO_{CDM}) for the CDM were also observed at 3.16-3.23 ppm. This group generated from AMDV was bonded to the chain end of the PMMA block. Signals attributed to the PTHF block were observed at 1.63 and 3.43 ppm for the methylene groups and at 3.34 ppm for the terminal methoxy (MeO_{PTHF}). The molar ratio of MeO_{CDM}/MeO_{PTHF} was estimated to be 1.00/0.83. The molecular weight of the PMMA block was Mn = 9,730 based on MeO_{CDM} and the signal at 3.61 ppm of the methoxy protons of the MMA units. The total molecular weight of the PMMA-*b*-PTHF copolymer was determined to be Mn = 10,900, since the molecular weight of the L-TEMPO was Mn = 1,170 by ^{1}H NMR [90]. This molecular weight of the copolymer was in good agreement with that estimated by GPC. The copolymers produced from the H-TEMPO provided spectra similar to those from the L-TEMPO.

The signals of the THF units and both chain ends relatively decreased as the conversion increased (Figure 8.1.2). The plots of the molar ratio of the MMA unit to the THF unit versus the conversion using their signal intensity are shown in Figure 8.1.3. The MMA/THF ratio linearly increased with the increase in the conversion, suggesting the living nature of the polymerization.

The living mechanism of the polymerization was confirmed on the basis of plots of the molecular weight of the copolymer vs. the initial concentration of AMDV ($[AMDV]_0$) using L-TEMPO. The polymerization was carried out at TEMPO/AMDV and BAI/TEMPO constant ratios. The results are listed in Table 8-1-1. The MWD of the copolymers was the same as or less than that of the L-TEMPO. The molecular weight decreased with an increase in the AMDV concentration, although there was a small difference in the conversion. The plots of the molecular weight vs. the reciprocal of the initial concentration of AMDV are shown in Figure 8.1.4. The linear correlation proved the living mechanism of the polymerization.

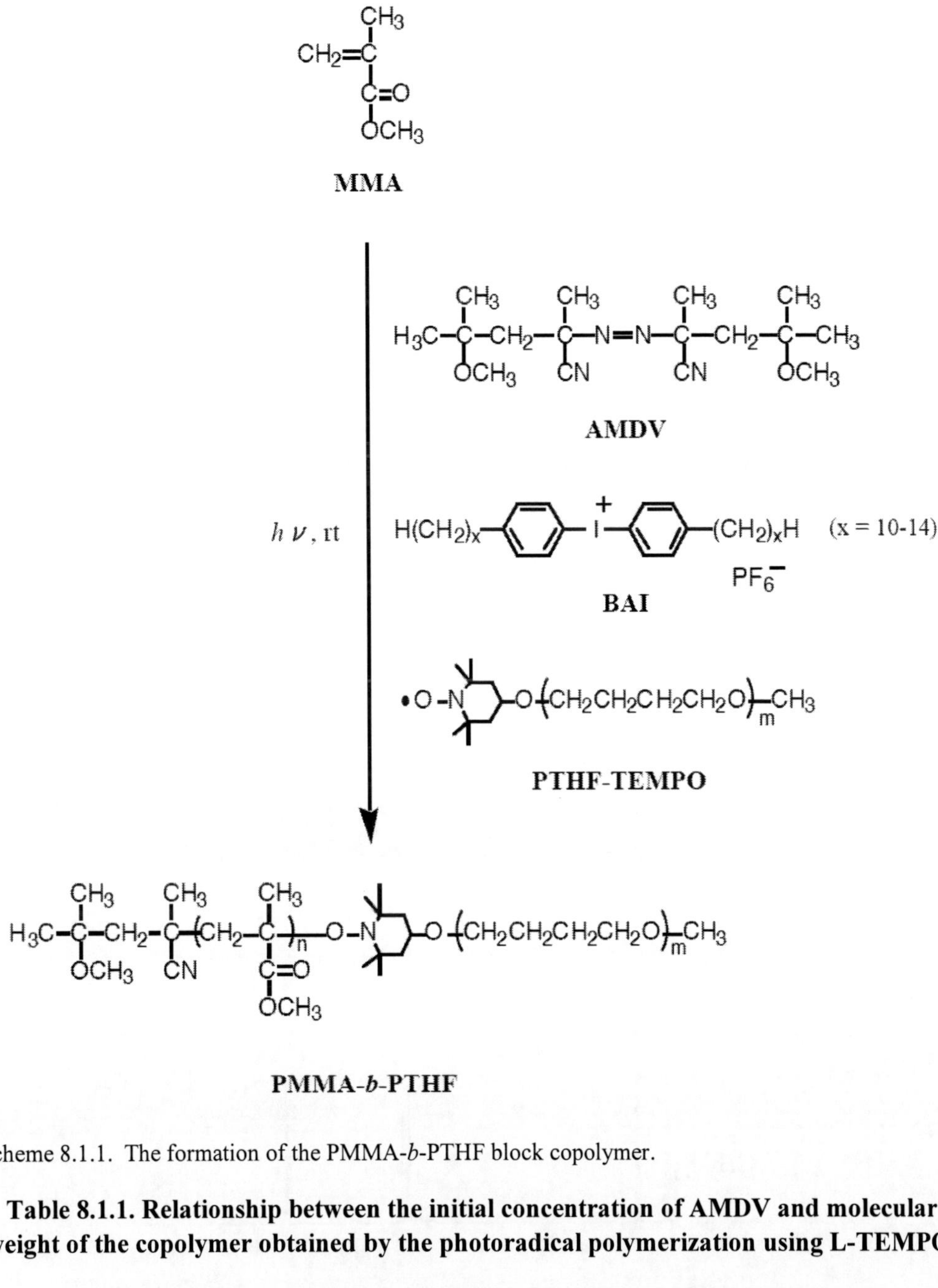

Scheme 8.1.1. The formation of the PMMA-*b*-PTHF block copolymer.

Table 8.1.1. Relationship between the initial concentration of AMDV and molecular weight of the copolymer obtained by the photoradical polymerization using L-TEMPO[a]

$[AMDV]_0$ ($X10^2$ mol/L)	Time (h)	Conversion (%)	Mn[b]	Mw/Mn[b]
1.62	12	51	29,700	1.55
2.27	8	62	25,400	1.62
2.92	8	56	22,700	1.65
4.54	3	69	19,500	1.88

[a] L-TEMPO/AMDV = 1.1 and BAI/L-TEMPO = 0.5. [b] Estimated by GPC based on PMMA standards.

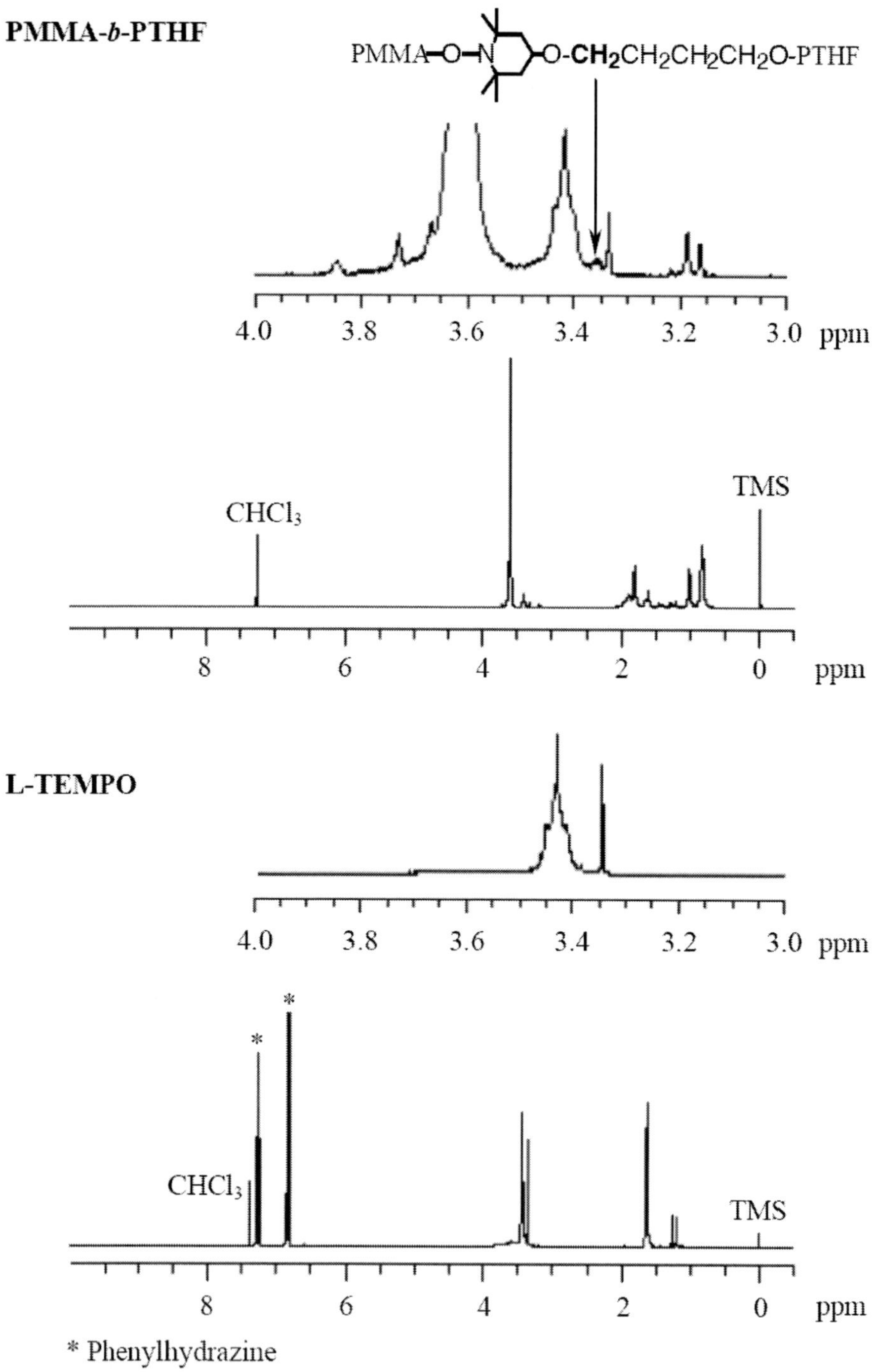

Figure 8.1.1. ^{1}H NMR spectra of the block copolymer and the L-TEMPO. PMMA-*b*-PTHF (Mn = 11,300 and Mw/Mn = 1.83) [41]. Solvent: $CDCl_3$.

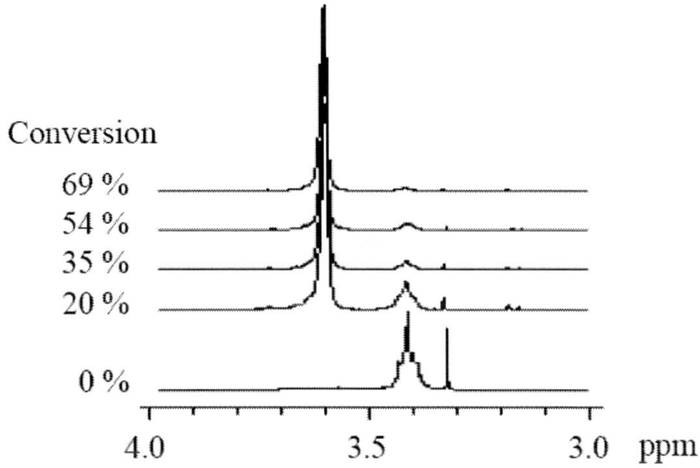

Figure 8.1.2. ^{1}H NMR spectra of the copolymers for each conversion. Solvent: $CDCl_3$ L-TEMPO/AMDV = 1.1, BAI/L-TEMPO = 0.5 $[AMDV]_0 = 4.54 \times 10^{-2}$ mol/L.

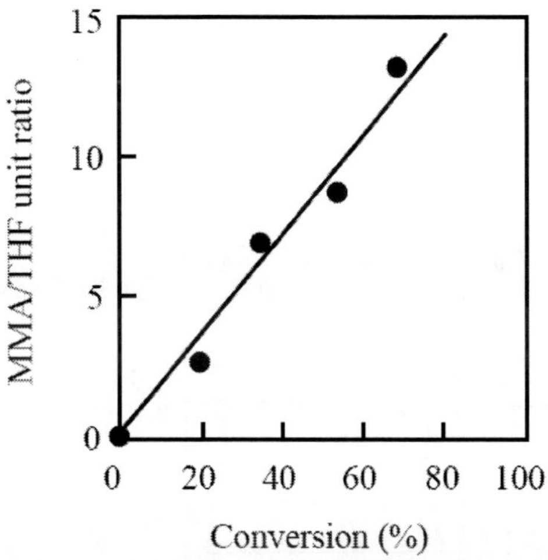

Figure 8.1.3. The plots of molar ratio of the MMA unit to the THF unit for the copolymers vs. the conversion. L-TEMPO/AMDV = 1.1, BAI/L-TEMPO = 0.5, $[AMDV]_0 = 4.54 \times 10^{-2}$ mol/L.

The first order time conversion plots of the polymerization also supported the living mechanism of the polymerization. Figure 8.1.5 shows the time-conversion plots of the polymerization using L-TEMPO. As the initial concentration of the initiator increased, the rate of the photopolymerization increased, while the thermal polymerization mediated by PTHF-TEMPO showed the opposite tendency [90]. It was difficult to raise the conversion over 80%, because the system was solidified at ca. 70% due to the bulk polymerization with the macromediator. The first order time conversion plots showed a linear correlation, independent of the initial concentration of the initiator. On the other hand, H-TEMPO did not provide a linear increase in the first order time-conversion plots (Figure 8.1.6).

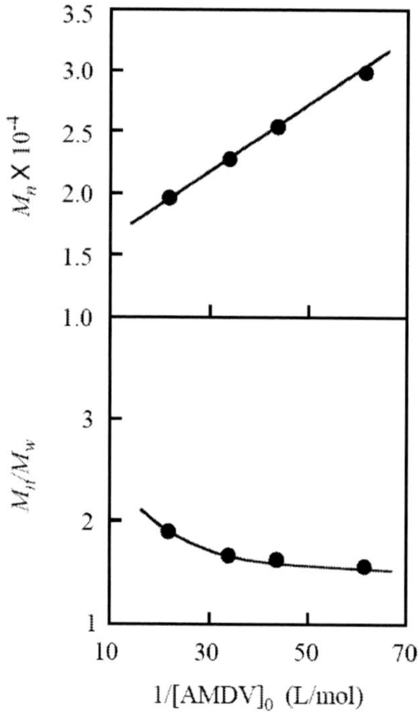

Figure 8.1.4. The plots of the molecular weight of the PMMA vs. the reciprocal of $[AMDV]_0$. L-TEMPO/AMDV = 1.1, BAI/L-TEMPO = 0.5.

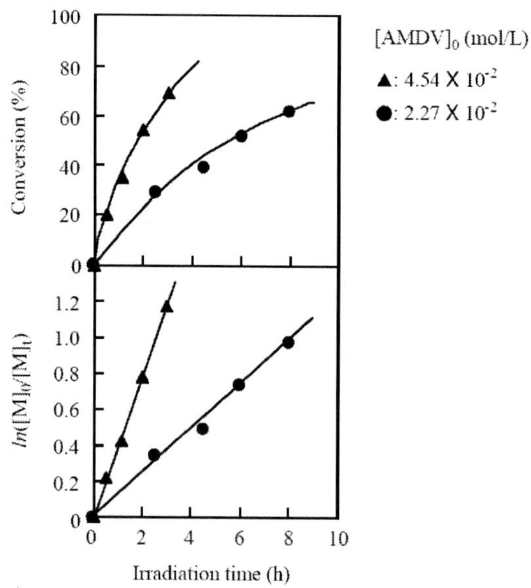

Figure 8.1.5. The first order time-conversion plots for the polymerization of MMA by L-TEMPO. L-TEMPO/AMDV = 1.1, BAI/L-TEMPO = 0.5.

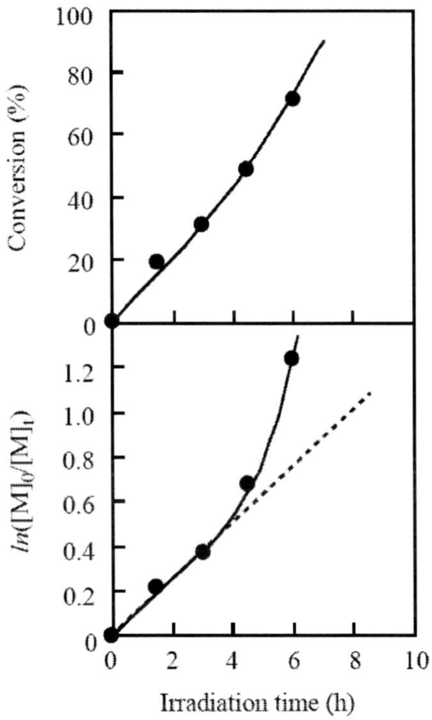

Figure 8.1.6. The first order time-conversion plots for the polymerization of MMA by H-TEMPO. H-TEMPO/AMDV = 1.1, BAI/H-TEMPO = 0.5, $[AMDV]_0 = 2.27 \times 10^{-2}$ mol/L.

The relationship between the molecular weight of the resulting copolymer and the conversion demonstrated that the polymerization proceeded in accordance with a living mechanism. Figures 8-1-7 and 8-1-8 show the plots of the molecular weight and MWD of the copolymers vs. the conversion for the polymerization with L-TEMPO and H-TEMPO, respectively. The molecular weight linearly increased with the conversion for both the L-TEMPO and H-TEMPO, supporting the living nature of the polymerization. The MWD of the copolymers produced from L-TEMPO decreased or retained the MWD of L-TEMPO, whereas the MWD from H-TEMPO showed an increase in the MWD over a 40% conversion. This MWD increase was based on the fact that the solution started clouding when over a 40% conversion. The uneven transmission of the light through the system caused the increase in the MWD.

A GPC analysis revealed that the increase in the MWD was caused by the partial propagation of the growing polymer chains. Figures 8-1-9 and 8-1-10 show the GPC profiles of the copolymers obtained from the L-TEMPO and H-TEMPO, respectively. The GPC curves obtained from the L-TEMPO were shifted to the higher molecular weight side, while retaining the MWD of the prepolymer. On the other hand, the GPC from H-TEMPO showed bimodal at a 71% conversion, although the curves were also shifted to the higher molecular weight side. The uneven transmission of the light produced the difference in the propagation rate.

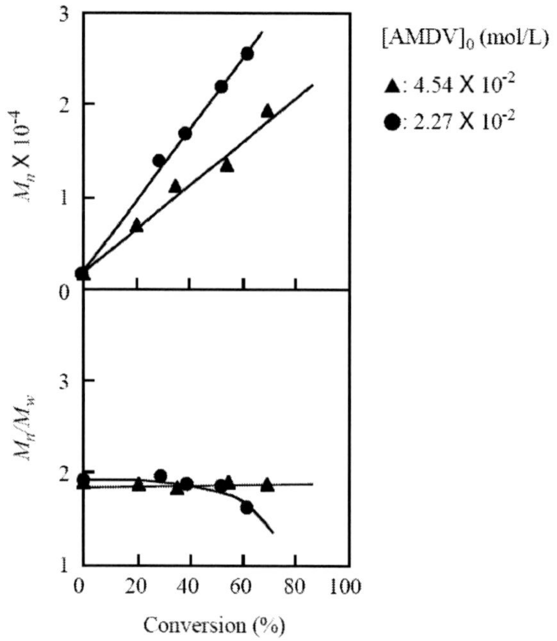

Figure 8.1.7. The plots of the molecular weight of the PMMA vs. the conversion for the polymerization by L-TEMPO. L-TEMPO/AMDV = 1.1, BAI/L-TEMPO = 0.5.

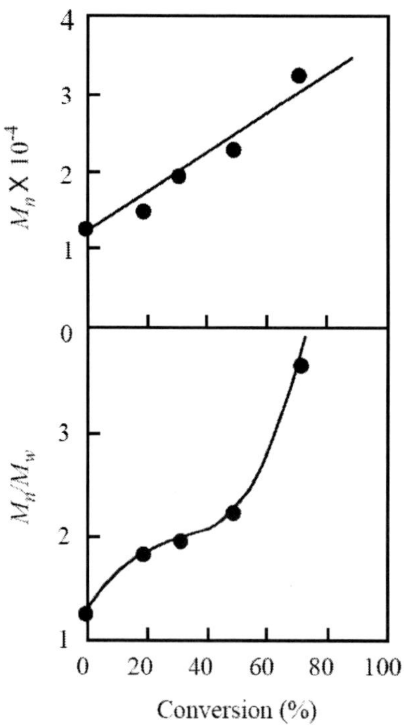

Figure 8.1.8. The plots of the molecular weight of the PMMA vs. the conversion for the polymerization by H-TEMPO. H-TEMPO/AMDV = 1.1, BAI/H-TEMPO = 0.5, [AMDV]$_0$ = 2.27 X 10^{-2} mol/L.

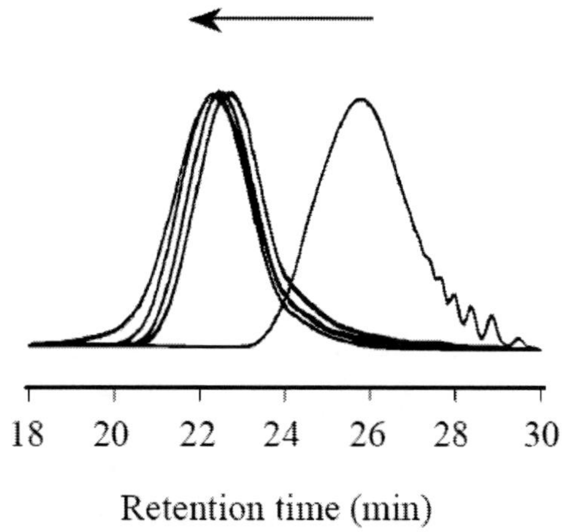

Figure 8.1.9. The GPC curves obtained by the polymerization using L-TEMPO at the conversions: 0% (0 h), 29% (2.5 h), and 39% (4.5 h), 52% (6 h), and 62% (8 h) from the *right*. L-TEMPO/AMDV = 1.1, BAI/L-TEMPO = 0.5, $[AMDV]_0 = 2.27 \times 10^{-2}$ mol/L.

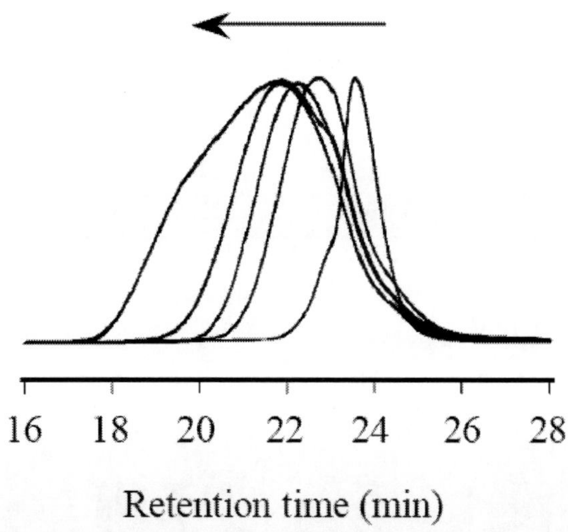

Figure 8.1.10. The GPC curves obtained by the polymerization using H-TEMPO at the conversions: 0% (0 h), 19% (1.5 h), and 31% (3 h), 49% (4.5 h), and 71% (6 h) from the *right*. H-TEMPO/AMDV = 1.1, BAI/H-TEMPO = 0.5, $[AMDV]_0 = 2.27 \times 10^{-2}$ mol/L.

The variation in the state of the polymerization using L-TEMPO and H-TEMPO is shown in Figure 8.1.11. The orange-colored monomer solution containing L-TEMPO became a colorless solid, while the solution containing H-TEMPO became cloudy as the polymerization proceeded and produced a white opaque mass. The increase in the molecular weight distribution of the copolymer obtained from H-TEMPO was based on uneven transmission of the light due to the miscibility of PMMA and the high molecular weight PTHF.

Although the molecular weight distribution increased during the late stage of the polymerization, the block copolymer was definitely formed from the H-TEMPO. Figure 8.1.12 shows the DSC spectra of the H-TEMPO, the resulting copolymers, and PMMA. The characterizations of these polymers are summarized in Table 8-1-2. The copolymers had a melting point based on the PTHF block around 20°C and a glass transition attributed to the PMMA block in the range of 100-120°C. The glass transition was shifted to a higher temperature with an increase in the molecular weight of the PMMA block. Furthermore, the proportion of the endotherm on the glass transition of PMMA to that on the melting point of PTHF increased as the molecular weight of the PMMA increased. The copolymers produced from L-TEMPO showed similar DSC spectra with the exception that L-TEMPO had a melting point around –4°C.

Table 8.1.2. The DSC analysis of the PMMA-*b*-PTHF block copolymers obtained from H-TEMPO

	Polymer	Polymerization time (h)	Mn[a]	Mw/Mn[a]	T_m (°C)	T_g (°C)
A	H-TEMPO	—	12,700	1.25	23.1	—
B	PMMA-*b*-PTHF	1.5	14,800	1.82	19.0	104.4
C	PMMA-*b*-PTHF	3	19,200	1.94	19.1	106.4
	Polymer	Polymerization time (h)	Mn[a]	Mw/Mn[a]	T_m (°C)	T_g (°C)
D	PMMA-*b*-PTHF	4.5	22,800	2.22	19.3	116.5
E	PMMA-*b*-PTHF	6	32,300	3.64	19.2	117.4
F	PMMA[b]	19	23,800	1.62	—	119.3

[a] Estimated by GPC based on PMMA standards. [b] Ref. 91.

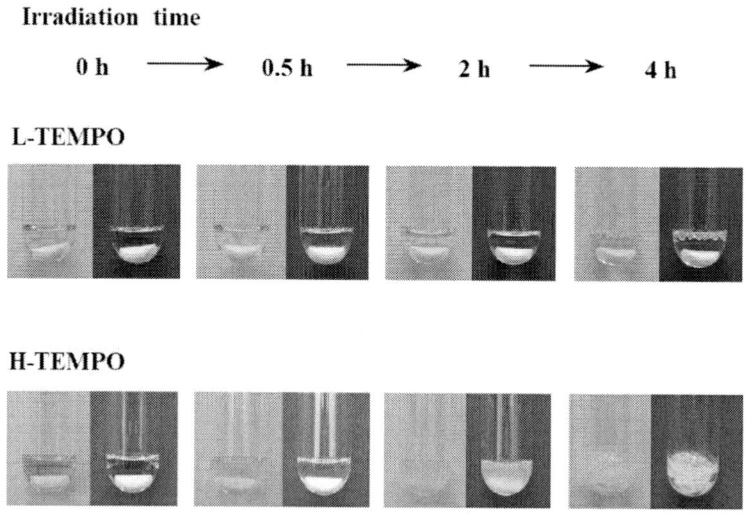

Figure 8.1.11. The variation in the state of the photopolymerization using L-TEMPO and H-TEMPO. L-TEMPO or H-TEMPO/AMDV = 1.1, BAI/L-TEMPO or H-TEMPO = 0.5, $[AMDV]_0 = 2.27 \times 10^{-2}$ mol/L.

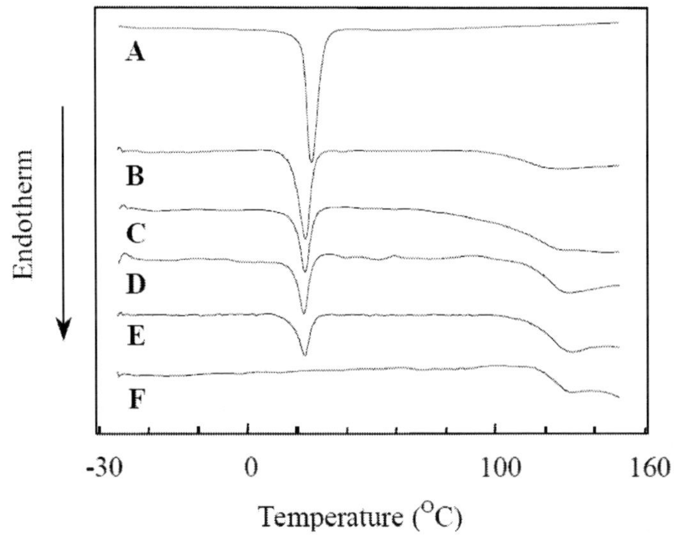

Figure 8.1.12. The DSC spectra of H-TEMPO, the resulting copolymers, and PMMA.

8.2. Graft Copolymerization Using TEMPO Supported on Polymer Side Chains [48]

The TEMPO can be supported not only on the polymer chain end, but also on side chains. In this section, the synthesis of a graft copolymer using the *Grafting from* method through the photopolymerization mediated by TEMPO supported on the side chains of a polystyrene backbone.

The poly(4-vinylbenzyl-4-oxy-2,2,6,6-tetramethylpiperidine-1-oxyl-*ran*-styrene) random copolymer (P(VTEMPO-*r*-St)) was prepared by the reaction of sodium 4-oxy-TEMPO and poly(4-chloromethylstyrene-*ran*-styrene) (P(ClMSt-*r*-St)) obtained through the random copolymerization ClMSt and St using MTEMPO as the mediator and BPO as the initiator at 125°C for 19 h. The molar ratio of the ClMSt and St units was estimated to be ClMSt/St = 0.40/0.60 by a ^{1}H NMR analysis using the signal intensity of the phenyl protons at 6.2-7.3 ppm and benzyl protons at 4.3-4.7 ppm. The degree of the polymerization (DP) of the ClMSt unit was 49 based on the signal intensity of these benzyl protons and the methoxy protons at 3.2-3.3 for MTEMPO attached to the polymer chain end. Based on this estimation, the DP of the styrene unit was 73. The total molecular weight of the copolymer was Mn = 15,100 by ^{1}H NMR. The molecular weight distribution was Mw/Mn = 1.43 by GPC based on the polystyrene standards. It was found by the ^{1}H NMR analysis that the ClMSt units were quantitatively converted into the VTEMPO units by the reaction of P(ClMSt-*r*-St) and sodium 4-oxy-TEMPO (Scheme 8-2-1). The ^{1}H NMR measurement was performed by the reduction of the TEMPO radicals by phenylhydrazine into the hydroxylamine derivatives. The molecular weight of P(VTEMPO-*r*-St) was Mn = 21,700 based on the DP of the ClMSt units.

The photoradical polymerization of MMA was performed at room temperature using P(VTEMPO-*r*-St) as the mediator and *r*-AMDV as the initiator in the presence of ^{t}BuS under the conditions in which the molar ratio of the VTEMPO unit to *r*-AMDV (VTEMPO/*r*-

AMDV) was 1.1 and the ratio of ^{t}BuS to the VTEMPO unit (^{t}BuS/VTEMPO) was 0.53. The orange solution turned colorless just after the irradiation. Figure 8.2.1 shows the first order time-conversion plots for the polymerization. The $ln([MMA]_0/[MMA]_t)$ linearly increased with time, suggesting that the number of growing polymer radicals was constant during the course of the polymerization. The plots of the molecular weight and MWD of the resulting graft copolymer vs. the conversion are shown in Figure 8.2.2. The molecular weight of the resulting graft copolymers linearly increased with an increase in the conversion, indicating that the photopolymerization proceeded by a controlled polymerization mechanism. The MWD of the resulting copolymers also increased with the increasing conversion. The reason for this increase was clarified by the GPC analysis (Figure 8.2.3). While the curves were shifted to the higher side of the molecular weight with the increasing conversion, the left side of the curve showed a much greater shift. This is accounted for because the polymerizations by the TEMPO supported on the polymers are accelerated due to lower mobility of TEMPO [39, 89, 90, 92]. In addition, the molecular weight of P(VTEMPO-*r*-St) estimated by GPC (*Mn* = 4,630) was still lower when compared to that by ^{1}H NMR (*Mn* = 21,700) due to the adsorption of the VTEMPO units within the columns, resulting in the estimated broader MWD of P(VTEMPO-*r*-St) (*Mw*/*Mn* = 3.31).

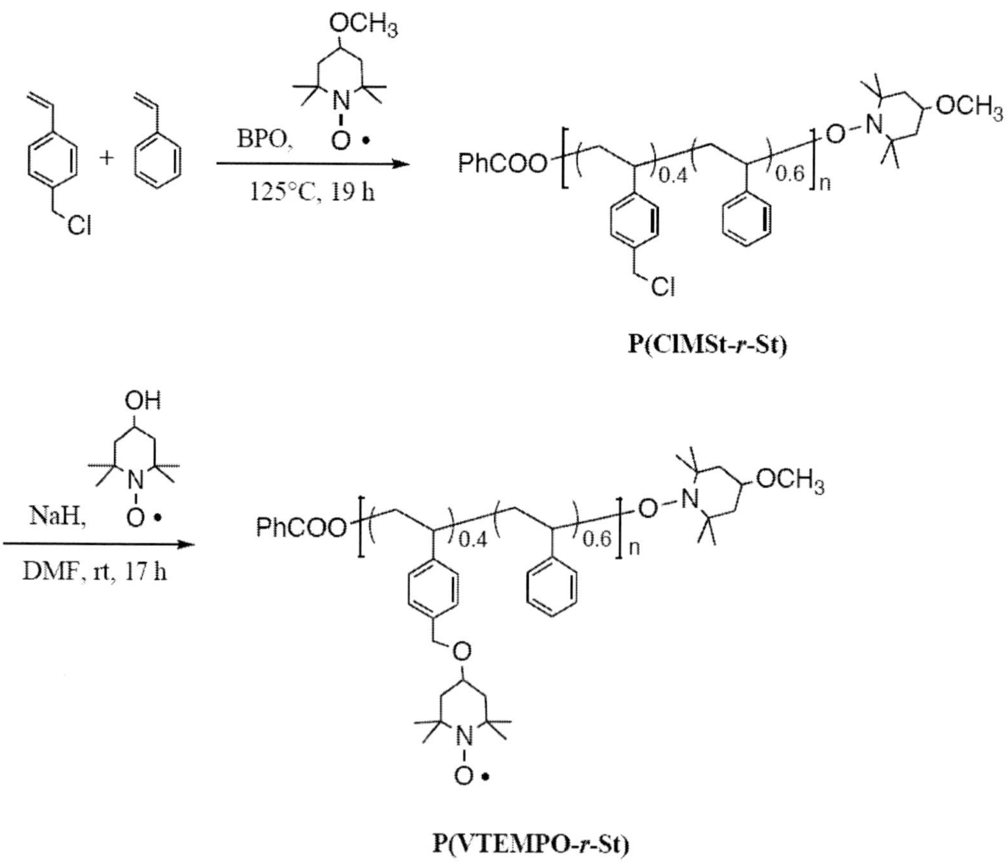

Scheme 8.2.1. The synthesis of P(VTEMPO-*r*-St).

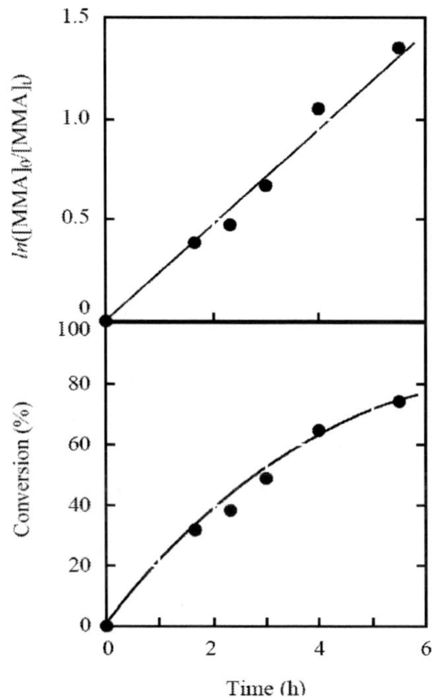

Figure 8.2.1. The time-conversion and its first order plots for the MMA polymerization. VTEMPO/*r*-AMDV = 1.1, ^{t}BuS/VTEMPO = 0.53.

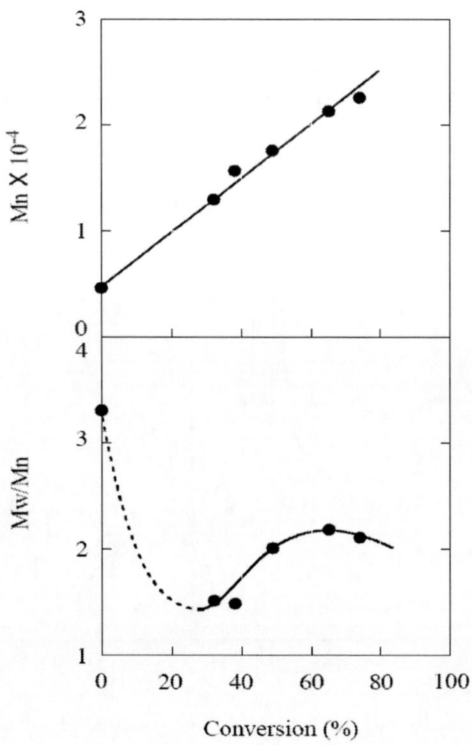

Figure 8.2.2. The plots of the molecular weight and molecular weight distribution vs. the conversion for the MMA polymerization. VTEMPO/*r*-AMDV = 1.1, ^{t}BuS/VTEMPO = 0.53.

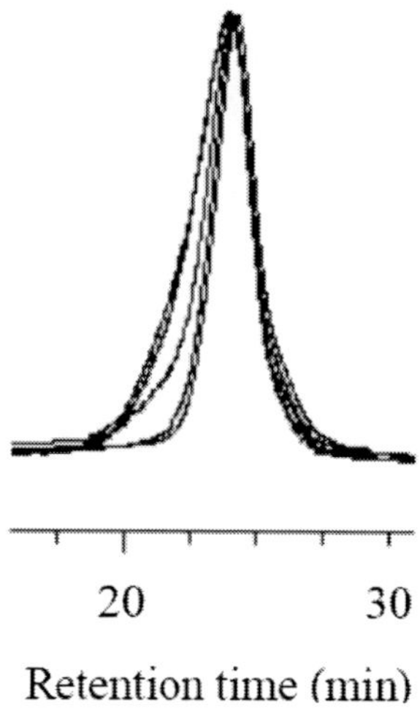

Figure 8.2.3. The GPC profiles of the resulting copolymer at each conversion: 32% (1.67 h), 38% (2.33 h), 49% (3.0 h), 65% (4.0 h), and 74% (5.5 h) from the right.

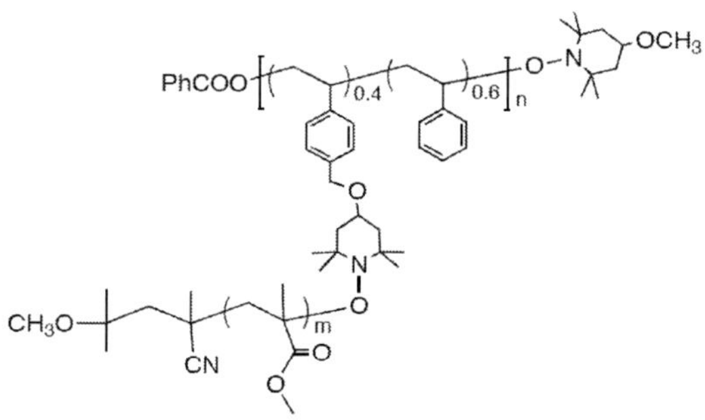

PSt-*graft*-PMMA

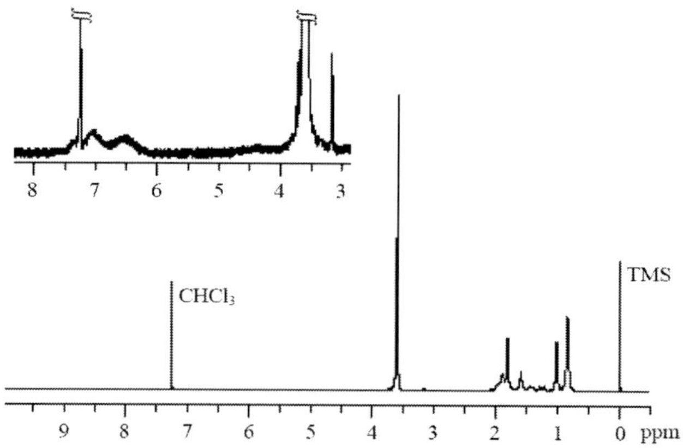

Figure 8.2.4. The ^{1}H NMR spectrum of the graft copolymer obtained by the photopolymerization (the irradiation time: 1.67 h, the conversion: 32%). Solvent: $CDCl_3$.

The ^{1}H NMR analysis confirmed that the grafted PMMA chains were controlled by the polymerization. Figure 8.2.4 shows the ^{1}H NMR spectrum of the graft copolymer obtained at a low conversion (32 %) through the polymerization by the irradiation for 1.67 h. Signals of the phenyl and benzyl protons based on the VTEMPO supporting the grafted PMMA were observed at 6.2-7.5 ppm and 4.0-5.0 ppm, respectively. Signals originating from the methoxyl protons of the CDM attached to the initiation chain end of the grafted PMMA were discerned at 3.1-3.2 ppm. The molar ratio of the VTEMPO at the terminal chain end of the grafted PMMA to the CDM at the initiation chain end (VTEMPO/CDM) was 1.05 based on the signal intensity for the phenyl and methoxy groups, the VTEMPO/St ratio, and their DPs. The almost unity of the VTEMPO/CDM ratio suggested that all the VTEMPO units support the controlled PMMA chains based on the previous results on the photo-living radical polymerization of MMA using 4-methoxy-TEMPO [38]. The molecular weight of the grafted PMMA chains was estimated to be Mn = 6,210 based on the CDM using the signal intensity of the methoxy protons and the methyl protons at 3.3-4.0 ppm for the methyl ester groups of the PMMA chains. The molecular weight estimated on the basis of the monomer conversion and the initial concentration of the VTEMPO units (0.0484 mmol) was Mn = 6,190. This good agreement in the molecular weights estimated on the basis of the initiation chain end and the monomer conversion also supports the conclusion that all the VTEMPO units were bonded to the PMMA chains. It was assumed that all the VTEMPO units support the PMMA chains, and the total molecular weight of the graft copolymer was Mn = 326,000 by ^{1}H NMR. Compared to the molecular weight estimated by GPC (Mn = 13,000), the molecular weight of the graft copolymer was estimated lower than that of its linear copolymer due to its graft structure.

CONCLUSIONS

The TEMPO-mediated photopolymerization is the novel method to control the molecular weight by photo irradiation. This polymerization also expanded the range of applicable monomers that could not be polymerized by the thermal NMP. The applicable monomers include vinyl acetate [41]. The MTEMPO-mediated photo controlled/living radical polymerization is expected to also expand the scope of molecular design by the NMP and to be applied for heterogeneous polymerizations and solid-state polymerization.

REFERENCES

[1] Odian, G. (2004). Principles of Polymerization, 4th Ed, pp.200, Wiley-Interface, New York.
[2] Otsu, T. (1956). *J. Polym. Sci.* 21, 559.
[3] Matyjaszewski, K. & Xia, J. (2001). *Chem. Rev. 101*, 2921.
[4] Uegaki, H., Kotani, Y., Kamigaito, M. & Sawamoto, M. (1997). *Macromolecules.* 30, 2249.
[5] Wayland, B. B., Poszmik, G. & Mukerjee, S. L. (1994). *J. Am. Chem. Soc.* 116, 7943.

[6] Ando, T., Kamigaito, M., Sawamoto, M. & Higashimura, T. (1995). *Macromolecules.* 30, 4507.

[7] Kato, M., Kamigaito, M., Sawamoto, M. & Higashimura, T. (1995). *Macromolecules.* 28, 1721.

[8] Percec, V., Barboiu, B., Neumann, A., Ronda, J. C. & Zhao, M (1996). *Macromolecules.* 29, 3665.

[9] Lecmote. Ph., Drapier, I., Dubois, Ph., Teyssie, Ph. & Jerome, R. (1997). *Macromolecules.* 30, 7631.

[10] Kotani, Y., Kamigaito, M. & Sawamoto, M. (1999). *Macromolecules.* 32, 2420.

[11] Moad, G., Rizzardo, E. & Thang, S. H. (2008). *Acc. Chem. Res.* 41, 1133.

[12] Tsarevsky, N. V. & Matyjaszewski, K. (2007). *Chem. Rev. 107*, 2270.

[13] Oka, M. & Tatemoto, M. (1984). Vinylidene fluoride –Hexafluoropropylene copolymer having terminal iodines, *Contemporary Topics in Polymer Science.* 4: 763, Plenum Press, New York, NY.

[14] Lacroix-Desmazes, P., Severac, R. & Boutevin, B. (2005). *Macromolecules.* 38, 6299.

[15] Pouget, E., Tonnar, J., Eloy, C., Lacroix-Desmazes, P. & Boutevin, B. (2006). *Macromolecules.* 39, 6009.

[16] Hawker, C. J., Bosman, A. W. & Harth, E. (2001). *Chem. Rev.* 101, 3661.

[17] Sciannamea, V., Jérôme, R. & Detrembleur, C. (2008). *Chem. Rev. 108*, 1104.

[18] Yousi, Z., Jian, L., Rongchuan, Z., Jianliang, Y., Lizong, D. & Lansun, Z. (2000). *Macromolecules.* 33, 4745.

[19] Devonport, W., Michalak, L., Malmstrom, E., Mate, M., Kurdi, B., Hawker, C. J., Barclay, G. G., Sinta, R. (1997). *Macromolecules.* 30, 1929.

[20] Mardare, D. & Matyjaszewski, K. (1994). *Macromolecules.* 27, 645.

[21] Granel, C., Jerime, R., Teyssie, P., Jasieczek, C. B., Shooter, A. J., Haddleton, D. M., Hastings, J. J., Gigmes, D., Grimaldi, S., Tordo, P., Greszta, D. & Matyjaszewski, K. (1998). *Macromolecules.* 31, 7133.

[22] White, D. & Matyjaszewski, K. (1997). *J. Macromol. Sci. -Pure Appl. Chem.* A34, 221.

[23] Hawker, C. J., Barclay, G. G., Orellana, A., Dao, J. & Devonport, W. (1996). *Macromolrcules.* 29, 5245.

[24] Dao, J., Benoit, D. & Hawker, C. J. (1998). J Polym Sci Part A: Polym Chem, 36, 2161.

[25] Banoit, D., Chaplinski, V., Braslau, R. & Hawker, C. J. (1999). *J. Am. Chem. Soc.* 121, 3904.

[26] Veregin, R. P. N., Georges, M. K., Hamer, G. K. & Kazmaier, P. M. (1995). *Macromolecules.* 28, 4391.

[27] Greszta, D. & Matyjaszewski, K. (1996). *Macromolecules.* 29, 5239.

[28] Braslau, R., Burrill II, L. C., Siano, M., Naik, N., Howden, R. K. & Mahal, L. K. (1997). *Macromolecules.* 30, 6445.

[29] Li, I. Q., Howell, B. A., Koster, R. A. & Priddy, D. B. (1996). *Macromolecules.* 29, 8554.

[30] Mercier, C. L., Acerbis, S., Bertin, D., Chauvin, F., Gigmes, D., Guerret, O., Lansalot, M., Marque, S., Moigne, F. L., Fischer, H. & Tordo, P. (2002). *Macromol. Symp.* 182, 225.

[31] Yoshida, E., Ishizone, T., Hirao, A., Nakahama, S., Takata, T. & Endo, T. (1994). *Macromolecules.* 27, 3119.

[32] Yoshida, E., Nakamura, K., Takata, T. & Endo, T. (1993). *J. Polym. Sci. Part A: Polym. Chem.* 31, 1505.
[33] Pyun, J., Tang, C., Kowalewski, T., Fréchet, J. M. J. & Hawker, C. J. (2005). *Macromolecules.* 38, 2674.
[34] Hawker, C. J. & Hedrick, J. L. (1995). *Macromolecules.* 28, 2993.
[35] Hawker, C. J. (1995). *Angew Chem. Int. Ed. Eng.* 134, 1456.
[36] Tsoukatos, T., Pispas, S. & Hadjichristidis, N. (2000). *Macromolecules.* 33, 9504.
[37] Yoshida, E. (2008). *Colloid Polym. Sci.* 286, 1663.
[38] Yoshida, E. (2009). *Colloid Polym. Sci.* 287, 767.
[39] Yoshida, E. (2009). *Colloid Polym. Sci.* 287, 1417.
[40] Yoshida, E. (2010). *Colloid Polym. Sci.* 288, 7.
[41] Yoshida, E. (2010). *Colloid Polym. Sci.* 288, 73.
[42] Yoshida, E. (2010). *Colloid Polym. Sci.* 288, 239.
[43] Yoshida, E. (2010). *Colloid Polym. Sci.* 288, 341.
[44] Yoshida, E. (2010). *Colloid Polym. Sci.* 288, 901.
[45] Yoshida, E. (2010). *Colloid Polym. Sci.* 288, 1027.
[46] Yoshida, E. (2010). *Colloid Polym. Sci.* 288, 1639.
[47] Yoshida, E. (2010). *Colloid Polym. Sci.* 288, 1745.
[48] Yoshida, E. (2011). *Colloid Polym. Sci.* 289, 837.
[49] Yoshida, E. (2011). *Colloid Polym. Sci.* 289, 1127.
[50] Lalevee, J., Allonas, X. & Fouassier, J. P. (2006). *Macromolecules.* 39, 8216.
[51] Nemoto, Y. & Nakayama, Y. (2008). J. Polym. Sci. Part A Polym. Chem. 46, 4505.
[52] You, Y., Hong, C., Bai, R., Pan, C. & Wang, J. (2002). *Macromol. Chem. Phys.* 203, 477.
[53] Ran, R., Yu, Y. & Wan, T. (2007). *J. Appl. Polym. Sci.* 105, 398.
[54] Baruah, S. R., Kakati, D. K. (2006). J. Appl. Polym. Sci. 100, 1601.
[55] Ajayaghosh, A. & Francis, R. (1999). *J. Am. Chem .Soc.* 121, 6599.
[56] Yang, W. & Ranby, B. (1996). *Macromolecules.* 29, 3308.
[57] Tasdelen, M. A., Durmaz, Y. Y., Karagoz, B., Bicak, N. & Yagci, Y. (2008). *J. Polym. Sci. Part A Polym. Chem.* 46, 3387.
[58] Ajayaghosh, A. & Francis, R. (1998). *Macromolecules.* 31, 1436.
[59] Yin, H., Zheng, H., Lu, L., Liu, P. & Cai, Y. (2007). *J. Polym. Sci. Part A Polym. Chem.* 45, 5091.
[60] Cohen, N. A., Tillman, E. S., Thakur, S., Smith, J. R., Eckenhoff, W. T. & Pintauer, T. (2008). *Macromol. Chem. Phys.* 209 (DOI: 10.1002/macp.200800481).
[61] Li, P. & Qiu, K. (2002). *Macromol. Rapid. Commun.* 23, 1124.
[62] Bevinton, J. C., Melville, H. W. & Taylor, R. P. (1954). *J. Polym. Sci.* 14, 463.
[63] Kita, Y., Gotanda, K., Murata, K., Suemura, M., Sano, A., Yamaguchi, T., Oka, M. & Matsugi, M. (1998). Org. Process Res. Dev. 2, 250.
[64] Talat-Erben, M. & Bywater, (1955). *J. Am. Chem. Soc.* 77, 3710.
[65] Talat-Erben, M. & Bywater, (1955). *J. Am. Chem. Soc.* 77, 3712.
[66] Hammond, G. S., Trapp, O. D., Keys, R. T. & Neff, D. L. (1959). *J. Am. Chem. Soc.* 81, 4878.
[67] Hammond, G. S., Wu, C. S., Trapp, O. D., Warkentin, J. & Keys, R. T. (1960). *J. Am. Chem. Soc.* 82, 5394.
[68] Wu, C. S., Hammond, G. S. & Wright, J. M. (1960). *J. Am. Chem. Soc.* 82, 5386.

[69] Kita, Y., Sano, A., Yamaguchi, T., Oka, M., Gotanda, K. & Matsugi, M. (1999). *J. Org. Chem.* 64, 675.
[70] The product information from Wako Pure Chemical Industries (http://www.wako-chem.co.jp/kaseihin/oilazo/VAm-110.htm).
[71] Yoshida, E. (2008). *Am. Chem. Soc. Polym. Prepr.* 49(2), 232.
[72] Bovey, F. A. & Tiers, G. V. D. (1960). *J. Polym. Sci.* 44, 173.
[73] Crivello, J. V. & Sangermano, M. (2001). *J. Polym. Sci. Part A: Polym. Chem.* 39, 343.
[74] Crivello, J. V. (1999). J. Polym. Sci. Part A: Polym. Chem. 37, 4241.
[75] Devoe, R. J., Sahyunn, M. R. V., Sepone, D. & Sharmac, D. K. (1987). *Can. J. Chem.* 65, 2342.
[76] Crivello, J. V., Lam, J. H. W., Moore, J. E., Schroeter, S. H. (1978). J. *Radiat. Curing.* 5, 2.
[77] Crivello, J. V. & Lam, J. H. W. (1980). *J. Polym. Sci. Polym. Chem.* 18, 2677.
[78] Crivello, J. V. & Lam, J. H. W. (1979). *J. Polym. Sci. Polym. Chem.* 17, 977.
[79] Meier, K. & Zweifel, H. (1986). *J. Rad. Curing.* 13, 26.
[80] Meier, K. & Zweifel, H. (1986). *J. Imag. Sci.* 30, 174.
[81] Schumann, V. H. (1984). *Chem-Ztg.* 108, 345.
[82] Lohse, F. & Zweifel, H. (1986). *Adv. Polym. Sci.* 78, 61.
[83] Pappas, S. P. (1989). Comprehensive Polymer Science, Pergamon Press, New York 4: 337.
[84] Liu, Y. C., Wu, L. M. & Chen, P. (1985). *Tetrahedron Lett.* 26, 4201.
[85] Varshney, S. K., Jacobs, C., Hautekeer, J. P., Bayard, P., Jerome, R., Fayt, R. & Teyssie, P. (1991). *Macromolecules.* 24, 4997.
[86] Yoshida, E. & Fujii, T. (1998). *J. Polym. Sci. Part A Polym. Chem. Ed.* 36, 269.
[87] Yoshida, E. & Okada, Y. (1997). *Bull. Chem. Soc. Jpn.* 70, 275.
[88] Li, I., Howell, B. A., Matyjaszewski, K., Shigemoto, T., Smith, P. B. & Priddy, D. B. (1995). *Macromolecules.* 28, 6692.
[89] Yoshida, E. & Sugita, A. (1998). *J. Polym. Sci. Polym. Chem. Ed.* 36, 2059.
[90] Yoshida, E. & Sugita, A. (1996). *Macromolecules.* 29, 6422.
[91] The PMMA was obtained by the photopolymerization of MMA by AMDV, 4-methoxy-TEMPO, and BAI. *See the Experimental. Ref.* 38.
[92] Yoshida, E. & Osagawa, Y. (1998). *Macromolecules.* 31, 1446.

In: Radical Polymerization: New Developments ISBN 978-1-62100-406-6
Editors: I. O. Paulauskas, L. A. Urbonas, pp. 149-173

Chapter 4

ALKALINE ANION-EXCHANGE MEMBRANES PREPARED BY PLASMA POLYMERIZATION: SYNTHESIS, STRUCTURAL CHARACTERIZATION AND APPLICATION IN DIRECT ALCOHOL FUEL CELLS

Jue Hu, Chengxu Zhang and Yuedong Meng
Institute of Plasma Physics, Chinese Academy of Sciences,
P. O. Box 1126, Hefei, P.R. China.

ABSTRACT

This chapter is focused on plasma polymerization method for the preparation of alkaline anion-exchange membranes. There are mainly two steps in fabrication of plasma-polymerized alkaline anion-exchange membrane: (1) plasma polymerization of monomer into polymer films; (2) quaternization of plasma polymer into alkaline anion-exchange membrane with quaternary ammonium functional groups. Among them, the plasma polymerization process is the most important step since the percentage of quaternary ammonium functional groups in alkaline anion-exchange membrane is determined by the preservation of functional groups in plasma polymerization. A brief introduction to the subject of plasma polymerization, and general characteristics of plasma polymers, is followed by an examination of recent literature on attempts to synthesize electrolyte membranes for fuel cells. The topics covered include effects of plasma polymerization conditions on the structures and properties of the films, the modern analytical techniques used for characterization of plasma polymer films and physical properties of the plasma-polymerized alkaline anion-exchange membranes. The application of the plasma-polymerized alkaline anion-exchange membranes in alkaline direct alcohol fuel cells is also presented.

1. FUNDAMENTAL ASPECTS OF PLASMA POLYMERIZATION

Plasma polymerization refers to the formation of polymeric materials under the influence on plasma. In the polymerization of monomers, plasma, which contains exciting species such as electrons, ions, excited molecules, can be considered a kind of "radiation" source. In plasma-state polymerization process, these exciting species bombard with monomers creating active species for polymerization, and as a result, depositing high cross-linked films on the substrate surface. In the plasma-induced polymerization, free radicals formed on the surface of polymers and other solid materials initiate graft polymerization [1]. Whereas, plasma polymerization is limited to the interface between materials plasma. There is a significant difference between radiation polymerization and plasma polymerization. In radiation polymerization the formation of the chain-carrying reactive specie is a consecutive process to the ionization of a monomer, whereas in plasma polymerization, the reactive species are not necessarily formed as the consequence of the ionization. These reactive species, especially ions, play a much more important role in radiation polymerization than in plasma polymerization. Because the energy levels of electrons are low in plasma polymerization, which are much lower than those involved in rays and high-energy electron beams, it is thought that these energetic species in plasma have less penetrating power. Whereas, compared with the more penetrating radiations, plasma bombarding is restricted to the surface [2, 3]. Chemical reactions that occur under plasma conditions are generally very complex. Although the concentration of free radicals in plasma is usually five to six orders of magnitude higher than that of ions and the reactive species might be free radicals, it should also be noted that the plasma polymerization is different from the free radical polymerization [4]. Although atmospheric plasma is becoming more and more popular for deposition plasma polymers, still, the focus on the usually plasma polymerization process carried on in a low pressure and low temperature is treated in this chapter. Among the many types of electric discharge, which characterized by the presence of free electrons or an electric field, glow discharge is by far the most frequently used in plasma polymerization. Consequently, the term glow discharge polymerization can be used nearly synonymously with plasma polymerization. Polymerization in a vacuum condition is likely to proceed by the rapid step-growth mechanism, which is based on reactions between reactive species, rather than the chain-growth addition mechanism, which is based on reactions between reactive species and monomer molecules. The rapid step-growth polymerization suggested that it is possible only by reactions of multifunctional reactive species or by the repeated reactivation of the reaction products. Plasma polymerization seems to occur by both mechanisms [3].

As far as plasma polymerization and plasma treatment of materials are concerned, plasma can be divided into three major groups: (1) chemically nonreactive plasma, (2) chemically reactive plasma, and (3) polymer-forming plasma. Chemically nonreactive plasma is mainly those of monoatomic inert gases, such as Argon plasma, which can ionize other molecules or sputter materials but is not consumed in chemical reactions. Chemically reactive plasma is those of inorganic and organic molecular gases. O_2, N_2, and CF_4 plasma are chemically reactive but do not form polymeric deposits in their pure gas plasma. Polymer-forming plasma is obviously chemically reactive but forms a polymeric solid deposit by itself [5]. From the viewpoint of plasma polymerization as a material production process, there are two opposing processes: polymer formation, which leads to the deposition of material, and ablation, which lead to the removal of material. The overall scheme of plasma polymerization encompasses the principle of the competitive ablation and polymerization mechanism [6]. Both polymer-forming species and species that cause the ablation of materials are created in the plasma. It may be, however, that many species in the plasma do not contribute to the plasma polymerization but rather contribute to the ablation. The nature and number of activated species formed are dependent on the conditions of plasma polymerization.

In plasma polymerization, the competitive nature of reactions is highly dependent on the conditions of plasma polymerization, in particular the energy input level. Plasma polymerization can be generally divided into two cases according to the rate-determining process, which are the discharge power-controlled case and the flow rate-controlled case. In the former case, the polymer deposition rate is dependent of the discharge power, on the contrary, the chemical etching rate increases with flow rate of the etching gas. But in the latter case, the polymer deposition rate is dependent of the dose rate, and the discharge wattage determines the etching rate [3]. The power input heavily influences the electron energy distributing function, the density of all the active species, as well as the relevant electric features (plasma potential, bias potential, etc.) of the discharge. Increasing the power results always in an increased fragmentation of the monomer in the plasma, thus in an increased production of both atoms and radicals [1].

2. General Characteristics of Plasma Polymer Films

The area of plasma polymerization is now well recognized as an important part of the material science. The materials formed by plasma polymerization are different from the conventional polymers. Polymers obtained by plasma polymerization have extremely important characteristics which are sought after in the modern technology, such as highly cross-linked and highly branched backbone structure, strong resistance to most chemicals, excellent adhesion to substrate materials, and so on. The properties of the plasma polymers depend on the conditions of plasma polymerization, the design of the reactor, the chemical properties of the monomer, and the chemical and physical characteristics of the substrate [3]. Plasma polymers structure is depends on the intensity and energy of the species bombarding the growing film. The slower the deposition and tighter the network of polymer, the larger is the internal stress. Consequently, the tighter the network structure, the more the application of such a plasma polymer should be limited to a thinner layer in order to capitalize on the characteristic properties of the plasma polymer [3]. Furthermore, better adhesion between plasma polymer films and substrate is often obtained by the slower deposition of plasma polymers. Surface characteristics, such as roughness, cleanness, degree of oxidation, and so on, are also very important factors for a given substrate. For example, adhesion of plasma polymers onto organic polymers of polytetrafluoroethylene (PTFE), polypropylene (PP) or polyethylene oxide (PEO) is generally good, but adhesion to smooth surfaces of metals, ceramics, silicon or glasses is relatively poor [7-9]. So it is necessary to make effort on improvement the adhesion to the smooth surface substrates. Because of the and high polarity and small size of water molecules, it is extremely difficult to stop the permeation of water vapor by conventional polymers. In contrast, the permeability of plasma polymers is governed by considerably different principles. Because of the absence of large-scale segmental mobility and the presence of a high degree of cross-linking, and a molecular-level “sieve” mechanism for such small permeants becomes evident. So plasma polymer films show an extremely low water permeability [3].

3. Alkaline Anion-Exchange Membranes Prepared by Plasma Polymerization for Application in Direct Alcohol Fuel Cells (DAFCs)

3.1. Introduction

Direct alcohol fuel cells (DAFCs) are expected to be promising power sources ranging from automotive to portable electronic device applications due to their high power density and low environmental pollution [10-14]. Proton exchange membrane direct alcohol fuel cells (PEMDAFCs) have been extensively explored thanks to the numerous advantages. However, there are still scientific and technological difficulties that impede their widespread commercialization: high cost for using noble metals as catalyst, fuel permeation and slow oxidation kinetics of alcohol [10, 15]. To overcome these problems, it is natural to consider an alkaline analogue of DAFC. Several merits for alkaline direct alcohol fuel cells (ADAFCs) have been suggested: (1) potential to forego noble metal catalysts [16, 17]; (2) lower fuel permeability [12]; (3) lower overpotentials associated with many electrochemical reactions at high pH [18, 19]; (4) less serious corrosion in alkaline environment [18]; and (5) a significant change in water management [18]. Alkaline anion-exchange membranes (AAEMs), which serve dual functions of hydroxide ion conducting and fuels separating, can seriously affect the performance of ADAFCs. The need for AAEMs with the necessary conductivity, alcohol permeability, thermal stability and chemical stability is a key challenge in the development of ADAFCs. Significant effort has been focused on the preparation of AAEMs by performing a chloromethylation reaction on the polymer to form benzylic chloromethyl groups and then converting the benzylic chloromethyl groups into benzyltrimethylammonium cationic groups in subsequent quaternization and alkalization steps [15, 20-23]. However, the chloromethyl ether, which was usually used in the chloromethylation reaction as a carcinogen, is potential harmful to human health [24]. Grafting of vinylbenzyl chloride (VBC) onto polymer matrix has been proved to be an effective way to avoid using chloromethyl ether [25-29]. Nevertheless, due to the high irradiation damage of the polymer matrix, AAEMs produced by radiation-grafting are promising only when fully fluorinated base films are used [25]. AAEMs with remarkable conductivity (in excess of 0.05 $S \cdot cm^{-1}$) have been synthesized by Cornelius [30], Coates [31], and Zhang [32], et al.. Although, previous studies have successfully fabricated AAEMs with high conductivity and excellent stability, it is difficult to synthesize AAEMs with several micrometers thickness due to the restriction of the traditional film-casting method [28, 30-33]. Considering the mobility of hydroxide ions is inherently slower than that of proton in dilute solution, the property of AAEMs with ultra-thin structure is also promising [34]. Herein, plasma polymerization was adopted to prepare AAEMs [35, 36]. The scheme of fabricating AAEM and novel membrane electrode assembly (MEA) for application in the ADAFC using plasma polymerization technique is shown in Figure 1. Plasma polymerization technique is promising to prepare ion-exchange membranes with a few microns thickness, highly cross-linked backbone structure, and good adhesion to the substrate. The highly cross-linked structure can improve the membrane chemical and mechanical properties. The strong adhesion to the electrode by directly depositing membrane onto catalyst layer will decrease the interfacial resistance between membrane and electrode, further improving the performance of ADAFCs. Vinylbenzyl chloride (VBC) is an excellent monomer for synthesis of AAEMs for having two functional groups: vinyl group for polymerization reaction and benzyl chloride group for quaternization. The VBC monomer has not been reported as having been used for preparing AAEMs by plasma polymerization so far. In this work, AAEMs were prepared from VBC monomer by forming poly(vinylbenzyl chloride) (PVBC) membranes in AGD-PP system, and then converting the benzylic chloromethyl groups into benzyltrimethylammonium cationic groups in subsequent quaternization and alkalization steps. Because of the low bond dissociation energy of the C-Cl bond in benzyl chloride groups, it is very important to investigate the effects of energy input level on the polymerization [37]. The objectives of this study are: (1) to prepare quaternized poly(vinylbenzyl chloride) (QPVBC) membranes with flat and uniform morphology, high thermal stability and high ion-exchange capacities for potential application in ADMFCs; (2) to establish the influence of plasma discharge power (energy input) on structure characteristics of the plasma-polymerized PVBC membranes to give information for further understanding the plasma polymerization process and improving the membranes performance.

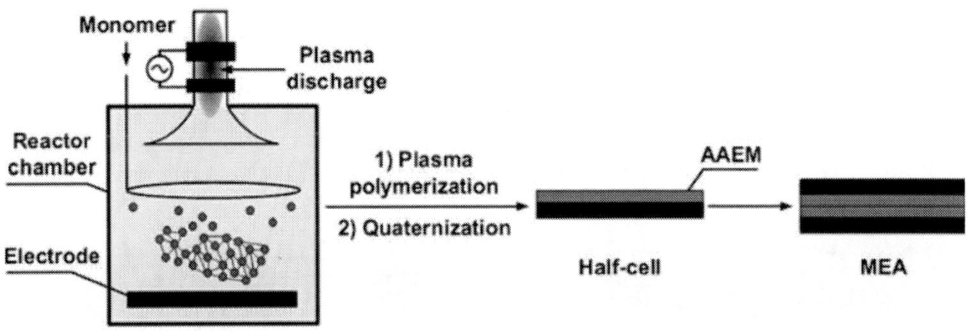

Figure 1. Schematic diagram for synthesis of alkaline anion-exchange membrane and membrane electrode assembly by plasma polymerization.

3.2. Synthesis of Plasma-Polymerized Alkaline Anion-Exchange Membrane

There are mainly two steps in fabrication of plasma polymerized alkaline anion-exchange membrane using VBC as monomer: (1) plasma polymerization of VBC monomer into poly(vinylbenzyl chloride); (2) quaternization of benzyl chloride groups into $-N^{+}(CH_3)_3OH^{-}$ groups. The plasma-polymerized membranes were synthesized in an after-glow discharge system, depicted in Figure 2, consisting of a stainless steel reactor, a radio frequency (RF) power supply with corresponding power coupling system, a pulse modulation, a substrate bias voltage supply, a monomer heater with a temperature control system, gas mass flow controllers with flow indicators, vacuum system (a mechanical booster pump, a rotary pump and liquid nitrogen trap) with vacuum gauge. The reactor was in a after-glow discharge configuration, meaning that the glow discharge was produced in the upper part of the reactor (in a Pyrex glass tube) and the plasma polymerization was performed in the lower part of it (stainless steel chamber). The plasma discharge was sustained by a RF power supply between two external electrodes (5 cm gap) in the Pyrex glass tube using Ar (supplied by Air liquid) as working gas. 4–Vinylbenzyl chloride monomer carried by H_2 (supplied by Air liquid), was introduced into the polymerization region through an air distribution ring. The flow rates of Ar, VBC monomer and H_2 were controlled by gas mass flow controllers. In order to avoid the condensation and polymerization of monomer on the inner walls of the gas lines, heating wires were wrapped around them, as shown in Figure 2. The liquid nitrogen trap was placed in front of the pumps to avoid erosion of the vacuum pumps by untreated monomer and byproducts of the plasma polymerization. There are different substrates used to support plasma-polymerized membrane: silicon wafers [P-doped Si (100)] for SEM observations and thermogravimetric analyses, stainless steel plates for chemical structural characterizations and polytetrafluoroethylene (PTFE) porous substrates for ionic conductivity and ethanol permeability measurements.

To quaternize of benzyl chloride groups into $–N^+(CH_3)_3OH^−$ groups, the plasma-polymerized vinylbenzyl chloride (PVBC) membranes were submerged in 33 wt % trimethylamine (TMA) aqueous solution for at least 48 h at room temperature. After soaking, the membranes ($Cl^−$ form) were washed with deionized water to remove excess TMA solution. The treated membranes were immersed in 2 mol $L^{−1}$ potassium hydroxide aqueous solution at room temperature for 48 h to convert the $–N^+(CH_3)_3Cl^−$ groups into $–N^+(CH_3)_3OH^−$ groups. Then, the quaternized poly(vinylbenzyl chloride) membranes (QPVBC) were washed by deionized water to remove any trapped potassium hydroxide and finally immersed in deionized water >48 h with frequent water changes.

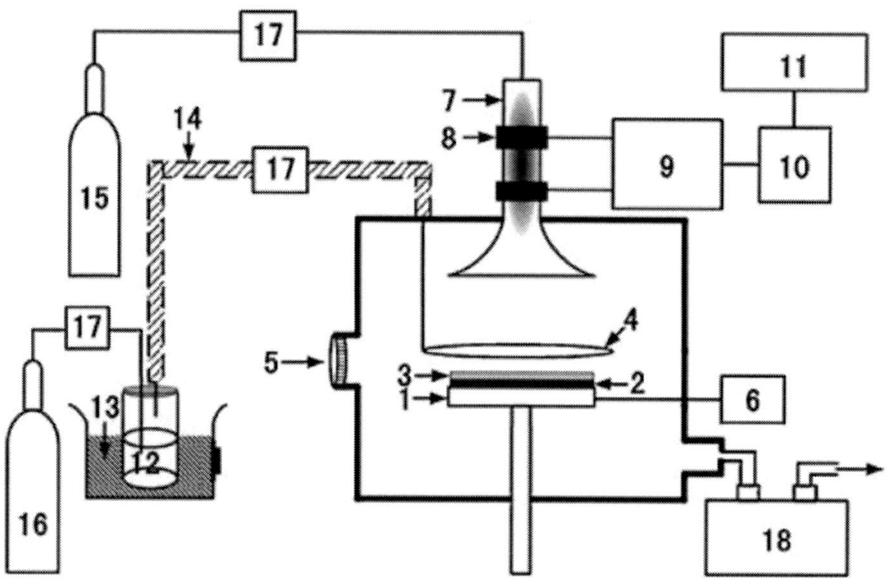

Figure 2. Schematic diagram of the apparatus used for plasma polymerization. 1-Substrate holder, 2-substrate, 3-plasma polymerized membrane, 4-air distribution ring, 5- watch window, 6-bias voltage power source, 7-Pyrex glass tube, 8-electrode, 9-mathching network, 10-RF power supply, 11-pulse modulation, 12-liquid monomer in glass test tube, 13-water bath, 14-heating wires, 15-argon liquid gas to pump, 16-hydrogen liquid gas, 17-gas mass flowmeter, 18-rotary pump.

3.3. Characteristic Measurements of Membranes

The morphology and microstructure of the plasma polymerized AAEM were observed by scanning electron microscope (SEM) (Sirion 200, FEI, USA) at operation voltage of 5.0 kV. In order to expose the cross-section, the QPVBC membrane with silicon substrate was frozen in liquid nitrogen and broken. Before observation, the membrane samples were sputtered with gold for several seconds since the plasma-polymerized membranes were nonconductive.

The chemical structure and composition of the PVBC and QPVBC membranes were analyzed by attenuated total reflection Fourier transform infrared spectroscopy (ATR-FTIR) and X-ray photoelectron spectroscopy (XPS). All the samples were dried in a vacuum oven at 60^0C for 12 h. The ATR-FTIR spectra of dried PVBC and QPVBC membranes were recorded using Nicolet NEXUS 870 spectrometer (Thermo Electron Corporation, USA) in the range of 4000-670 cm^{-1}. The spectra were obtained after 256 scans at 2 cm^{-1} resolution with subtracting the contributions from CO_2 and H_2O (gas). The XPS analysis was carried out using a Thermo ESCALAB 250 spectroscopy (Thermo Electron Corporation, USA) at a power of 150 W with a monochromatic Al Kα radiation at 1486.6 eV. The photoelectrons were detected with a hemispherical analyzer positioned at an angle of 90° with respect to the sample plane. XPS of PVBC and QPVBC membranes were recorded at pass energies of 70 eV for survey spectra and 20 eV for core level spectra. An additional electron gun was used to allow for surface neutralization during the measurements since the plasma polymerized membranes were nonconductive. The energy resolution was about 0.6 eV. The spectrometer energy scale calibration was checked by setting Ag $3d_{5/2}$ = 368.26 eV and the spectra were calibrated with respect to the C 1s peak at 284.6 eV. The curves were fitted with symmetrical Lorentz-Gauss functions.

Thermal degradation and stability of the membranes were investigated using a thermogravimetric analyzer (TGA) (DTG-60H, SHIMADZU, Japan). ~2 mg dried membrane sample, placed in a crucible, was heated from ambient temperature to 800 ^{0}C under flowing nitrogen at the scanning rate of 10^0C min^{-1}.

In order to evaluate the capability of hydroxyl ion transport, the ion-exchange capacities (IECs) of PVBC and QPVBC membranes were measured by classical back titration method [38]. For IECs measurement, dry OH^- form samples were accurately weighed. Three pieces of dried membrane samples prepared in the same time were accurately weighed and then respectively equilibrated with 25 ml 0.005 mol L^{-1} HCl solution for 48 h and then back titrated by 0.005 mol L^{-1} NaOH solution. IEC values of the samples were calculated as the following relation

$$\mathrm{IEC}\ (\mathrm{mmol\ g^{-1}}) = \frac{n_{1,\mathrm{HCl}} - n_{2,\mathrm{HCl}}}{m_{\mathrm{dry}}} \tag{1}$$

where m_{dry} is the mass (g) of the dried sample, $n_{1,\mathrm{HCl}}$ and $n_{2,\mathrm{HCl}}$ are the amount (mmol) of hydrochloric acid required before and after equilibrium, respectively. The average value of the three samples calculated from Equation 1 is the IEC value of the measured membrane.

The water uptake of the PVBC and QPVBC membranes were carried out by measuring the change of weight between the membrane before and after immersion in deionized water.

Firstly, the membrane was soaked in deionized water at room temperature and equilibrated for more than 48 h. The weight of the wet membrane was recorded as a benchmark by measuring after removing excess surface water. The wet membrane was then dried under vacuum at 60 °C until a constant weight was obtained. The water uptake could be calculated using Equation 2

$$\text{Water uptake }\% = \frac{m_{\text{wet}} - m_{\text{dry}}}{m_{\text{dry}}} \times 100\% \tag{2}$$

where m_{wet} is the mass (g) of a wet membrane and m_{dry} is the mass of a dry membrane.

The hydroxide ionic conductivity of the obtained membranes was measured by three-electrode AC impedance spectroscopy using an Autolab potentiostat/galvanostat (IM6e, Zahner, Germany) over a frequency ranging from 0.1 Hz to 1 MHz with oscillating voltage of 10 mV. Before conductivity measurements, the OH^- form of plasma polymerized membrane (QPVBC) samples (2.0 cm × 4.0 cm) and commercial AAEM (AHA, 2.0 cm × 4.0 cm) were fully hydrated in deionized water for at least 48 h until neutral pH was obtained. After removing the surface water, the hydrated membrane was then rapidly placed between two polytetrafluoroethylene (PTFE) plates, with the QPVBC membrane side contacted with three parallel platinum wires. The reference electrode (RE) was connected to the inner platinum wire, and the counter electrode (CE) and working electrode (WE) were connected to the outer two platinum wires, respectively. The testing cell was placed in a chamber with deionized water to keep the water content of the membrane constant during the measurements. The hydroxide ion conductivity σ_{OH^-} could be calculated by

$$\sigma_{\text{OH}^-}\ (\text{S cm}^{-1}) = \frac{l}{R_{\text{m}} A} \tag{3}$$

where A was the cross-sectional area (cm^2) of the membrane, l was the distance (cm) between the working electrode and reference electrode, R_{m} was the membrane resistance (Ω) obtained from the AC impedance data.

3.4. Morphology of PVBC Membranes [39]

The cross-section of plasma membranes deposited on silicon wafer at plasma discharge power of 10 W to 50 W, total pressure of 60 Pa and bias voltage of -10 V can be recorded by SEM image, shown in Figure 3. The plasma-polymerized PVBC membranes are thin and flat with uniform structure at the level of the SEM observation. The thicknesses of the PVBC membranes deposited from 10 W to 50 W, based on SEM image, are 1740 nm, 2880 nm, 608 nm, 400 nm and 787 nm, respectively. From the cross-sectional view in Figure 3 (a), membrane deposited at 10 W peeled off from the substrate because of its poor practical adhesion to the substrate, suggesting hard preservation of membrane in the quaternization and alkalization steps and application in the ADMFC. However, as shown in Figure 3 (b), (c), (d) and (e), it is observed that the operation of higher discharge power (20 W to 50 W) improved the practical adhesion between membrane and substrate, and brought better morphology performance of plasma-polymerized membranes for potential application in ADMFC. Therefore, the discharge power of 20 W, 30 W, 40 W and 50 W was chosen for the further investigation.

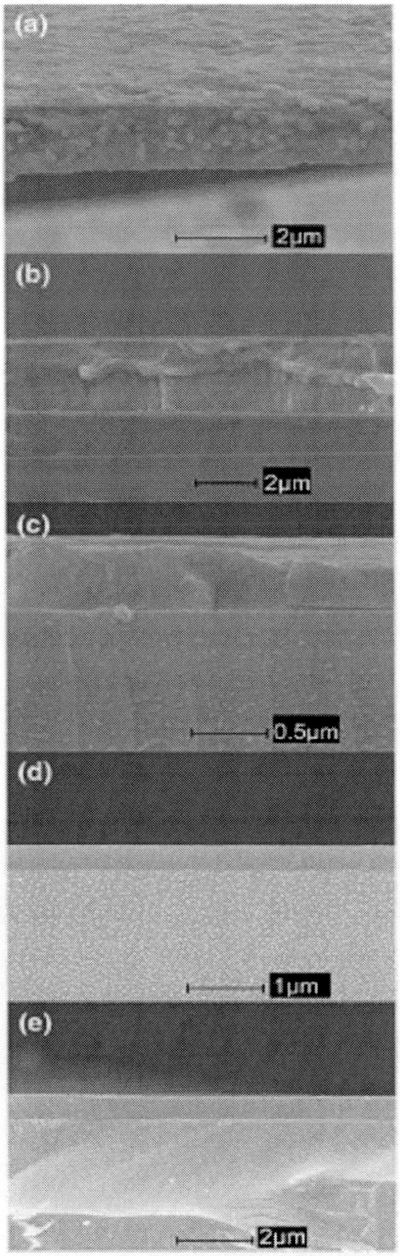

Figure 3. SEM images of the cross-section morphology of PVBC membranes on silica wafer at total pressure of 60 Pa, bias voltage of -10 V and discharge power of: (a) 10 W, (b) 20 W, (c) 30 W, (d) 40 W and (e) 50 W.

3.5. IEC Analysis [39]

IEC is performed to determine the capability of hydroxyl ion transport, and of course, to evaluate the applicability in ADMFCs of obtained membranes. The IEC of QPVBC membrane deposited at discharge power of 20 W is 1.24 mmol/g, which is higher than those of VBC grafted FEP AAEM reported before [26]. Figure 4 shows the influence of plasma discharge power on the IECs of QPVBC membranes deposited at the total pressure of 60 Pa and bias voltage of -10 V. It can be observed that the IECs linearly decrease with increasing plasma discharge power from 20 W to 50 W. The operation of higher discharge power enhances the energy level of plasma, leading to more destruction in functional benzyl chloride groups, and thus decreases the IECs which associate with the contents of quaternary ammonium groups of QPVBC membranes. It can also be indicated that the contents of quaternary ammonium groups in QPVBC membranes are greatly influenced by plasma discharge power due to the damage of benzyl chloride groups in plasma polymerization. In this regard, the effects of plasma discharge power on the preservation of benzyl chloride groups of PVBC membranes should be elucidated.

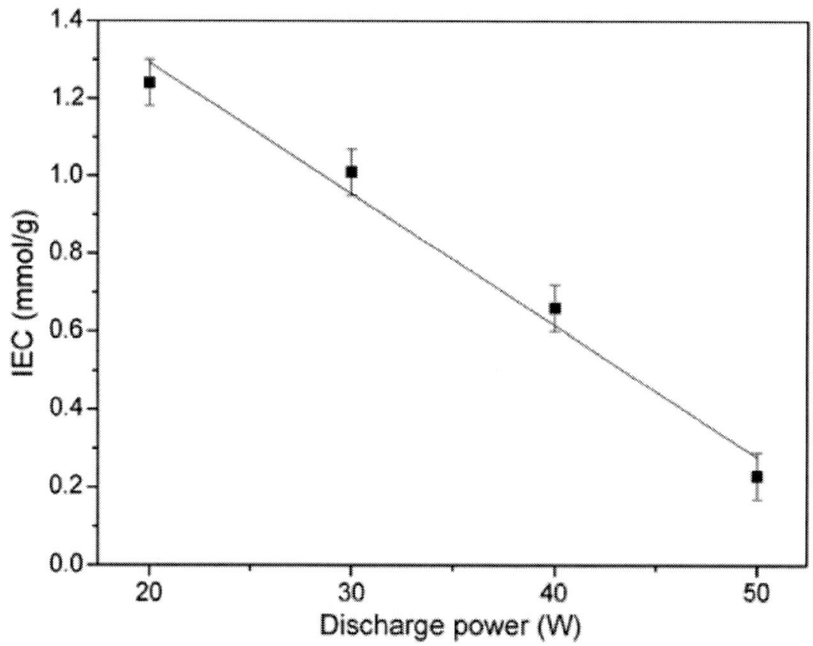

Figure 4. Influence of plasma discharge power on the IECs of QPVBC membranes deposited on PTFE substrate at the total pressure of 60 Pa and bias voltage of -10 V.

3.6. Chemical Structure Characterization of the Membranes [35, 39]

The chemical structures of the PVBC membranes deposited on stainless steel substrate at the total pressure of 60 Pa and bias voltage of -10 V, as a function of plasma discharge power, were analyzed by ATR-FTIR spectra, shown in Figure 5. The absorption at 826 cm^{-1} and 3020 cm^{-1} related to the C–H deformation for para-substituted aromatics and aromatic C–H stretches, respectively, indicates the existence of the benzene ring structure in PVBC membranes at the discharge power of 20 W to 50 W [40, 41]. The peaks at 1265 cm^{-1} and 844 cm^{-1} are assigned to CH_2–Cl wag of benzyl chloride groups and C–Cl stretching vibrations, respectively, suggesting the preservation of the benzyl chloride groups in AGD-PP system[42, 43]. The decrease in the absorption at 1265 cm^{-1} at discharge power from 20 W to 50 W indicates greater polymer damage under the higher discharge power. The absorption at 961 cm^{-1} and 1654 cm^{-1} attributed to the *trans* CH wagging and C=C stretching, respectively, proves the existence of C=C bonds in C–C backbone [43].

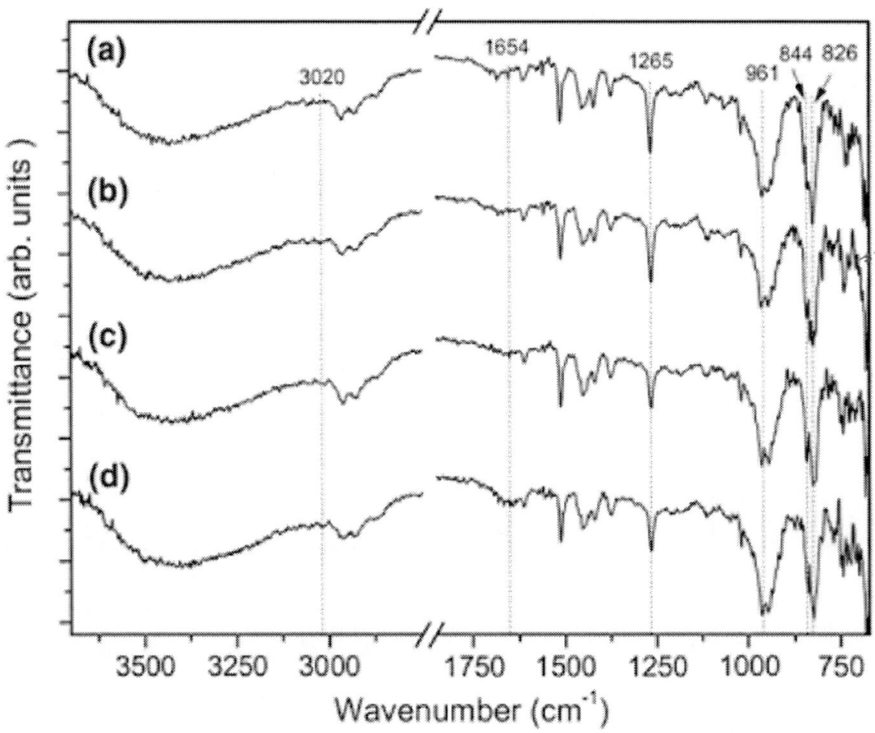

Figure 5. Influence of plasma discharge power on ATR-FTIR spectra of the PVBC membranes deposited at total pressure of 60 Pa, bias voltage of -10 V: (a) 20 W, (b) 30 W, (c) 40 W and (d) 50 W.

In order to characterize the changes in the contents of benzyl chloride groups, XPS data were conducted. Table 1 shows the XPS results of PVBC membranes with the contents of O, N and Cl expressed as atom %, as a function of discharge power. The Cl 2p core-level spectrum of the sample can be curve-fitted with two spin-orbit-split doublets, shown in Figure 6, with the binding energy for Cl $2p_{3/2}$ peak components located at about 199.9 ± 0.2 eV attributable to the covalently bonded chlorine species in benzyl chloride groups and polymer backbone, and 197.4 ± 0.2 eV related to the ionic chloride (Cl^-) in HCl [44, 45]. The quantitative analysis confirms the high preservation of benzyl chloride groups in PVBC membranes at discharge power of 20 W. The huge decrement of covalently bonded chlorine species under the higher discharge power indicates an effect of plasma ablation [6]. From Table 1 and Figure 6, it can be seen that the plasma ablation of PVBC membranes at discharge power of 30 W compared with 20 W does not lead to large changes in the content of Cl, but changes in the content of Cl^c (covalently bonded chlorine species in benzyl chloride groups and polymer backbone) obviously. The decrease in the C–Cl fraction at discharge power from 20 W to 30 W indicates that most of the covalently bonded chlorine species might be destroyed by plasma ablation. Moreover, the increase in Cl^- fraction when the discharge power is up to 30 W from 20 W also proves the dissociation of C–Cl bonds and the formation of HCl on the surface of PVBC membranes. At the discharge power from 30 W to 50 W, there are slight changes in the content of Cl^c, whereas, the content of Cl^- decreases. Considering the air contamination during and after plasma polymerization process due to the trapped free radicals within the membrane network and on the membrane surface, the influence of oxygen and nitrogen on membrane chemical structure during plasma polymerization process is hard to estimate [1].

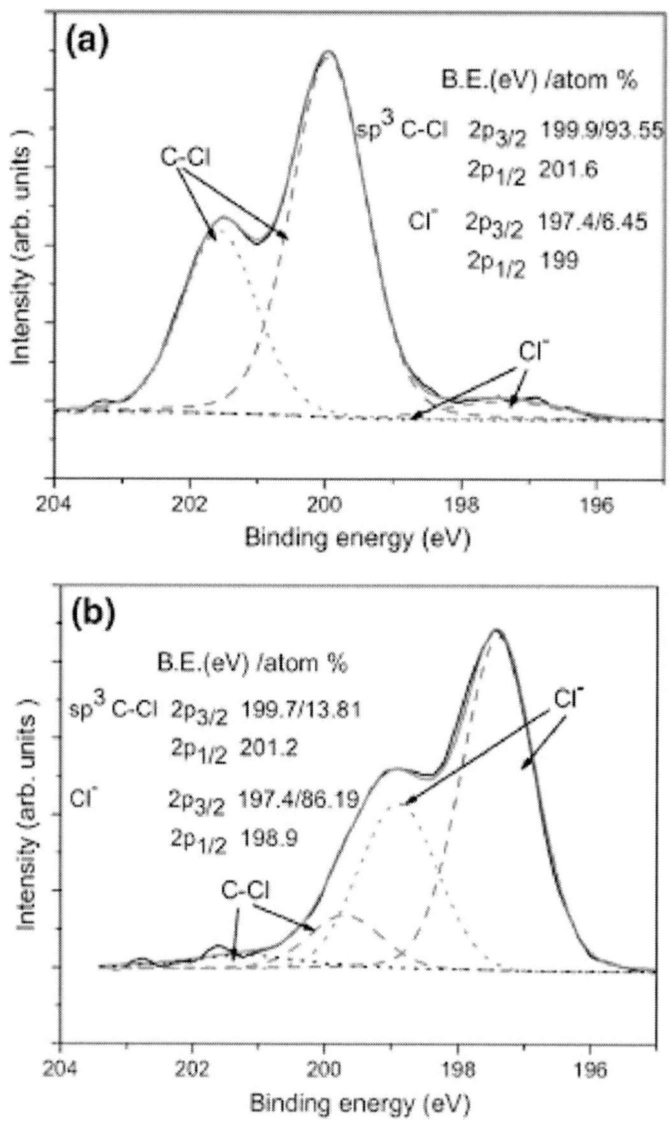

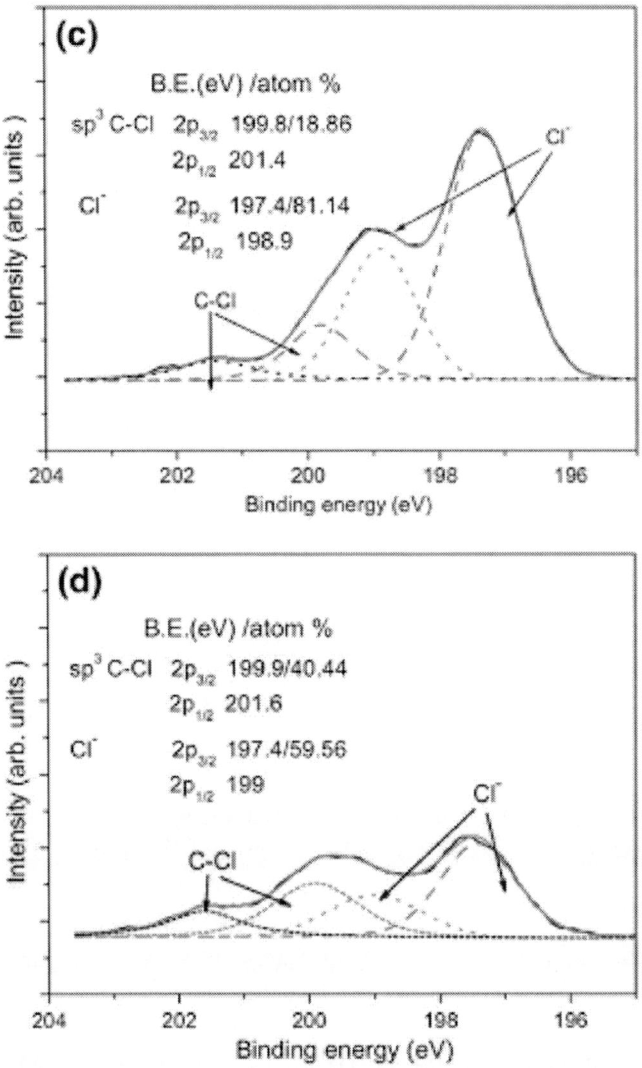

Figure 6. Influence of plasma discharge power on XPS Cl 2p spectra of the PVBC membranes deposited at total pressure of 60 Pa, bias voltage of -10 V: (a) 20 W, (b) 30 W, (c) 40 W and (d) 50 W.

Table 1. XPS elemental analysis of plasma-polymerized PVBC membranes as a function of plasma discharge power deposited at total pressure of 60 Pa and bias voltage of -10 V

PVBC membranes	C atom %	N atom %	O atom %	Cl atom %	Cl^{c*} atom %	Cl^- atom %
20 W	92.00	0.72	4.13	2.98	2.79	0.19
30 W	83.23	4.51	8.58	2.73	0.38	2.35
40 W	83.72	3.37	10.82	2.09	0.39	1.70
50 W	89.54	1.77	7.49	1.19	0.48	0.71

[*] Cl^c: covalently bonded chlorine species in benzyl chloride groups and polymer backbone.

To acquire more information on the chemical structure characteristics of plasma-polymerized membranes influenced by the plasma discharge power, C 1s spectra were analyzed. As shown in Figure 7, the C 1s peak is deconvolved into four peaks [45-48]: peak (1) at 284.3 eV attributed to the sp^2 carbon atoms (C=C); peak (2) at 284.8 eV corresponded to the sp^3 carbon atoms (C–C and C–H); peak (3) at 286.1 eV related to sp^3 carbon carrying one chlorine atom (chlorine fixed by a free radical process on two nonconjugated sp^2 carbon atoms) or to sp^3 carbon bonded to one oxygen atom (C–O); peak (4) at 287.2 eV assigned to the oxygenated groups (carbonyl). The initial decrease in the C=C fraction at discharge power from 20 W to 30 W is likely due to the occurrence of polymerization. Afterward, the C=C fraction increases at discharge power from 30 W to 50 W, whereas Cl^- content decreases and Cl^c content slightly increases. These changes suggest the occurrence of elimination reaction in C–C backbone with participation of chloridion through generating covalently bonded chlorine. A possible reaction mechanism based on XPS analysis might occur during the plasma polymerization process from VBC monomer, as shown in Figure 8. Most studies of plasma polymerization mechanisms suggest that free radicals are the most likely reactive species involved in polymer formation under plasma conditions. It is believed that free radicals are generated firstly on the dissociated π bonds in C=C because of its low bond dissociation energy [37]. Dissociation can also be occurred in C–Cl bonds of benzyl chloride groups when the impinging particles energy is higher than its bond dissociation energy. Free radicals can further react with monomers and active species to form polymerized membranes. Chlorine radicals existing in plasma atmosphere can also react with active species. This process may produce C–Cl bonds in C–C backbone, as shown in Figure 8 (a). The benzyl chloride groups in polymer membranes might be further destroyed due to the high-energy particles bombardment (as an effect of plasma ablation on polymer), which indicates the decreasing of benzyl chloride groups contents by turning up the discharge power, as shown in Figure 8 (b). The mechanism of C=C fraction formation may be more complicated. A possibility is the occurrence of elimination reaction through eliminating covalent chlorine and beta hydrogen by the high-energy particles bombardment shown in Figure 8 (c). This gives reasons for the decrease of Cl^c content while Cl^- content increases. Figure 8 (d) shows another possibility that the C=C bonds may be generated by a series of reactions including the formation of carbonium ion by high-energy particles bombardment, nucleophilic attack by chloridion and elimination reaction. This can be certified by the changes of C=C fraction and Cl^- content and the slight increase of Cl^c content at the discharge power from 30 W to 50 W.

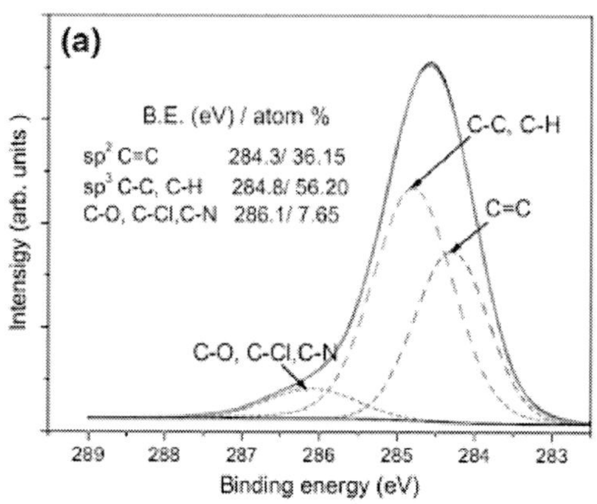

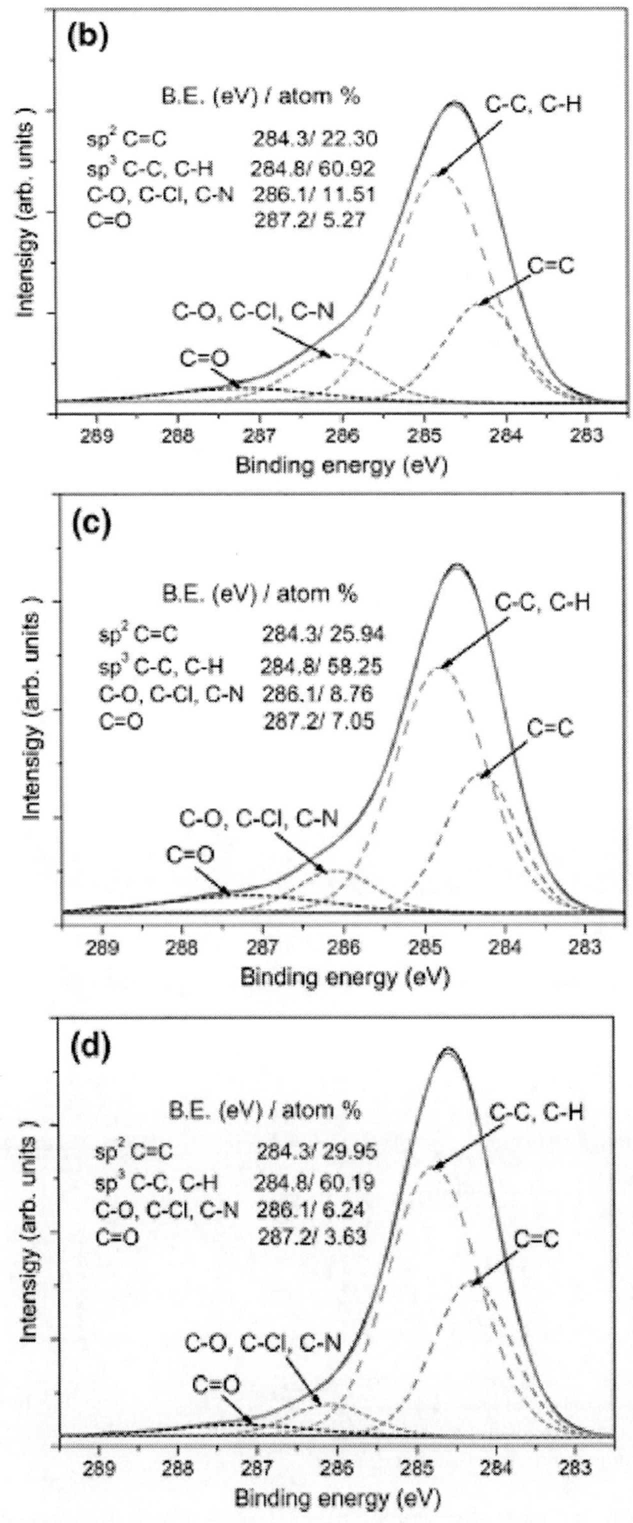

Figure 7. Influence of plasma discharge power on XPS C 1s spectra of the PVBC membranes deposited at total pressure of 60 Pa, bias voltage of -10 V: (a) 20 W, (b) 30 W, (c) 40 W and (d) 50 W.

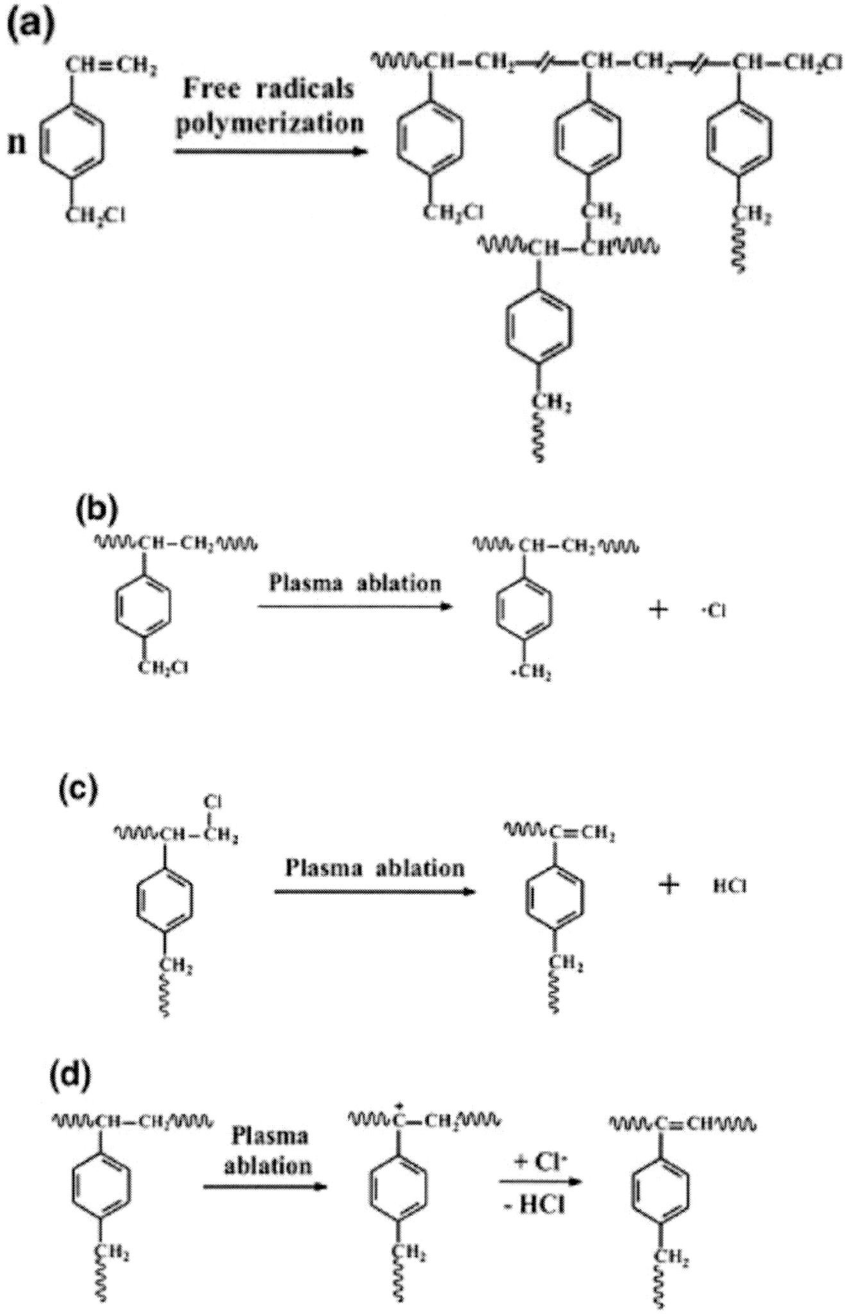

Figure 8. Possible mechanism of plasma polymerization from VBC monomer. (a) formation of cross-linked PVBC membranes; (b) destruction of functional groups in PVBC membranes; (c) and (d) generation of C=C bonds in polymer backbone.

This mechanism indicates the competitive free radicals polymerization and plasma ablation in plasma polymerization process which affect the polymerization of monomers, and the structure of the functional membrane, respectively [6, 49]. It is believed that most of the chemical reactions occur between the plasma-generated radicals and the active species in the plasma atmosphere [50]. It may be, however, that many species in the plasma do not contribute to the plasma polymerization but rather contribute to the ablation [3]. The competitive nature of reactions is highly dependent on the condition of plasma polymerization, in particular the energy input level. The density of free radicals and other reactive species increases in various degrees as the plasma discharge power increases. In the range of lower discharge power, the effect of plasma ablation is less obvious. In the range of higher discharge power, however, the effect of plasma ablation becomes a vital element to affect the membrane chemical structure when other plasma polymerization conditions keep constant. This result is crucial because the plasma discharge power as one of the most important parameters for the industrial application of plasma-polymerized electrolyte synthesis can directly influence the electron energy distributing function, the density of all the active species of the discharge, and then the membrane chemical structure [1]. Except for the reactions mechanism mentioned in Figure 8, many other reactions may be involved in the plasma polymerization process due to the complexity of plasma polymerization mechanism [3]. Herein, we take the free radicals generated on the C=C bonds and benzyl chloride groups, C=C bonds generated by the reactions of covalent chlorine and chloridion as examples and illustrate the possible reactions on the plasma polymerization and plasma ablation.

The existence of quaternary nitrogen groups is further demonstrated by the decomposition of N 1s spectrum, shown in Figure 9. The peak at 401.7 eV for QPVBC membrane was attributed to the quaternary or protonated nitrogen (N^+), confirming that plasma polymerization and quaternization produced the expected functional groups [51-53]. Quantitatively analysis of the N 1s spectrum of QPVBC membrane, shown in Table 2 and Figure 9, reveals that the percentage of quaternary ammonium groups with respect to the total atom quantity is 1.85% for plasma-polymerized alkaline anion-exchange membrane, which is higher than that of sulfonic acid groups for Nafion 117 membrane, shown as 1.23% in ref. [54]. The high percentage of quaternary ammonium functional groups indicates high ion conductivity of the plasma-polymerized AAEM and, as a result, a good performance in ADAFCs application. The peaks appearing at 399.2 eV and 401.2 eV in Figure 9 of PVBC membrane are assigned with C–N groups and amide nitrogen, respectively, as a result of air contamination [55-57]. In order to characterize the changes in the contents of benzyl chloride groups, the Cl 2p core-level spectrum of the samples were curve-fitted with two spin-orbit-split doublets, shown in Figure 10, with the binding energy for Cl $2p_{3/2}$ peak components located at 199.9 ± 0.2 eV attributable to the covalently bonded chlorine species in benzyl chloride groups, and 197.4 ± 0.2 eV related to the ionic chloride (Cl^-) [44, 45]. The results of Cl 2p spectra analysis are in good agreement with what shown by N 1s spectra. The decrease of the chloride and covalently bonded chlorine species contents verify the successful quaternization of benzyl chloride groups into $-N^+(CH_3)_3OH^-$ groups. However, the residual benzyl chloride groups and ionic chloride in QPVBC membrane, according to Figure 4f, indicates the incomplete quaternization and alkalization. In the quaternization process, plasma polymerized membrane was soaked in TMA solution for 48 h, which was much longer than 4 h described as the maximum quaternization levels in the literature [28]. This difficulty for complete quaternization may be due to the high cross-linked C-C backbone and dense structure of the plasma-polymerized membrane.

Table 2. XPS elemental analysis of plasma-polymerized PVBC and QPVBC membranes deposited at total pressure of 60 Pa, discharge power of 20 W and bias voltage of -10 V

Membranes	C atom %	O atom %	Cl atom %	N atom %	N^+ atom %
PVBC	92.00	4.13	2.98	0.72	0
QPVBC	76.12	18.74	0.71	4.43	1.85

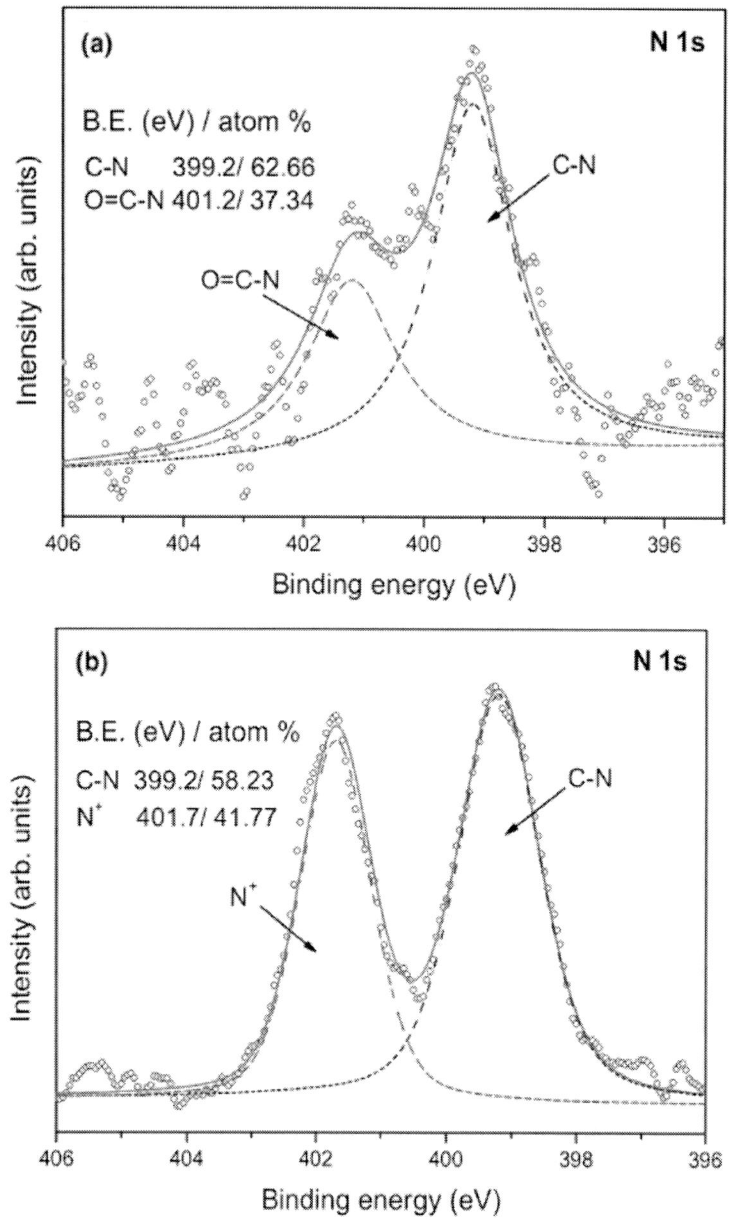

Figure 9. XPS N 1s spectra for PVBC membrane (a) and QPVBC membrane (b).

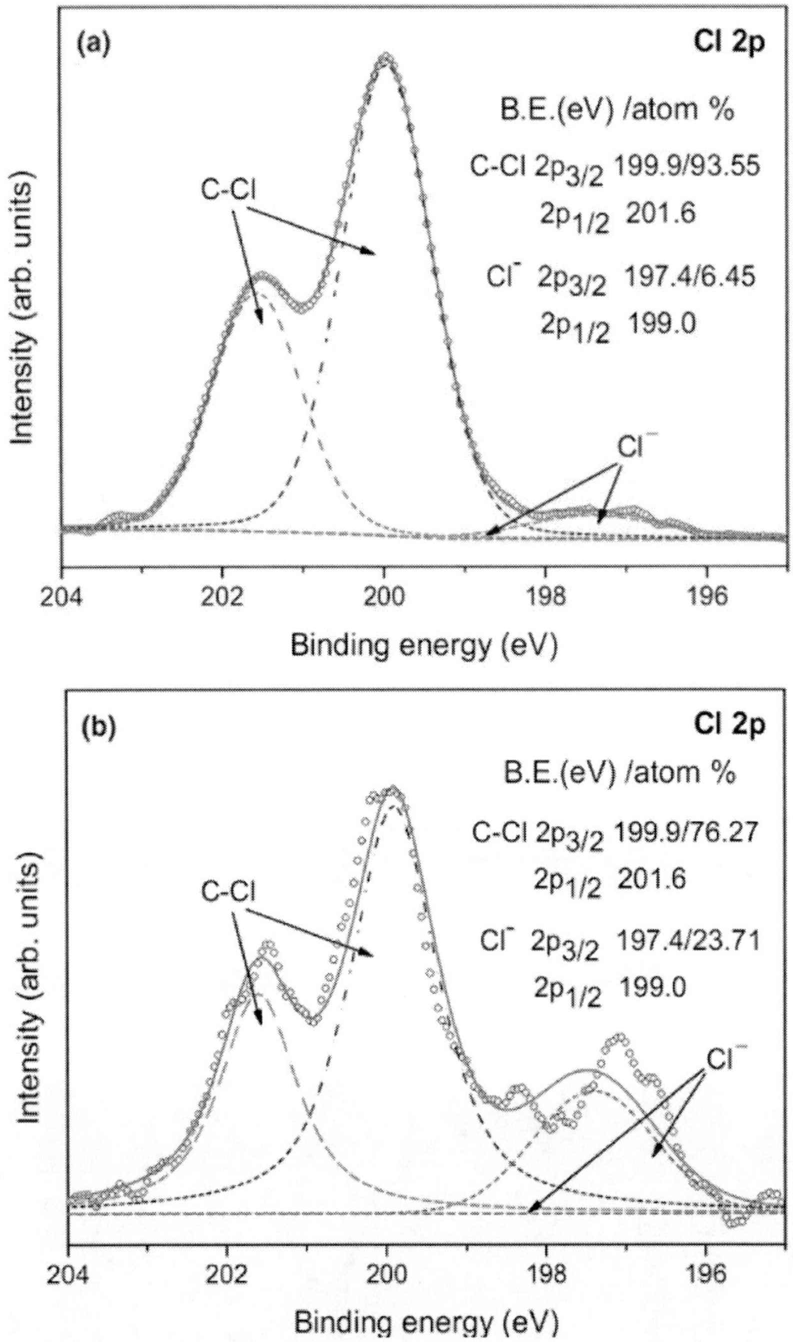

Figure 10. XPS Cl 2p spectra for PVBC membrane (a) and QPVBC membrane (b).

3.7. Ionic Conductivity [35]

Hydroxide ion conductivity of the AAEM is a crucial criterion for evaluating its performance in ADAFCs. To evaluate the intrinsic nature of the polymer electrolyte, electrolyte solution was not used during ionic conductivity measurements in this work. Figure

11 shows the Nyquist plot of ($-Z''$ vs. Z') for the QPVBC membrane measured in deionized water, fitted by the equivalent circuit. The OH^- ion conductivity of the membrane (σ_{OH^-}) can be calculated by Equation 3 from the AC impedance data. The resistance of QPVBC membrane R_m is 47 862 Ω in the hydrated membrane thickness of 3.16 μm at 20 ℃. The OH^- ion conductivity of QPVBC membrane, according to Equation 3, is 0.0331 S cm^{-1} which is in the order of 0.01 S cm^{-1}, indicating the satisfactory for application of the membrane in fuel cells [25]. However, our membrane with dual properties of high conductivity and high cross-linked structure seems not agree with the inference in the literature that cross-linked structure can restrict the mobility of charged sites and decrease the free volume of the membrane, resulting in less and smaller hydrophilic pathway for ion mobility and water absorption [58, 59]. This observation can be attributed to: (1) the high content of quaternary ammonium functional groups (1.85 atom%) fixed in the QPVBC membrane matrix indicating the excellent hydrophility and the great potential for promoting conductivity; (2) the satisfactory IEC value suggesting the high activity of the quaternary ammonium functional groups; (3) the high water uptake (66.67 wt%) leading to adequate water absoption in ion transport; and (4) the cross-linked structure resulting in the increase of ion incorporation [31, 60].

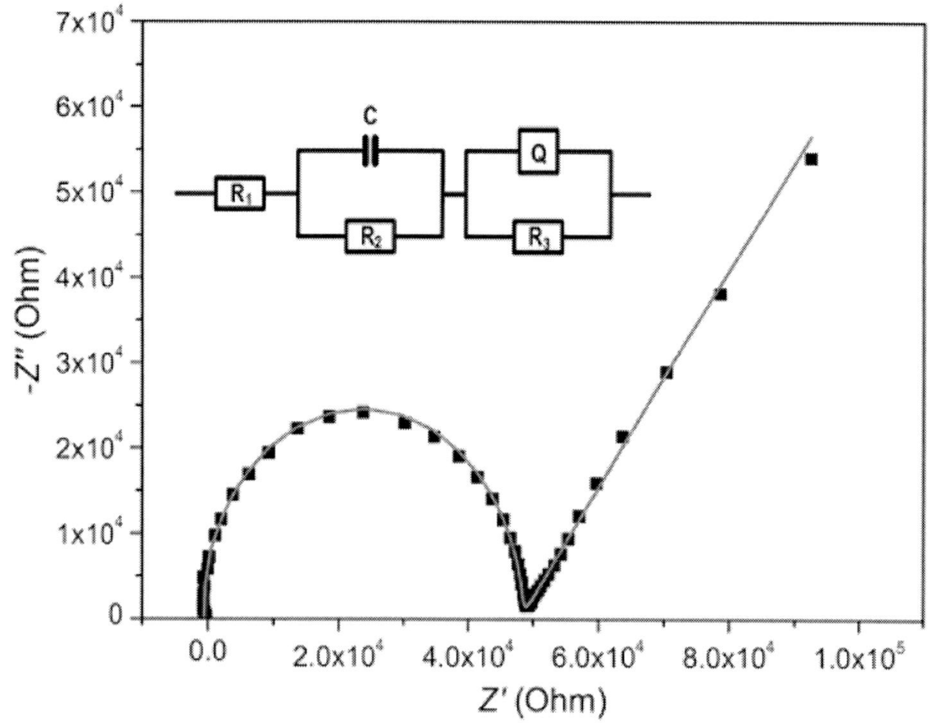

Figure 11. Nyquist plot for the QPVBC membrane and an equivalent circuit. The dots are the experimental results. The curves are the fitting of the experimental data using the equivalent circuit.

Previous studies have demonstrated that the ionic conductivities of ion-exchange membranes are temperature dependence [61-64]. Figure 12 shows the correlation between conductivity and temperature for QPVBC membrane. The conductivity of our membrane increases with temperature and exhibits conductivities up to 60℃ when immersed in deionized water. Linear regression of $\ln \sigma_{OH^-}$ vs. $1000/T$ (T is the absolute temperature in Kelvins) was performed assuming an Arrhenius relationship. The anion transport activation energy (E_a) of the QPVBC membrane, which corresponds to the energy barrier for carrier transfer from one free site to another, can be obtained according to the Arrhenius equation:

$$E_a \ (\text{kJ mol}^{-1}) = RT^2 \frac{d \ln \sigma_{OH^-}}{dT} \tag{4}$$

where R is the pure gas constant (8.314 J K^{-1} mol^{-1}). The activation energy value for alkalinized QPVBC membrane is 7.632 kJ mol^{-1}, which is much lower than the value for the radiation-grafted AAEMs [65]. The high σ_{OH^-} and low E_a values might be attributed to the high content of tetraalkylammonium ions, water uptake and IEC of the membrane, which conduce to forming continuous ion transferring channel and make the movement of ion easily [22]. The high conductivity and low ion transport activation energy of the QPVBC membrane will facilitate low temperature fuel cell operation.

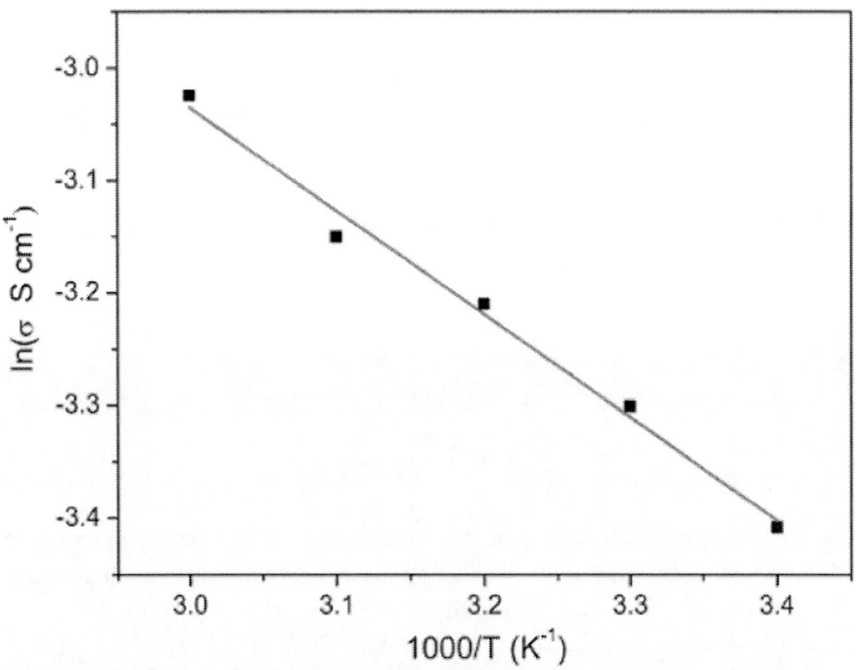

Figure 12. Arrhenius plot showing the temperature dependence of hydroxide conductivity for the QPVBC membrane.

3.8. Thermal Stability [35]

AAEMs with high thermal stability are desirable since operation of ADMFCs at elevated temperature would not only reduce thermodynamic voltage losses but also improve the electrokinetics [18]. The thermal stability of PVBC membranes deposited at discharge power of 20 W to 50 W was investigated by DTA in flowing nitrogen from ambient temperature to 800℃. The sample for DTA is PVBC membranes combined with PTFE substrate. In DTA trace, as shown in Figure 13, the PVBC membranes are degraded mainly in two steps, with the first step ascribed to degradation of PVBC membranes in the range of 310-366 °C, indicating the high thermal stability of PVBC membranes due to the highly cross-linked structure, and the second step related to the degradation of PTFE substrate in the range of 460-624 °C [66].

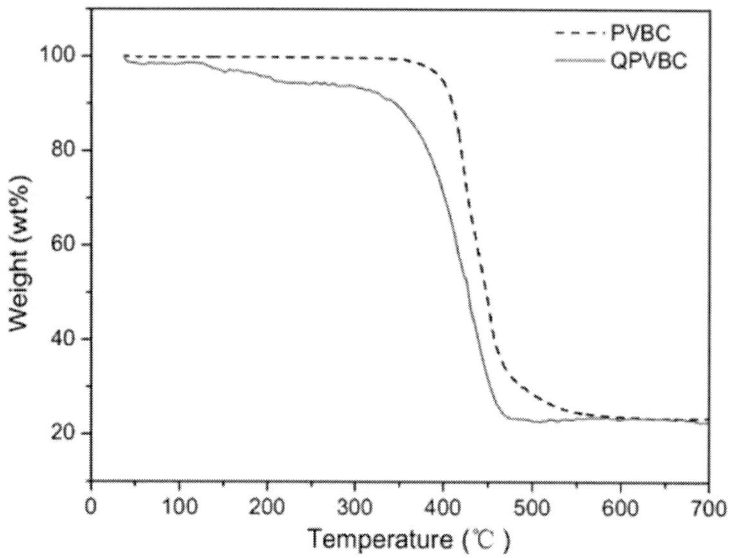

Figure 13. Thermo gravimetric analysis of PVBC and QPVBC membranes.

4. SUMMARY

In this work, we presented an approach to prepare quaternized poly(vinylbenzyl chloride) (QPVBC) membranes by after-glow discharge plasma polymerization, quaternization and alkalization process. SEM images showed that the plasma-polymerized membranes were flat and uniform. The operation of higher discharge power (20 W to 50 W) improved the practical adhesion between membrane and substrate, brought better morphology performance of plasma-polymerized membranes for potential application in ADMFC. XPS and ATR-FTIR data showed the variation of the chemical structure properties of plasma-polymerized membranes as a function of plasma discharge power. A mechanism of plasma polymerization using VBC as monomer was proposed, indicating the competitive effects of free radicals polymerization and plasma ablation in the plasma polymerization process. Because of the enhancement of plasma ablation with increasing energy input, plasma-polymerized membranes deposited at lower discharge power have higher contents of functional groups and less structure damage. As a result, it seems to be an effective way to fabricate QPVBC membranes by after-glow discharge plasma polymerization at total pressure of 60 Pa, bias voltage of -10 V and plasma discharge power of 20 W. The content of the quaternary ammonium groups in QPVBC membrane deposited at 20 W, based on XPS results, is up to 1.85 atom%. TGA analysis indicates that the plasma-polymerized AAEMs can be stably used below 100 °C. The high content of functional groups, high water uptake (66.67 wt%) and satisfactory IEC provide sufficient fixed cations and water molecules to transport OH^- ions, leading to an excellent hydroxide ion conductivity (0.0331 S cm^{-1} in deionized water at 20 °C) and low activation energy (7.632 kJ mol^{-1}). This performance criterion expects that such kind of plasma-polymerized AAEMs have great potential for application in alkaline directly alcohol fuel cells.

REFERENCES

[1] Biederman, H. Plasma Polymer Films; Imperial College Press: Covent Gardent, London, 2004.

[2] Lentz, R. W. null; Wiley: New York, 1967.

[3] Yasuda, H. Plasma Polymerization; Acadamic Press, INC: Orlando, Florida, 1985.

[4] Denaro, A. R.; Owens, P. A. and Crawshaw, A. *Eur. Polym. J.* 1968, 4, 93.

[5] Yasuda, H. and Hirotsu, T. *J. Polym. Sci. Pol. Chem.* 1978, 16, 2587-2592.

[6] Yasuda, H. and Yasuda, T. *J. Polym. Sci. Pol. Chem.* 2000, 38, 943-953.

[7] Yasuda, H.; Sharma, A. K. and Yasuda, T. *J. Polym. Sci., Polym. Chem. Ed.* 1981, 19, 1285.

[8] Morosoff, N.; Newton, W. and Yasuda, H. *J. Vac. Sci. Technol.* 1978, 15, 1815.

[9] Sharma, A. K. and Yasuda, H. *J. Adhes.* 1982, 13, 201.

[10] Wang, Y.; Li, L.; Hu, L.; Zhuang, L.; Lu, J. and Xu, B. *Electrochem. Commun.* 2003, 5, 662-666.

[11] Sevilla, M.; Sanchís, C.; Valdés-Solís, T.; Morallón, E. and Fuertes, A. B. *Electrochim. Acta.* 2009, 54, 2234-2238.

[12] Yu, E. H. and Scott, K. J. *Power Sources.* 2004, 137, 248-256.

[13] Antolini, E. J. *Power Sources.* 2007, 170, 1-12.

[14] Borup, R.; Meyers, J.; Pivovar, B.; Kim, Y. S.; Mukundan, R.; Garland, N.; Myers, D.; Wilson, M.; Garzon, F.; Wood, D.; Zelenay, P.; More, K.; Stroh, K.; Zawodzinski, T.; Boncella, J.; McGrath, J. E.; Inaba, M.; Miyatake, K.; Hori, M.; Ota, K.; Ogumi, Z.; Miyata, S.; Nishikata, A.; Siroma, Z.; Uchimoto, Y.; Yasuda, K.; Kimjima, K. C. and Iwashita, N. *Chem. Rev.* 2007, 107, 3904-3951.

[15] Lu, S. F.; Pan, J.; Huang, A. B.; Zhuang, L. and Lu, J. T. P. *Natl. Acad. Sci. U.S.A.* 2008, 105, 20611-20614.

[16] Varcoe, J. R.; Slade, R. C. T.; Wright, G. L. and Chen, Y. L. *J. Phys. Chem. B* 2006, 110, 21041-21049.

[17] Coutanceau, C.; Demarconnay, L.; Lamy, C. and Leger, J. M. *J. Power Sources.* 2006, 156, 14-19.
[18] Varcoe, J. R. and Slade, R. C. T. *Fuel Cells.* 2005, 5, 187-200.
[19] Kordesch, K. and Oliveira, J. C. T. *Int. J. Hydrogen Energ.* 1988, 13, 411-427.
[20] Hwang, G.-J. and Ohya, H. J. *Membr. Sci.* 1998, 149, 163-169.
[21] Komkova, E. N.; Stamatialis, D. F.; Strathmann, H. and Wessling, M. *J. Membr. Sci.* 2004, 244, 25-34.
[22] Xiong, Y.; Fang, J.; Zeng, Q. H. and Liu, Q. L. *J. Membr. Sci.* 2008, 311, 319-325.
[23] Wang, G. G.; Weng, Y. M.; Chu, D.; Xie, D. and Chen, R. R. *J. Membr. Sci.* 2009, 326, 4-8.
[24] Xu, T.; Liu, Z. and Yang, W. *J. Membr. Sci.* 2005, 249, 183-191.
[25] Danks, T. N.; Slade, R. C. T. and Varcoe, J. R. *J. Mater. Chem.* 2003, 13, 712-721.
[26] Herman, H.; Slade, R. C. T. and Varcoe, J. R. *J. Membr. Sci.* 2003, 218, 147-163.
[27] Varcoe, J. R. *Phys. Chem. Chem. Phys.* 2007, 9, 1479-1486.
[28] Varcoe, J. R.; Slade, R. C. T.; Yee, E. L. H.; Poynton, S. D.; Driscoll, D. J. and Apperley, D. C. *Chem. Mater.* 2007, 19, 2686-2693.
[29] Hu, J.; Zhang, C.; Cong, J.; Toyoda, H.; Nagatsu, M. and Meng, Y. J. *Power Sources.* 2011, 196, 4483-4490.
[30] Hibbs, M. R.; Fujimoto, C. H. and Cornelius, C. J. *Macromolecules.* 2009, 42, 8316-8321.
[31] Robertson, N. J.; Kostalik, H. A.; Clark, T. J.; Mutolo, P. F.; Abruna, H. D. and Coates, G. W. *J. Am. Chem. Soc.* 2010, 132, 3400-3404.
[32] Wang, J.; Zhao, Z.; Gong, F.; Li, S. and Zhang, S. *Macromolecules.* 2009, 42, 8711-8717.
[33] Pan, J.; Lu, S.; Li, Y.; Huang, A.; Zhuang, L. and Lu, *J. Adv. Funct. Mater.* 2010, 20, 312-319.
[34] Hibbs, M. R.; Hickner, M. A.; Alam, T. M.; McIntyre, S. K.; Fujimoto, C. H. and Cornelius, C. J. *Chem. Mater.* 2008, 20, 2566-2573.
[35] Zhang, C.; Hu, J.; Nagatsu, M.; Meng, Y.; Shen, W.; Toyoda, H. and Shu, X. Plasma Process. Polym. 2011, in press.
[36] Zhang, C.; Hu, J.; Cong, J.; Zhao, Y.; Shen, W.; Toyoda, H.; Nagatsu, M. and Meng, Y. *J. Power Sources.* 2011, 196, 5386-5393.
[37] D.R. Lide; G. Baysinger; L.I. Berger; R.N. Goldberg; H.V. Kehiaian; K. Kuchitsu; G. Rosenblatt; D.L. Roth and Zwillinger, D. *Handbook of Chemistry and Physics;* CRC Press: Boca Raton, Florida, 2006.
[38] Nagarale, R. K.; Gohil, G. S.; Shahi, V. K. and Rangarajan, R. Macromolecules 2004, 37, 10023-10030.
[39] Hu, J.; Meng, Y.; Zhang, C. and Fang, S. *Thin Solid Films.* 2011, 519, 2155-2162.
[40] Sundaraganesan, N.; Saleem, H.; Mohan, S.; Ramalingam, M. and Sethuraman, V. *Spectrochim. Acta A* 2005, 62, 740-751.
[41] Reyes-Labarta, J.; Herrero, M.; Tiemblo, P.; Mijangos, C. and Reinecke, H. *Polymer.* 2003, 44, 2263-2269.
[42] Teare, D. O. H.; Barwick, D. C.; Schofield, W. C. E.; Garrod, R. P.; Ward, L. J. and Badyal, J. P. S. *Langmuir.* 2005, 21, 11425-11430.
[43] Rajendran, S. and Uma, T. J. *Power Sources.* 2000, 88, 282-285.

[44] Roscoe, S. B.; Yitzchaik, S.; Kakkar, A. K.; Marks, T. J.; Xu, Z. Y.; Zhang, T. G.; Lin, W. P. and Wong, G. K. *Langmuir.* 1996, 12, 5338-5349.
[45] Papirer, E.; Lacroix, R.; Donnet, J.-B.; Nanse, G. and Fioux, P. Carbon 1995, 33, 63-72.
[46] Briggs, D. and Beamson, G. *Anal. Chem.* 1992, 64, 1729-1736.
[47] Okpalugo, T. I. T.; Papakonstantinou, P.; Murphy, H.; McLaughlin, J. and Brown, N. M. D. *Carbon.* 2005, 43, 153-161.
[48] Xu, F. J.; Kang, E. T. and Neoh, K. G. *Macromolecules.* 2005, 38, 1573-1580.
[49] Denaro, A. R.; Owens, P. A. and Crawshaw, A. *Eur. Polym. J.* 1968, 4, 93-106.
[50] Chen, C.; Liang, B.; Lu, D.; Ogino, A.; Wang, X. and Nagatsu, M. *Carbon.* 2010, 48, 939-948.
[51] Tan, K. L.; Tan, B. T. G.; Kang, E. T. and Neoh, K. G. *J. Mater. Sci.* 1992, 27, 4056-4060.
[52] Welle, A.; Liao, J. D.; Kaiser, K.; Grunze, M.; Mäder, U. and Blank, N. *Appl. Surf. Sci.* 1997, 119, 185-198.
[53] Shi, Z.; Neoh, K. G. and Kang, E. T. *Biomaterials.* 2005, 26, 501-508.
[54] Jiang, Z.; Jiang, Z.; Yu, X. and Meng, Y. *Plasma Process. Polym.* 2010, 7, 382-389.
[55] Chen, C.; Liang, B.; Lu, D.; Ogino, A.; Wang, X. and Nagatsu, M. *Carbon.* 2010, 48, 939-948.
[56] Kelemen, S. R.; Afeworki, M.; Gorbaty, M. L.; Kwiatek, P. J.; Solum, M. S.; Hu, J. Z. and Pugmire, R. J. *Energ. Fuel.* 2002, 16, 1507-1515.
[57] Wang, Y.; Shao, Y.; Matson, D. W.; Li, J. and Lin, Y. *ACS Nano.* 2010, 4, 1790-1798.
[58] Qiao, J.; Hamaya, T. and Okada, T. *Polymer.* 2005, 46, 10809-10816.
[59] Feng, S.; Shang, Y.; Xie, X.; Wang, Y. and Xu, J. *J. Membr. Sci.* 2009, 335, 13-20.
[60] Kumar, M.; Singh, S. and Shahi, V. K. *J. Phys. Chem.* B 2010, 114, 198-206.
[61] Yang, C. C. *J. Membr. Sci.* 2007, 288, 51-60.
[62] Li, L. and Wang, Y. J. *Membr. Sci.* 2005, 262, 1-4.
[63] Tripathi, B. P.; Kumar, M. and Shahi, V. K. *J. Membr. Sci.* 2010, 360, 90-101.
[64] Xiong, Y.; Liu, Q. L.; Zhu, A. M.; Huang, S. M. and Zeng, Q. H. *J. Power Sources.* 2009, 186, 328-333.
[65] Slade, R. C. T. and Varcoe, J. R. *Solid State Ionics.* 2005, 176, 585-597.
[66] Yu, T.; Lin, H.; Shen, K.; Huang, L.; Chang, Y.; Jung, G. and Huang, J. J. *Polym. Res.* 2004, 11, 217-224.

In: Radical Polymerization: New Developments ISBN 978-1-62100-406-6
Editors: I. O. Paulauskas, L. A. Urbonas, pp. 175-198

Chapter 5

A Review on Radical Polymerization Used for Design and Development of Biomaterials

K. S. V. Krishna Rao[1]* and K. Madhusudana Rao[2]
[1] Department of Chemistry, Yogi Vemana University, Kadapa, India.
[2] Department of Chemistry, Sri Krishnadevaraya University, Anantapur, India.

Abstract

Recent mechanistic developments of radical polymerization in the field of biomaterials will be reviewed. Carbohydrate-based polymers are used in the field of biomedical applications, such as drug delivery devices, contact lenses, bioseparation, biosensors systems tissue engineering scaffolds, cell culture supports, etc. Advances in this field are particularly relevant to applications in the areas of drug delivery and regenerative medicine etc. Particular emphasis is placed on stimuli-responsive (pH and temperature) polymers, micro/nano gels, membranes, and films. The aim of present review analyses and summarizes recent developments in the field of graft modification of carbohydrates by controlled/living radical polymerization.

Keywords: Radical Polymerization, Carbohydrate Polymer, Drug Delivery Devises, Biomaterials.

Abbreviations

Monomers

AA	Acrylic acid
AAP	4-aminoantipyrene

* Corresponding author: K.S.V. Krishna Rao; E-mail: drksvkrishna@yahoo.com.

AcM	acrylomorpholine
BA	Butyl acrylate
DEAEMA	2-(diethylamino) ethyl methacrylate
DMAEMA	(2-dimethyl amino) ethylmethacrylate
EMO	3-ethyl-3-methacryloyloxy-methyloxytane
HAEAPMA	2-hydroxy-3-(2-aminoethyl)amino]propyl methacrylate
HDMA	Hexadecylmethacrylate
MA	Methyl acrylate
MAm	Methacrylamide
MEO_2MA	2-(2-methoxyethoxy) ethyl methacrylate
MMA	Methyl methacrylate
MPC	2-Methacryloxyethyl phosphorylcholine
NIPA	N-isopropylacrylamide
PEGEEMA	Poly (ethylene glycol)ethyl ether methacrylate
PEGMA	Poly (ethylene glycol)methyl ether methacrylate
SPMA	3-sulfopropyl methacrylate
SStS	Sodium 4-styrene sulphonate
St	Styrene
TBA	tert-butyl acrylate
VP	4-vinylpyridine

Polymers

ST	Starch
JF	Jute fiber
CLDA	Cellulose diacetate
HPC	Hydroxypropyl cellulose
ECL	Ethyl cellulose
CS	Chitosan
DX	Dextran
GG	Guar gum
β-CD	β-cyclodextrin
PSt	Poly styrene

Ligands

BDAT	S,S'-Bis (a,a'-dimethyl-a"-acetic acid) trithiocarbonate
Bpy	2,2-bipyridyl
DDACT	S-1-Dodecyl-S'-(a,a'-dimethyl-a"-acetic acid) trithiocarbonate
DMAP	4-(N,N-dimethylamino) pyridine
HMTETA	1,1,4,7,10,10-hexamethyltriethylenetetraamine
Me6-TREN	tris(2-(dimethylamino)ethyl)amine
n-Pr-PMI	N-(n-propyl)-2-pyridylmethanimine
Phen	1,10-phenanthroline

PMDETA	N,N,N',N",N"-pentamethyldiethylenetriamine
TEMDA	N,N,N',N'-tetramethylethylenediamine
MCPDB	S-Methoxycarbonylphenylmethyl Dithiobenzoate
BSCTPAcCl	3-Benzylsulfanylthiocarbonylsulfanylpropionicacid chloride

1. INTRODUCTION

Polymeric materials are used in medical implants directly or as coatings to improve biocompatibility of underlying materials, as scaffolds in tissue and drug delivery systems. Since controlled/living radical polymerization (CLRP) is tolerant of functional groups it could be used to synthesize traditionally used polymers in medical applications, the control over the structure and composition of polymer chains could be exploited to get optimal performance of these materials. Polymers could be conjugated with biomolecules such as carbohydrate polymers to give interesting materials capable of interacting with the biological systems. CLRP of hydrophilic and hydrophobic monomers can be applied to generate biocompatible surfaces by incorporating functional groups such as NIPA, PEG, AA, phosphorylcholine, MMA, St. It can also be applied to prepare stimuli responsive polymers that respond to external stimuli such as temperature and pH.

Carbohydrate polymer based macro/micro/nano biomaterials have recently attracted a great deal of interest in drug delivery and tissue engineering applications. They have all the properties of synthetic counterparts as well as being intrinsically abundant in nature, biodegradable, renewable, nontoxic and relatively cheap. In addition, biomaterials possess a high content of functional groups including hydroxyl, amino, and carboxylic acid groups. These functional groups are utilized in crosslinking with additional functional crosslinkers and furthermore, for further bioconjugation with cell targeting agents. In addition, this review will also describe various methods to modify carbohydrate polymers, including methacrylation, surface initiation grafting, and covalent grafting by free radical polymerization and CLRP.

2. FREE RADICAL POLYMERIZATION

Free radical polymerization (FRP) is one of the most useful methods for polymer synthesis, partly because it can be applied to a large of monomers and requires relatively moderate conditions. However, one of the major problems of conventional free radical polymerization is a lack of control over the structure of polymer molecules; it is not possible to obtain polymers with precisely defined end-groups and narrow distribution of molecular weights. Advantages are the simple experimental setup, the use of inexpensive or easy to prepare and purify reagents, in addition to tolerance toward functional groups, solvents and impurities. Various initiation methods for FRP of vinyl monomers on carbohydrate polymers were examined, including persulphate, ceric ion, Fenton's reagent, γ-radiation, thermal and photolysis [1,2]. Persulphate and ceric ion in tetravalent state (Ce^{4+}) is a versatile oxidizing agents that through a redox reaction with various organic compounds creates free radicals capable of initiating radical polymerization [3-5]. Fenton's reagent involves the redox reaction of ferrous ion (Fe^{2+}) with hydrogen peroxide, yielding hydroxy radicals [6]. In addition, γ-radiation utilizes high energy radiation to generate macroradicals on carbohydrate polymers [7]. These initiation methods create macroradicals on carbohydrate polymers that can initiate FRP of vinyl monomers, producing CL, CS, and CT derivatives grafted with vinyl polymers. Several examples include ceric ion initiation to graft DX with PMMA in aqueous nitric acid medium [8] and thermal initiation of NIPA in the presence of CS to yield PNIPA-g-CS consisting of temperature-sensitive PNIPA core with pH-sensitive CS shell [9,10]. However, FRP method had several drawbacks. They include (1) inadequate control over molecular weight and molecular weight distribution ($Mw/Mn > 2.0$) of grafted vinyl polymers and (2) production of homopolymer as a side product with graft copolymer [11]. The intermediates involved in free radical polymerization are highly energetic and therefore, very reactive. In order to control the molecular structure of polymer chains, it is necessary to curb those unwanted reactions of the radicals that do not contribute to the growth of a polymer chain. However, because of extensive research during the last 10-15 years, several procedures for controlling radical polymerization have been developed [12-15].

3. Controlled/Living Radical Polymerization

Controlled/living radical polymerization (CLRP) has emerged as an elegant and versatile method of creating polymers with controlled molecular weight, polydispersity, composition, chain architecture, and site-specific functionalities (Figure 1) [16] which cannot be synthesized by conventional free radical chemistries. These types of polymers are useful in a variety of advanced materials including the ability to prepare bioconjugates, organic/inorganic composites, and surface-tethered copolymers. In CLRP, all of the reaction mechanisms of free radical polymerization still apply; however, a mediating species is employed to control the polymerization and can aid in the creation of block copolymers, very short oligomers and polymers with narrow molecular weight distributions. CLRP has been successfully implemented in bulk or solution polymerization, but to truly produce these specialty polymers at an industrially viable scale and to reduce the use of volatile organic compounds, successful transfer of these polymerization reactions into aqueous dispersed phase systems, such as emulsion, miniemulsion or microemulsion is necessary. In the following part will be described how this control over the radical polymerization process has been achieved. However, we will first briefly focus on the basic concepts of controlled polymerization.

3.1. Basic Concepts of Controlled/Living Radical Polymerization

A CLRP was defined by Swarc [17] as a polymerization reaction that proceeds without the occurrence of any irreversible termination or transfer reactions. In combination with some side conditions such as

1) Fast initiation in comparison to propagation, and

2) Fast exchange between species of various reactivity's in comparison to propagation, polymers with the following characteristics are obtained:[18-21]
   i. Controlled molecular weights, with the degree of polymerization (DP_n) predetermined by the ratio of the concentrations of consumed monomer to the introduced initiator. $DP_n = \Delta$ [Monomer] / [Initiator]$_0$.
   ii. Polydispersities close to Poisson distribution. $DP_w / DP_n \approx 1 + 1/DP_n$.
   iii. All chains end-functionalized.

Experimentally, the best way to evaluate a polymerization technique for its livingness is to follow the kinetics of the polymerization and the evolution of the molecular weight (M_n), polydispersity (PDI) and functionalities with conversion. Well-controlled systems should provide:

a) Linear kinetic plots in semi logarithmic coordinates ($\ln([M]_0/[M])$ vs time), if the reaction is first order in monomer concentration. Acceleration on such plots may indicate slow initiation whereas deceleration may indicate termination or deactivation of the catalyst.
b) Linear evolution of molecular weights (DP_n) with conversion. Molecular weights lower than predicted indicate transfer, higher molecular weights indicate inefficient initiation or chain coupling.
c) Polydispersities should decrease with conversion for systems with slow initiation and slow exchange.

3.2. Types of Controlled/Living Radical Polymerization

From the past two decades a number of CLRP methods have been developed for narrow molecular weight distribution (Mw/Mn < 1.5), designed architectures, and useful end-functionalities [196–198] and the most successful CLRP methods include atom transfer radical polymerization (ATRP), reversible addition-fragmentation chain transfer (RAFT) polymerization, stable free radical polymerization (SFRP) and nitroxide mediated polymerization (NMP). A schematic representation of the polymerization mechanism of the three CLRP is given in Figure 2.

In order to extend the life time of the propagating chains, each of these controlled radical polymerization methods relies on establishing a dynamic equilibrium between a low concentration of active propagating chains ($P_n^{\bullet}$) and a predominant amount of dormant chains (P_n-X) that are unable to propagate (addition of monomer M-k_p) or terminate (k_t). The total small number of dead chains (k_t) produced in a CRP can be neglected in comparison with the total amount of chains which are "living". In the case of NMP, the concentration of active propagating chains is kept low by spontaneous reversible homolytic cleavage of a dormant chain end (activation/deactivation-k_a/k_d). ATRP involves a catalytic reversible homolytic cleavage of a covalent bond via a redox process. RAFT is based on a bimolecular exchange between growing radicals and a dormant species.

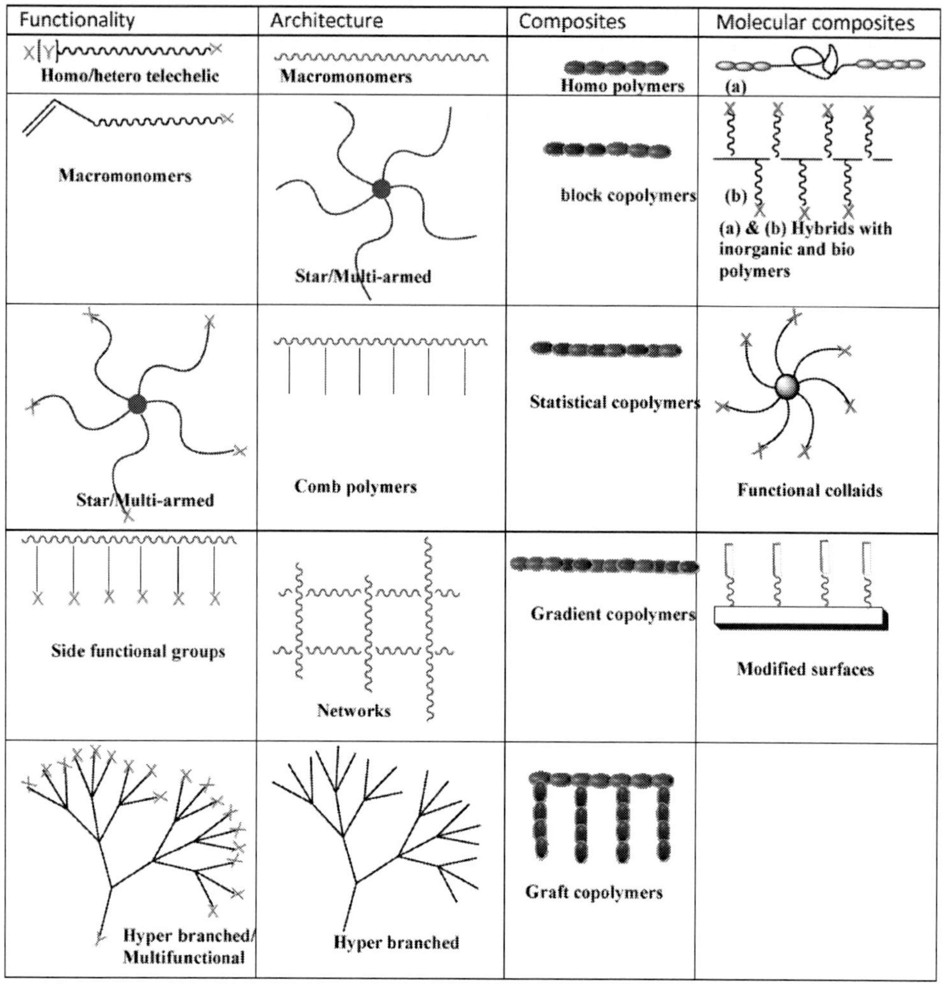

Figure 1. Some examples of molecular structures attained through the use of controlled/living radical polymerization methods.

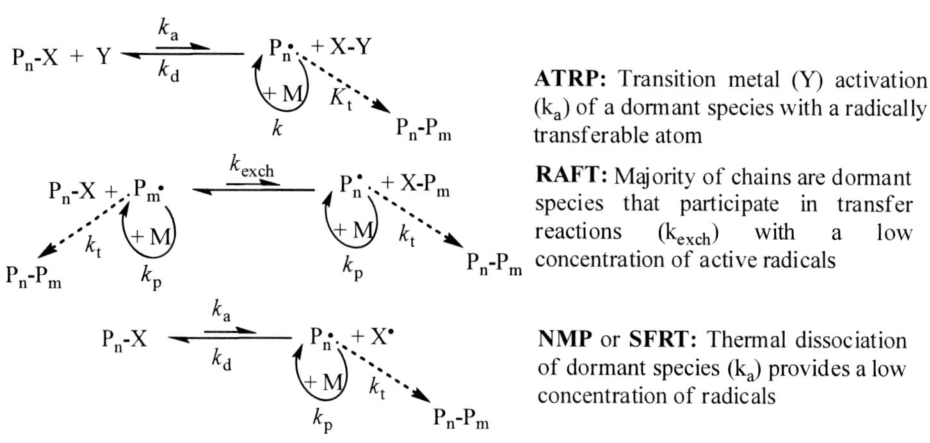

Figure 2. Three main methods for controlled/living radical polymerization.

3.2.1. Atom Transfer Radical Polymerization: Kinetics and Mechanism

Atom transfer radical polymerization (ATRP) was developed in 1995 and further refined by several research groups including the ones of Sawamoto [12-15], Matyjaszewski [16-24] etc. The polymerization mechanism of ATRP (Figure 3) is based on the cleavage of a halogen atom of the initiator R-X (k_a^o) or a dormant polymer chain P_n-X (k_a) by a transition metal complexed with a ligand (L) in its lower oxidation state (M_t^n/L). An alkyl radical $R^•$ or an active polymer chain $P_n^•$ is generated and the transition metal complex is transformed to its higher oxidation state (X-M_t^{n+1}/L). In the propagation step, monomer is added to grow a polymer chain (k_p, k_i when monomer is added to $R^•$) until the dormant species P_n-X is formed again by abstraction of a halogen atom of X-M_t^{n+1}/L with formation of M_t^n/L (k_d). Control over chain length (molecular weight), molecular weight distribution and functionality is thus obtained by a dynamic equilibrium (k_a/k_d) between a dormant chain P_n-X and an active chain $P_n^•$ by an exchange of electrons (redox process) between the transition metal complex and the radical chain ends of the active species. However, termination can result in a small amount of coupled polymer chains P_n-P_m in case of combination ($k_{t,c}$), or can result in disproportionated chains P_n = or P_mH in case of disproportionation ($k_{t,d}$).

ATRP can be applied for the polymerization of a wide range of monomers such as (meth)acrylates, (meth)acrylamides, styrenes, acrylonitriles, and acid functional monomers as well in bulk as in solvent [22-24]. The initiator has to be selected carefully, in accordance with the structure and reactivity of the monomer and the used metal complex. Initiation has to be quantitative and the initiation step has to be fast in comparison to propagation in order to obtain a controlled polymerization. The initiator role is to provide a radical via the first activation/deactivation cycle of the polymerization. All most of all the initiators such as organic halides can successfully employed with a carbon-halogen bond, which can easily generate a radical species through electronic and steric effects of their substituents. And also sulfonyl halides have been used as initiators for ATRP by Percec *et al.*[25-30].

There are many transition elements have been used for ATRP, among frequently used metals are Cu, Ru, Ni, Fe [31-35] and less frequently used Re, Pd, Mo [36-39]. Cu-based metal complexes are used as catalyst due to its good reactivity, commercially availability, easy to synthesize, and copper complexes show a high selectivity for atom transfer (they posses a low affinity for e.g. alkyl radicals) [23]. Due to the toxicity of the catalyst and the intense colour of the resulting polymers, various post polymerization purification methods have been developed for removal of the catalyst [40]. As removal of the catalyst is a rather expensive and time-consuming process, research concerning the development of solid-phase catalyst systems has been reported by many groups [41-45].

The choice of ligand influences the activity of the transition metal complex, and thus the radical concentration in the polymerization system. These ligands have the following purposes: they solubilise the transition metal (homogeneous reaction medium) and they have an influence on the ATRP equilibrium by means of their electronic and steric effects. A more reducing catalyst complex (with lower redox potential) usually shows a higher catalytic activity (higher k_a/k_d value). For Cu-catalyzed ATRP, most commonly amine ligands are used, which can be classified according to their number of nitrogen atoms i.e Bidentate ligands, Tridentate ligands and Quadridentate ligands. Some examples are shown in Table 1.

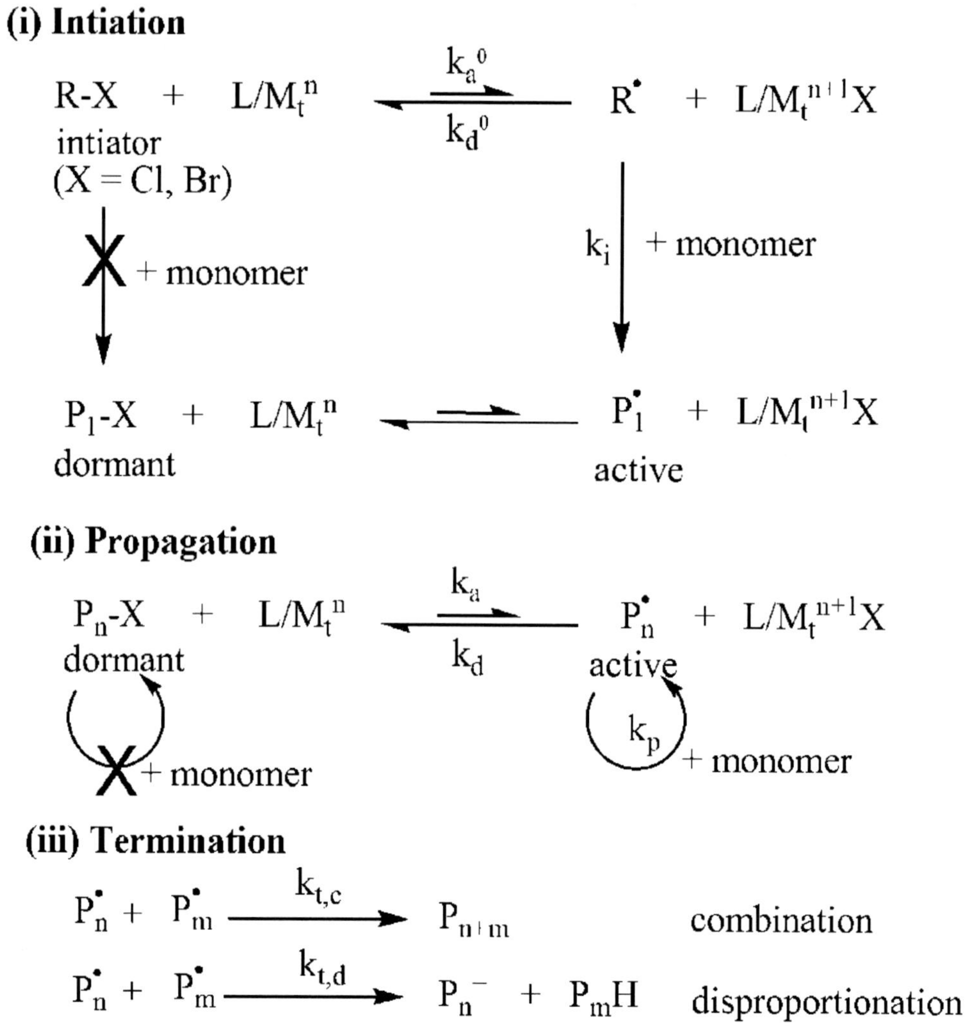

Figure 3. Detailed reaction mechanism of atom transfer radical polymerization.

The typical examples for naturally occurring carbohydrate polymers are; chitosan (CS), hyaluronan (HA), dextran (DX), cellulose (CL), guar gum (GG) , pullulan (PL), chondroitin sulfate (CDS), and alginate (ALG) [46,47]. Figure 4 shows their chemical structures of the above carbohydrate polymers. CS is a polymeric β(1→4) linked 2- amino-deoxy-D-glucan. It is commercially produced by the hydrolysis of aminoacetyl groups of chitin (CT), polymeric (1→4) linked N-acetyl-D-β glucosamine. CT is the main component of shells, crabs, shrimp, and krill [48,49]. HA is a mucopolysaccharide composing of N-acetyl-D-glucosamine and D-glucuronic acid. It is mainly found in the extracellular tissue matrix of vertebrates. DX is a bacterial-derived polysaccharide consisting of α(1→6) linked D-glucopyranose residues [50]. CL is a β(1→4) linked homopolymer of D-glucose. It is the most abundant polysaccharide and the building element of plants, providing shape and structure. PL is a linear polysaccharide consisting of α(1→6) linked maltotriosyl units. CDS is a sulfated CT. ALG is a linear unbranched polysaccharide containing β-D-mannuronic acid and α-L-guluronic acid residues. It coats the surfaces of many forms of seaweed and marine kelp. GG is a branched polygalactomannan, isolated from the seeds of leguminous herbs [51]. It is a linear β (1→4) mannose to which α (1→6) galactopyranoside single subunits are attached as side chains [52].

Table 1. Some ATRP system components

Initiator	Metal halide	Ligand
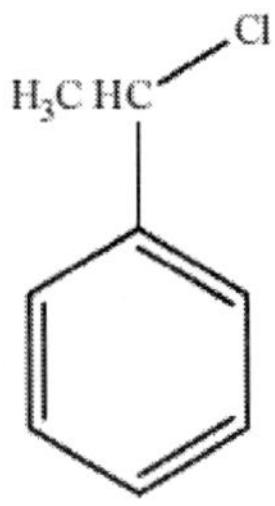	CuCl	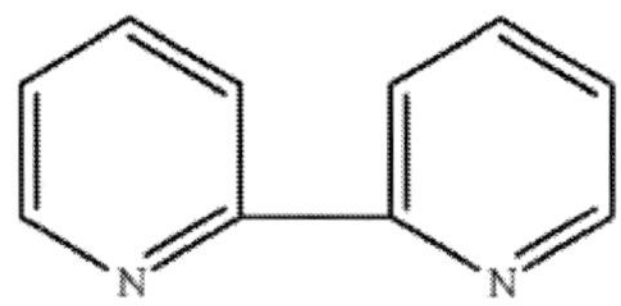
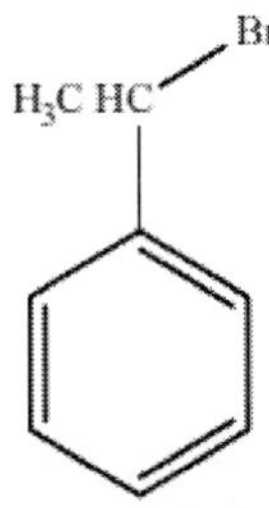	CuBr	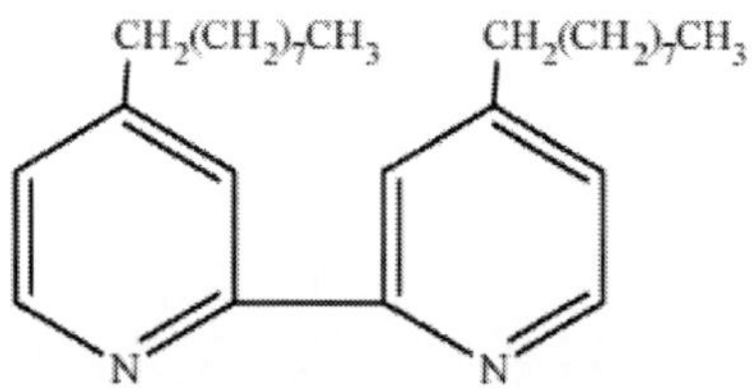
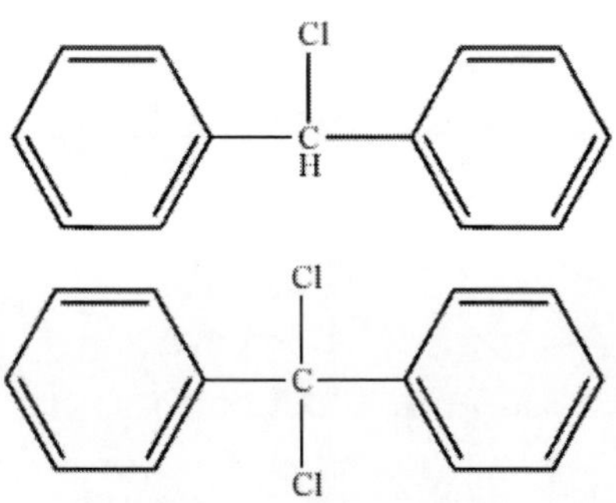		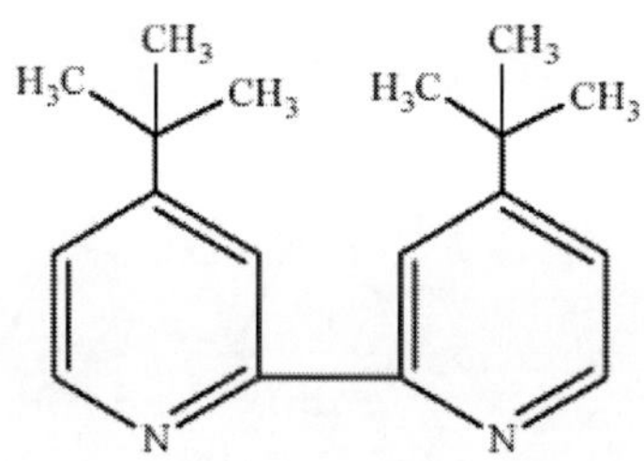

The ATRP method was extensively explored to modified carbohydrate polymers such as ALG, CL, CD, CDS, CS, CT, DX, and HA derivatives with well controlled hydrophilic and hydrophobic polymers. This is due to a facile modification of carbohydrates with ATRP initiating species. The resulting ATRP macroinitiators of carbohydrate polymer can initiate ATRP polymerization of vinyl monomers in various solutions, producing polymer grafted carbohydrate polymers. Figure 5 shows a schematic illustration of the use of the ATRP method to prepare CL grafted with well-controlled polymers in a solution, as an example.

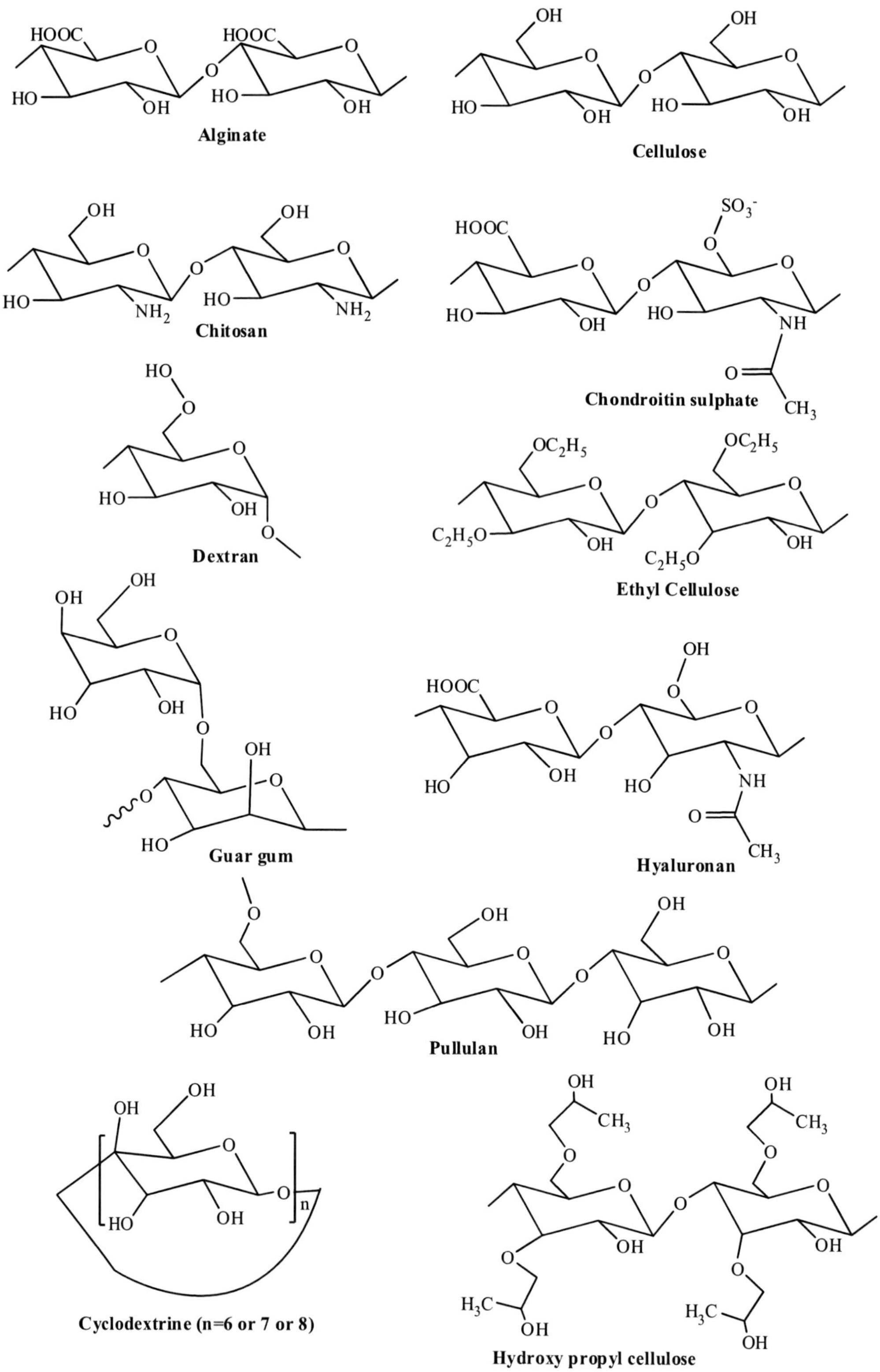

Figure 4. Chemical structures of some carbohydrate polymers.

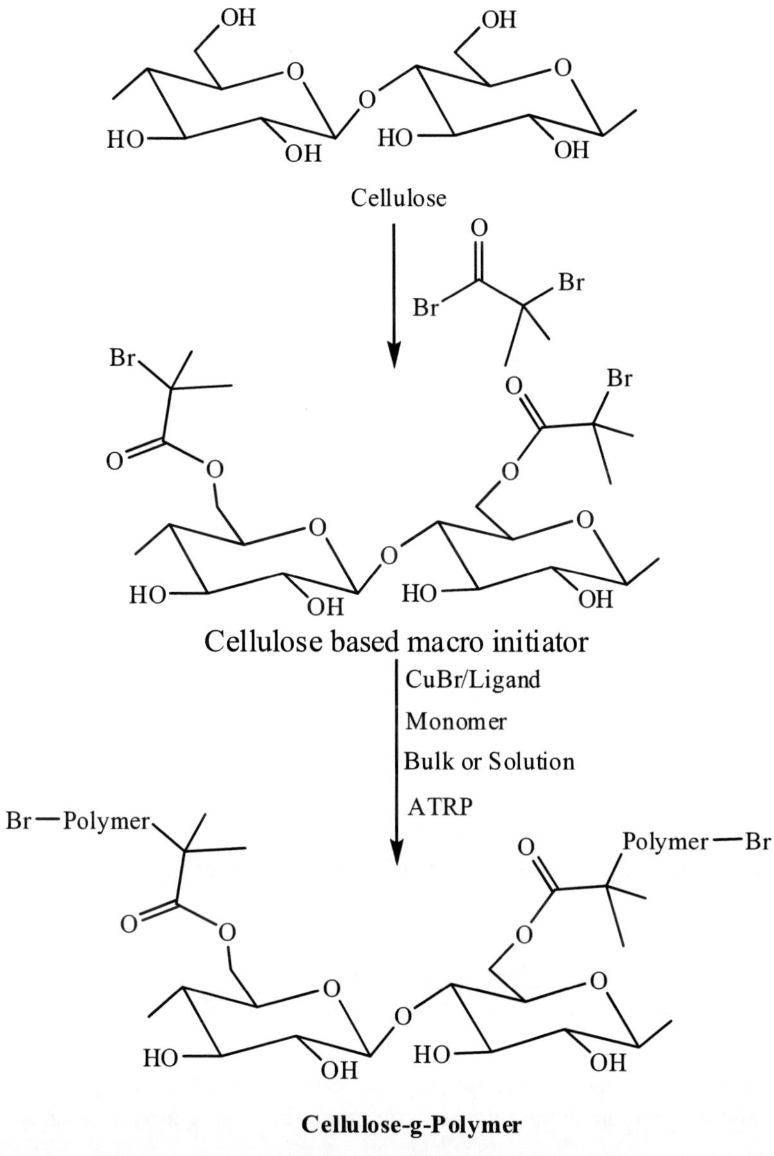

Figure 5. Schematic representation of cellulose-g-synthetic polymer by ATRP.

3.2.1.1.Carbohydrate Based Graft Copolymers

A CS grafted with methoxy capped (PEG 350) by ATRP in the presence of pyridine as a base. The kinetic study revealed a first order polymerization reaction and polydispersity of about 1.25 was obtained [53]. CDA has been modified by reacting the residual hydroxyl groups with 2-bromoisobutyryl bromide. The modified CDA then was grafted with methyl methacrylate through ATRP [54]. The graft copolymers of cellulose with MMA, AcM and MAm, were prepared by means of these groups using the Cu(I)-1,2-dipiperidinoethane complex as a catalyst in DMF at 130 °C, that is, by ATRP [55]. The well defined densely grafted copolymers of EC were synthesized with St and MMA through ATRP. Copper(I) bromide (CuBr) was used as a catalyst, and PMDETA was used as a complexing reagent [56].

CDA functionalized with 2-bromoisobutyryl or dichloroacetyl groups as initiating sites for ATRP of St, MMA, BA and some block copolymers [57]. NIPA, MMA, HEMA grafted onto PL and DX by ATRP was synthesized. These results can be very interesting for biomedical applications. DLS results showed that the average hydrodynamic diameter of the aggregates is about 92 nm [58]. A bottle brush-like graft copolymer with CDA backbone and chemically different grafts, prepared in a two-step grafting process, combining techniques of ROP (ε-caprolactone, CL) and ATRP (St, MMA and BA) [59].

Comb type polymers were produced by HPC as a backbone for "grafting from" polymerization under ATRP condition. Moreover, dendronization of the HPC using 2,2-bis(methylol)-propionic acid (bis-MPA) dendrons and subsequent ATRP grafting from the surface of the dendronized polymer was studied. The advantage of using HPC is the available hydroxyl groups, which show high reactivity with bis-MPA derivatives combined with the improved solubility HPC exhibits as compared to native cellulose. Furthermore, bis-MPA and HPC are both biocompatible, which could make these dendronized polymers promising candidates for biomedical applications [60]. Commercial cottons based on CL were purified with water and methanol and then modified with Br-iBBr in pyridine. The bromine-functionalized CL cottons were used as macro initiators for ATRP of MPC in DMSO. The resulting PMPC-grafted cottons had good protein adsorption resistance for the development of the hemopurification membrane system [61]. CL-ClAc was synthesized by the acylation reaction of cellulose in a homogeneous medium of a DMAc/LiCl mixture and pyridine was used as the acid scavenger in the acylation reaction. ATRPs of EMO and MMA were performed respectively by using Cell-ClAc as initiator. Furthermore, the second ATRP of EMO and MMA were conducted by using Cell-PMMA and Cell-PEMO as initiators, respectively [62].

BiBB groups were successfully linked to acetylated DX and ATRP of MMA was then carried out from the resulting derivative used as a macroinitiator [63]. Regioselective CL macroinitiator (6-O-bromoisobutyryl-2,3-di-O-methyl cellulose) was performed with NIPA by ATRP and copolymers with polyNIPA side-chains of DP up to 46.3 were synthesized [64]. A versatile multifunctional carbohydrate polymer, CL, was easily converted to an ATRP macroinitiator through direct homogeneous acylation of CL in ionic liquid. Stimuli-responsive CL-g-PDMAEMA copolymers were synthesized by grafting polymerization of DMAEMA monomers onto the ATRP macroinitiator in homogeneous solution under mild conditions [65].

High molecular weight comb block copolymers, consisting of a HPC backbone, inner blocks of PCL and outer blocks of different lengths of PtBA were successfully synthesized using a combination of ROP and ATRP. These amphiphilic polymers are shell cross-linked in dilute water solution to produce nanoparticles with hydrogel like shell. The nanoparticles encapsulated with pyrene for an initial investigation [66]. ECL was modified with POEOMA by the ATRP method under bulk conditions. Moreover, the self-assembly and thermosensitive properties of the resultant copolymers in water were characterized. The resultant amphiphilic graft copolymers have the potential application in drug delivery [67]. The tertiary bromo ester groups on the CL are efficient for ''graft from'' copolymerization. Graft copolymers of CL have obtained by ATRP of MMA or/and St under mild controllable conditions. The obtained comb graft copolymer CL–PMMA could aggregate and self-assemble in solution. The average diameters of sphere-like particulates derived from the solution of good solvent DMF and selective solvent acetone are roughly 50 and 200 nm, respectively [68].

ATRP was used to synthesize the graft copolymer of CL with PMMA using an ionic liquid BMIMCl as a reaction medium. In the synthesis process, the CL modified to serve as a macroinitiator that initiates the ATRP of MMA [69]. CS-g-poly(OEGMA) copolymers prepared by two synthetic routes, "grafting-from" and "grafting-to" from ATRP. A comparison of the two synthetic methods showed differences in the ratio of synthetic and natural polymers present in the resulting combs. The "grafting-to" synthetic route was preferable as the polymers were of higher purity and more defined [70]. Well-defined star-shaped cationic CDPD, consisting of a β-CD core and P(DMAEMA) arms, and CDPDPE, consisting of CDPD and P(PEGEEMA) end blocks, were prepared as gene carriers via ATRP from the bromoisobutyryl-terminated β-CD core [71].

Utilization of a mixed solvent of 1-methyl-2-pyrrolidone and water to facilitate the living polymerization of a series of novel cationic star polymers with 21 arms (21ACSPs) through ATRP using 21Br-b-CD initiator were studied. In addition, monomers bearing primary, tertiary amino and quaternary ammonium groups, which have been polymerized into linear polymers as DNA vectors and with suitable pKa values, were all used in an attempt to identify the polymer structure bioactivity relation in terms of physicochemical properties, cell transfection efficiency and cytotoxicity [72]. A series of new cationic copolymers prepared from biocompatible CS and P(DMAEMA) via ATRP. By controlling P(DMAEMA) side chain length, these chitosan-based cationic copolymers are expected to provide much flexibility to condense pDNA with suitable particle sizes, exhibit strong buffering capacity, and mediate efficient gene transfection with minimum cytotoxicity [73].

Pyrene used as the fluorescent probe and model of the poor water-soluble drug and loaded in EC-g-PPEGMA. These copolymers have well-defined structure different graft density and graft length were investigated. The EC-g-PPEGMA copolymers showed a potential application in controlled drug delivery systems [74]. Hyper branched poly(β-CD) core was synthesized to accomplish a so-called selective encapsulation, where two types of guest molecules (Levofloxacin lactate and Phenolphthalein) can be encapsulated into two types of molecular cavities from β-cyclodextrin (β-CD) and topography structure of molecular cavities from β-CD and topography structure of hyper branched polymer, respectively [75]. The comb-shaped copolymer conjugates composed of HPC backbones and short P(NIPA) side chains (HPC-g-P(NIPA) or HPN) synthesized via ATRP from the bromoisobutyryl-functionalized HPC backbones. The HPN with average P(NIPA) content of above 53 wt % exhibited a LCST below the body temperature of 37 °C. The MTT assay from the HEK293 cell line indicated that HPNs possess reduced cytotoxicity. Some of the remaining unreacted hydroxyl groups of HPNs were used as cross-linking sites for the preparation of stable HPN hydrogels [76].

With the versatility of ATRP, properly grafting a low molecular weight side chains from a natural backbone is an effective means for designing novel polysaccharide based nano biomaterials. ATRP was used to prepare well-defined comb-shaped copolymers (DPDs) consisting of biocompatible DX backbones and P(DMAEMA) side chains for nonviral gene delivery. The P(DMAEMA) side chains can be further partially quaternized to produce the QDPDs with quaternary ammonium groups. DPDs and QDPDs as gene carriers can effectively bind pDNA to form nanoparticle complexes of 100 to 150 nm in sizes [77]. The polymerization of NVP by ATRP using copper (I)/bpy complex process was successfully carried out to get glucose based 5-arms star polymer. The synthesized polymer shows narrow molar mass distribution between 1.41 and 1.11 [78]. Amphiphilic CS graft copolymers with double side chains of PCL and P(MEO_2MA-co-OEGMA) were synthesized via combination of ROP, ATRP and click chemistry. The ratio of PCL and P(MEO_2MA-co-OEGMA) was varied to alter the hydrophilic/hydrophobic balance. The terminal pyrene group of PCL endowed the self-assembled thermosensitive CS graft copolymer micelles with fluorescence and the fluorescent intensity was characterized. Controlled drug release behavior of CS graft copolymer micelles at buffer solution was investigated at different temperatures [79].

A simple, rapid, accurate and suitable method developed for the application in the investigations monomer conversion in CL graft MMA by ATRP using a full evaporation head space gas chromatography (FE HSGC). The data showed that a near-complete mass transfer of MMA from the very small liquid sample size (<30mg) to the vapour phase (head space) was achieved within 5 min at 105 ^{o}C [80]. ATRP graft copolymerization of AAP was successfully carried out using "grafting from" methodology in aqueous medium at ambient temperature on GG. Pseudo-first order kinetics was observed and a polydispersibility as low as 1.28 was obtained. High percentage of grafting is achieved without the formation of homopolymer [81]. Different amphiphilic oligosaccharide-functionalized well-defined polySt synthesized in three steps combining ATRP of St and chain ends modifications (nucleophilic substitution followed by reductive amination with glucose, maltose, maltotriose or dextran) [82]. Details of carbohydrate based graft copolymers are given in Table 2.

3.2.1.2.Carbohydrate Based Surface Graft Copolymers by ATRP

ATRP of MA from initiators immobilized on CL fibers were reported first time [83]. PolySt grafted on jute (*Corchorus capsularis*) using an ATRP technique to improve mechanical properties of plant fibres [84]. PBA grafted onto starch prepared via a SI-ATRP of BA from the surfaces of the starch bromo-acetic ester as macroinitiator in presence of *phen* and Cu(I)Br as catalyst in toluene system [85]. MA and St grafted from various solid polysaccharide substrates in terms of filter paper, microcrystalline CL, dialysis membrane, Lyocell fibers, and CS film via ATRP [86]. PolySt grafted CS particles were prepared by the surface-initiated (SI) ATRP of St from the macroinitiators, bromo-acetyl CSs in toluene system with the catalysts of *phen* and Cu(I)Br [87]. Core-shell type "Spherical polyelectrolyte brush" of silica- g-PSStNa prepared by ATRP and coated CS supports were treated with glutaraldehyde for stability in both alkaline and acidic media. The pectinase immobilized on the silica-coated CS support exhibited improved resistance against pH denaturation. This immobilized enzyme could be applied in a wide pH range (3.0-4.5) without significant loss in its activity [88].

Stimuli-responsive CL surfaces prepared via grafting of NIPA and 4VP from filter paper by ATRP. The PNIPA-grafted CL surface showed hydrophilic and hydrophobic character below and above the LCST of PNIPA respectively. Stimuli-responsive CL surfaces were prepared via grafting of PNIPA-*b*-P4VP and PVP-*b*-PNIPA for biomedical applications as drug carriers or sensors [89]. CS/CNT nanocomposites were prepared using PSSA-functionalized CNTs (CNT-PSSA). The sulfonic acid groups of PSSA provide chemical reactivity and interfacial compatibility between CNT-PSSA and CS. Moreover, the PSSA-modified CNTs were readily soluble in acidic aqueous solution of CS, to warrant the well dispersion of CNT-PSSA in CS. In addition to physical and mechanical properties, the electric conductivities of the prepared CS/CNT nanocomposites were also examined to probe their potential applications in electrodes and sensors [90]. Butyl acrylate was grafted on the surface cellulose micro fibrils (CMF) by SI-ATRP and these hydrophobic-modified CMF biocomposits with tailored soft corona suitable for a variety of potential applications [91]. Details of carbohydrate based surface graft copolymers are given in Table 3.

Table 2. Details of carbohydrate based graft copolymers by ATRP

S.No.	Polymer	Monomer	Metal:ligand	RT	Ref.
1	CS	MPEG	CuBr/bpy	----	53
2	CLDA	MMA	CuBr/TMDETA	-----	54
3.	CL	MMA,St, Mam, AcM	CuBr/1,2-Dipiperidinoethane	20 h	55
4	ECL	St, MMA	Cu(I)Br, PMDETA	----	56
5	CLDA	St, MMA, BA	$CuCl_2$/HMTETA	2.5 h	57
6	PL & DX	HEMA, MA, SPMA, NIPA,	CuCl and $CuCl_2$/bpy	----	58
7	HPC	MMA, HDMA, TBA	Cu (I)Br, Cu(II)Br/PMDETA	----	59
8	CLDA	MMA, BA	Cu salts (CuCl, CuBr, $CuCl_2$)/ Poly amines (HMTETA and PMDETA)	7 h	60
9	CL	MPC	Cu(I)Br/ bpy	----	61
10	CL	EMO, MMA	Cubr/TEMDA	----	62
11	DX	MMA	Cu(I)Br and Cu(II)Br_2/n-Pr-PMI	----	63
12	CL	NIPA	Cu(I)Br/PMDETA	12	64
13	CL	DMAEMA	CuBr/PMDETA	----	65
14	HPC	TBA	Cu(I)Br and Cu(II)Br_2/PMDETA	----	66
15	ECL	PEGMA	$CuCl_2$/4,4'-denonyl- bpy	----	67
16	CL	MMA, St	Cu(I)Cl/bpy	----	68
17	CL	MMA	CuBr/ bpy	----	69
18	CS	OEGMA	CuBr/bpy	----	70
19	β-CD	PEGEEMA, DMAEMA	CuBr/HMTETA	0.5-8	71
20	β-CD	HAEAPMA	Cu(I)Br: bpy	----	72
21	CS	DMAEMA	Cu(I)Br/HMTETA	0.3-8	73
22	ECL	PPEGMA	$CuCl_2$/4,4'-denonyl- bpy	----	74
23	β-CD	DMAEMA	CuCl/PMDETA	----	75
24	HPC	NIPA	CuBr: HMTETA	8-24 h	76
25	DX	DMAEMA	Cu(I)Br/HMTETA	5-12	77
26	Glu	VP	CuBr/bpy		78
27	CS	MEO_2MA-co-OEGMA	CuCl/bpy	----	79
28	CL	MMA	CuBr2/TEMED/Ascorbic acid	----	80
29	GG	AAP	Cu(I)Br/ bpy	----	81
30	D(+) Glu, D(+) Mal, DX	St	Cu(I)Br/PMDETA	----	82

Table 3. Details of surface graft copolymers by ATRP

S.No.	Polymer	Monomer	Metal:ligand	RT	Ref.
1	CL	MA	Cu(I)Br/Me_6-TREN	18 h	83
2	JF	St	Cu(I)Br/PMDETA	1.5h	84
3	ST	BA	CuBr/phen	----	85
4	CS	MA, St	Cu(I)Br/Me_6-TREN	18h	86
5	CS	St	Cu(I)Br/phen	----	87
6	CS	SStS	Cu(I)Br/bpy	3h	88
7	CL	NIPA VP	Cu(I)Cl, Cu(II)Cl_2/ Me_6-TREN	0.15-1h, 2-6h	89
8	CS	SStS	Cu(I)Br/PMDETA	30h	90
9	CL	BA	CuBr/bpy	----	91

3.2.1.3. Carbohydrate Based Self-Assembled Copolymers by ATRP

Self-assembly and pH- and thermo-sensitivities properties of well-defined HPC-g-PDMAEMA copolymers were synthesized by ATRP. The LCST of $HPC_{0.1}$-g-$PDMAEMA_{20}$ shifts to the lower temperature with the increase in pH values because of the deprotonation of the side chains PDMAEMA at higher pH value [92]. PStSS grafted CS was synthesized by the controlled polymerization in the presence of CS-TEMPO macroinitiator and the graft copolymers would be anticipated to have the improved capacity of cation removal [93]. Stimuli-responsive CS-g-PDMAEMA-g-PNIPA terpolymer synthesized based on the CS derivative via the combination of ATRP and click reaction. This unique dual hydrophilic graft CS copolymer is believed to have a great potential in application for gene/drug delivery and controlled release [94].

Novel amphiphilic CS-g-(PDEAEMA-b-PEO) copolymer was synthesized by the combination of ATRP and click chemistry. The pH-responsive CS amphiphilic copolymers have potential applications in biomedical field [95]. ECL-g-PDEAEMA micelles were synthesized via ATRP. The developed ECL-g-PDEAEMA graft copolymers have hydrophobic EC backbones and pH-responsive PDEAEMA side chains. ECL-g-PDEAEMA copolymeric micelles were tested for drug loading and release properties of Rifampicin as the model drug [96]. Details of carbohydrate based self-assembled copolymers are given in Table 4.

3.2.2. Reversible Addition-Fragmentation Chain Transfer Polymerization

Reversible addition-fragmentation chain transfer polymerization (RAFT) was first reported in 1998 by Rizzardo *et al.* [97-99] and Charmot *et al.*[100]. The mechanism of the RAFT process is believed to involve a series of reversible addition-fragmentation steps as shown in Figure 6. Addition of a propagating radical $P_n^{\bullet}$ to the thiocarbonylthio compound gives an adduct radical that fragments to a polymeric thiocarbonylthio compound and a new radical $R^{\bullet}$. The radical $R^{\bullet}$ then re-initiates polymerization to give a new propagating radical $Pm^{\bullet}$. Subsequent addition-fragmentation steps set up equilibrium between the propagating radicals $P_n^{\bullet}$ and $P_m^{\bullet}$ and the dormant polymeric chains by the way of an intermediate radical. This equilibration of the growing chains gives rise to a narrow molecular weight distribution. Throughout the polymerization (and at the end) the vast majority of the polymer chains are end capped by a thiocarbonylthio group (dormant chains), so the concentration of active (free) radicals remains low and termination reactions are minimized. The RAFT method allows the synthesis of well-defined random, block copolymers, gradient copolymers as well as star copolymers [98, 101-105]. For the initiation, thus for the generation of radicals, conventional free radical polymerization initiators are used (e.g. AIBN). However, there is often an optimal ratio for the amount of initiator to RAFT agent [106].

Table 4. Details of self-assembled copolymers by ATRP

S.No.	Polymer	Monomer	Metal:ligand	RT	Ref.
1	HPC	DMAEMA	Cu(I)Cl/HMTETA	24 h	92
2	CS	SStS		-----	93
3	CS	DEAEMA, NIPA	CuBr/HMTETA	6 h	94
4	CS	DMAEMA	CuBr/PDMETA	-----	95
5	EC	DEAEMA	CuBr/bpy	-----	96

The general structure of the RAFT agent is Z-C(=S)S-R (Figure 6). Z is an activating group [45] (e.g. aryl, alkyl, NR′, SR′,…), while R is a radical leaving group [107] (e.g. CH_2Ph, $C(CH_3)_2Ph$, $CH(CH_3)Ph$, $C(CH_3)(CN)CH_2CH_2C(O)OH$, $C(CH_3)_2CN$,…). For the MADIX process, Z = Z′-O (Z′ is e.g. C_2H_5, $C_{12}H_{25}$, CH_2CF_3, $CH(CF_3)[P(O)(OEt)_2]$) [108,109]. For an optimal control over the polymerization, the RAFT agent (Figure 7) has to be fine-tuned by the design of the Z and R substituents for the polymerization of different monomers.

Daoben Hua et al. reported the controlled graft modification of CS via RAFT polymerization. The CS was modified with (DDACT) to serve as RAFT agent, and then the graft polymerization of acrylic acid was performed at 80 °C [110]. And also they reported that "one-pot" polymerization of AA/CS may be a facile method to prepare functionalized CS-based nanoparticles by omitting the N-phthaloylation process of CS [111]. In addition, the thermosensitive CS was first successfully prepared by RAFT polymerization of NIPA in the presence of CS-RAFT agent. The "well-defined" grafted PNIPA and the marked thermally responsive property would positively enable CS to designate for the drug delivery applications [112]. A hydrophobic CL surface was prepared by grafting polySt onto a cellulose substrate (Whatman No. 1 filter paper) via the RAFT polymerization method with a view to using the hydrophobic cellulose fibers in composites [113]. Regular porous honeycomb structured films were prepared with carbohydrates, such as α-D-glucose, β-CD and CL, modified by polySt via RAFT polymerization. The polymers were allows the incorporation of a range of monomers with functional groups leading to films with active sites on the surface [114]. Details of carbohydrate based copolymers by RAFT are given in Table 5.

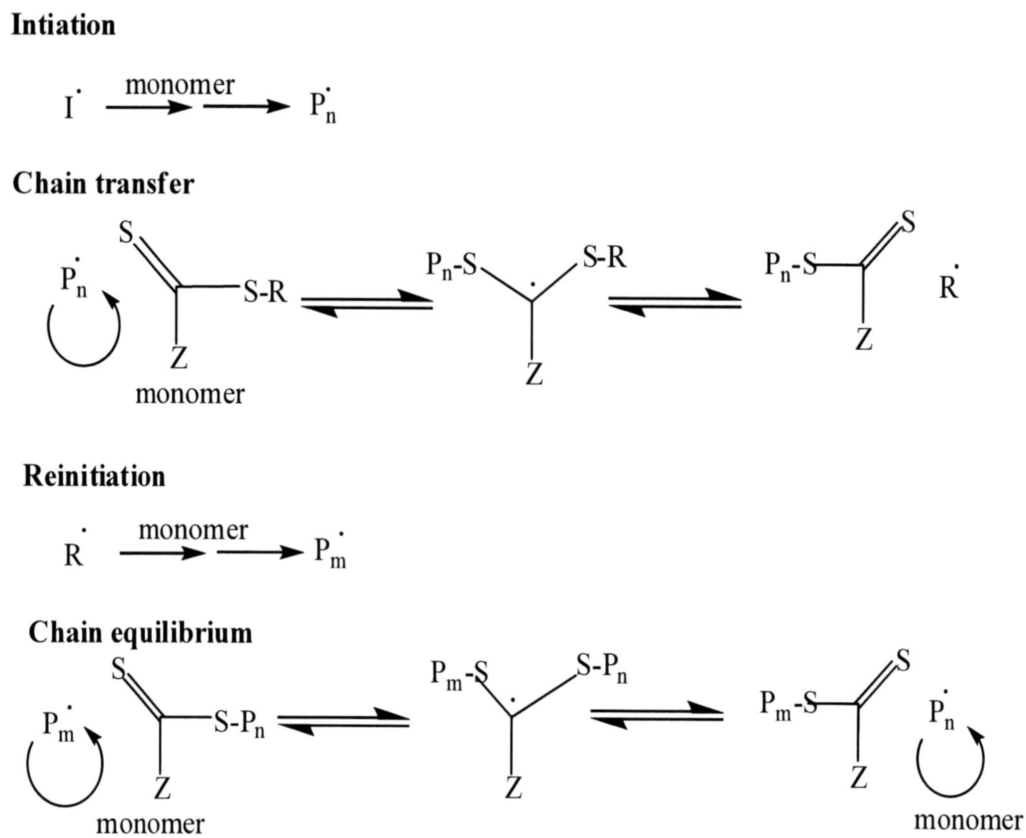

Figure 6. Mechanism of RAFT polymerization.

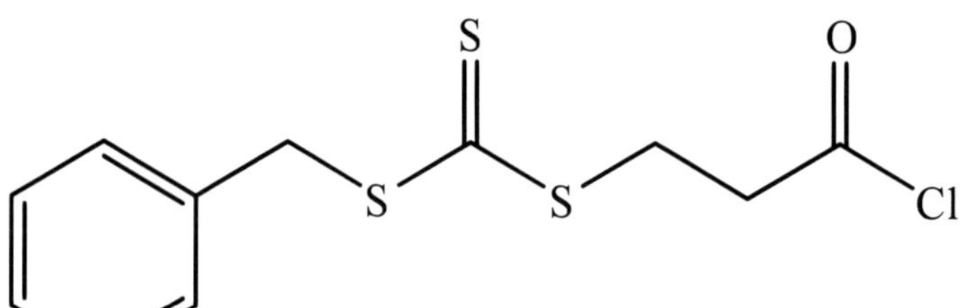

BSCSP-Cl

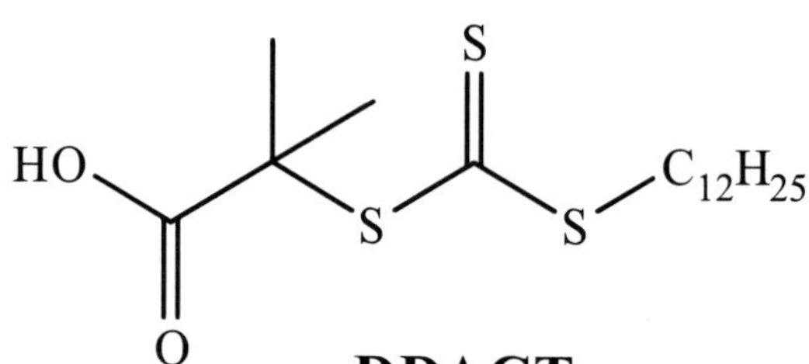

DDACT

Figure 7. Structures of some RAFT Agents.

Table 5. Details of graft copolymers by RAFT

S.No.	Polymer	Monomer	RAFT agent:Ligand	RT	Ref.
1	CS	AAc	DDACT/DMAP	48h	110
2	CS	NIPA, AAc	BDAT	24h	111
3	CS	NIPA	DDACT/DMAP	48h	112
4	CL	St	MCPDB	24 h	113
5	Glucose, β-CD, CL	PSt	BSCTPAcCl	----	114

3.2.3. Nitroxide Mediated Polymerization (NMP)

Nitroxide mediated polymerization (NMP) belongs to the class of stable free radical polymerizations. It was first reported by Rizzardo *et al.* in 1985 [115] and also Georges *et al.* made a lot of efforts in the field [116-118]. The stable free radicals, which originate from spontaneous homolytic cleavage (thermal dissociation) of a nitroxide compound at elevated temperature, combine with the radical at the end of the growing polymer chains with the formation of a covalent bond in the dormant species (Figure 2). In this way, the concentration of active radicals in the system is kept low and control over the polymerization is established. For NMP, a number of nitroxides have been used. In the first reports, the use of cyclic nitroxides such as TEMPO (2,2,6,6-tetramethyl-1-piperidinoxyl) is described [116-118]. Polymerization of methacrylates remains uncontrolled due to disproportionation of the nitroxide and the growing radical when TEMPO is used [119,20], while the use of non-cyclic nitroxides such as SG1 leads to a high concentration of radicals because of a high k_a/k_d constant, which results in irreversible termination[121]. However, Charleux *et al.* proved that a better control can be obtained by copolymerizing a limited amount of styrene, which lowers the k_a/k_d equilibrium constant [122].

CONCLUSION

Carbohydrate polymer based macro/micro/nano biomaterials have high water content, adjustable chemical and mechanical properties, controlled architecture, tunable size from submicrons to tens of nanometers, a large surface area for multivalent bioconjugation, and an interior network for the incorporation of therapeutics. Carbohydrate polymers are basically abundant in nature, nontoxic, biodegradable, renewable and relatively cheap. They also possess a high content of functional groups that can be utilized for crosslinking with additional functional crosslinkers as well as for further bio-conjugation with cell-targeting agents. These unique properties offer great potential for the utilization of carbohydrate polymer based macro/micro/nano biomaterials in tissue engineering, biomedical implants, bio-nanotechnology, and particularly, drug delivery. The present carbohydrate polymer based macro/micro/nano biomaterial applications still require a high degree of control over the properties for in vivo applications. In the design and development of effective micro/nano-based drug delivery systems for targeting cancer cells, several criteria are required. The last important criterion is novel biodegradability under acidic conditions for the sustained release of drugs in cancer cells and facile removal of the empty device. Finally, it is clear that the development of CLRP methods and all advances made in the field have provided synthetic chemists with the ability to prepare materials that were impossible to synthesize a decade ago. Controlled synthesis and processing now allows the properties of materials to be characterized and readily manipulated on the nanoscale. As a result, numerous corporations in a broad range of markets are preparing a spectrum of new materials. They now have the capability to tailor the properties of their products to their customer's needs.

ACKNOWLEDGEMENT

The authors Dr. K.S.V. Krishna Rao gratefully acknowledge the support by the Department of Science and technology (DST), New Delhi, India, (DST No.SR/FT/CS-047/2009) for a financial support.

REFERENCES

[1] Jenkins, D. W.; Hudson, S. M. *Chem Rev.* 2001, 101, 3245-3273.
[2] Berlin, A. A.; Kislenko, V.N. *Prog. Polym. Sci.* 1992, 17, 765-825.
[3] McDowall, D. J.; Gupta, B. S.; Stannett, V. T. Prog Polym Sci. 1984, 10, 1-50.
[4] Lagos, A.; Yazdani-Pedram, M.; Reyes, J.; Campos, N. *J. Macromol. Sci. Pure Appl. Chem.* 1992, A29, 1007-1015.
[5] Gaborieau, M.; De Bruyn, H.; Mange, S.; Castignolles, P.; Brockmeyer, A. *J. Polym. Sci. Part A Polym. Chem.* 2009, 47, 1836-1852.
[6] Lagos, A.; Reyes, J. *J. Polym. Sci. Part A Polym. Chem.* 1988, 26, 985-991.
[7] Singh, D.K.; Ray, A.R. *J. Appl. Polym. Sci.* 1994, 53, 1115-1121.
[8] Derkaoui, S.M.; Avramoglou, T.; Barbaud, C.; Letourneur, D. *Biomacromolecules.* 2008, 9, 3033-3038.
[9] Leung, M. F.; Zhu, J.; Harris, F.W.; Li, P. *Macromol. Rapid Commun.* 2004, 25, 1819-1823.
[10] Leung, M. F.; Zhu, J.; Harris, F. W.; Li, P. *Macromol. Symp.* 2005, 226, 177-185.
[11] Shantha, K. L.; Bala, U.; Rao, K. P. *Eur. Polym. J.* 1995, 31, 377-382.
[12] Matyjaszewski, K., Controlled Radical Polymerization; ACS Symposium Series 685, American Chemical Society: Washington DC, 1997.
[13] Matyjaszewski, K., Controlled/Living Radical Polymerization: Progress in ATRP, NMP and RAFT; ACS Symposium Series 786, American Chemical Society: Washington DC, 2000.
[14] Matyjaszewski, K., Advances in Controlled/Living Radical Polymerization; ACS Symposium Series 854, American Chemical Society: Washington DC, 2003.
[15] Matyjaszewski, K., Controlled/Living Radical Polymerization: From Synthesis to Materials; ACS Symposium Series 944, American Chemical Society: Washington DC, 2006.

[16] Matyjaszewski, K.; Spanswick, J. *Materials Today.* 2005, 8, 26-33.
[17] Szwarc, M. *Nature.* 1956, 178, 1168-1169.
[18] Matyjaszewski, K. *J. Phys. Org. Chem.* 1995, 8, 197-207.
[19] Matyjaszewski, K., Cationic Polymerization: Mechanisms, Synthesis and Applications; Marcel Dekker: New York, 1996.
[20] Muller, A. H. E.; Zhuang, R. G.; Yan, D. Y.; Litvinenko, G. *Macromolecules.* 1995, 28, 4326-4333.
[21] Quirk, R.; Lee, B. *Polym. Int.* 1992, 27, 359-367.
[22] Coessens, V.; Pintauer, T.; Matyjaszewski, K. *Progr. Polym. Sci.* 2001, 26, 337-377.
[23] Patten, T. E.; Matyjaszewski, K. *Adv. Mater.* 1998, 10, 901-915.
[24] Matyjaszewski, K.; Xia, J. H. *Chem. Rev.* 2001, 101, 2921-2990.
[25] Matyjaszewski, K.; Wang, J. L.; Grimaud, T.; Shipp, D. A. *Macromolecules.* 1998, 31, 1527-1534.
[26] Percec, V.; Barboiu, B.; Neumann, A.; Ronda, J. C.; Zhao, M. Y. *Macromolecules.* 1996, 29, 3665-3668.
[27] Percec, V.; Barboiu, B. *Macromolecules.* 1995, 28, 7970-7972.
[28] Percec, V.; Barboiu, B.; Kim, H. J. *J. Am.* Chem. *Soc.* 1998, 120, 305-316.
[29] Percec, V.; Kim, H. J.; Barboiu, B. *Macromolecules.* 1997, 30, 6702-6705.
[30] Percec, V.; Kim, H. J.; Barboiu, B. *Macromolecules.* 1997, 30, 8526-8528.
[31] Kato, M.; Kamigaito, M.; Sawamoto, M.; Higashimura, T. *Macromolecules.* 1995, 28, 1721-1723.
[32] Wang, J. S.; Matyjaszewski, K. J. *Am.* Chem. *Soc.* 1995, 117, 5614-5615.
[33] Granel, C.; Dubois, P.; Jerome, R.; Teyssie, P. M*acromolecules.* 1996, 29, 8576-8582.
[34] Matyjaszewski, K.; Coca, S.; Gaynor, S. G.; Wei, M. L.; Woodworth, B. E. *Macromolecules.* 1997,30, 7348-7350.
[35] Matyjaszewski, K.; Wei, M. L.; Xia, J. H.; McDermott, N. E. *Macromolecules.* 1997, 30, 8161-8164.
[36] Kotani, Y.; Kamigaito, M.; Sawamoto, M. *Macromolecules.* 1999, 32, 2420-2424.
[37] Lecomte, P.; Drapier, I.; Dubois, P.; Teyssie, P.; Jerome, R. *Macromolecules.* 1997, 30, 7631-7633.
[38] Le Grognec, E.; Claverie, R.; Poli, R. *J. Am.* Chem. *Soc.* 2001, 123, 9513-9524.
[39] Brandts, J. A. M.; van de Geijn, P.; van Faassen, E. E.; Boersma, J.; van Koten, G. *J. Organomet. Chem.* 1999, 584, 246-253.
[40] Shen, Y. Q.; Tang, H. D.; Ding, S. *J. Progr. Polym. Sci.* 2004, 29, 1053-1078.
[41] Shen, Y. Q.; Zhu, S. P.; Zeng, F. Q.; Pelton, R. H. *Macromolecules.* 2000, 33, 5427-5431.
[42] Kickelbick, G.; Paik, H. J.; Matyjaszewski, K. *Macromolecules.* 1999, 32, 2941-2947.
[43] Haddleton, D. M.; Kukulj, D.; Radigue, A. P. *Chem. Commun.* 1999, 99-100.
[44] Duquesne, E.; Labruyere, C.; Habimana, J.; Degee, P.; Dubois, P. *J. Polym. Sci., Part A: Polym. Chem.* 2006, 44, 744-756.
[45] Duquesne, E.; Degee, P.; Habimana, J.; Dubois, P. *Chem. Commun.* 2004, 640-641.
[46] Payne, G. F.; Raghavan, S. R. *Soft Matter.* 2007, 3, 521–27.
[47] Stern, R.; Jedrzejas, M. *J. Chem Rev.* 2008, 108, 5061–85.
[48] Berger, J.; Reist, M.; Mayer, J. M.; Felt, O.; Peppas, N. A.; Gurny, R. *Eur. J. Pharm. Biopharm.* 2004, 57, 19–34.

[49] Laurent, T. C, editor. The chemistry, biology and medical applications of hyaluronan and its derivatives. Wenner-Gren Int Se, vol. 72. LondonPortland Press; 1998.
[50] Lindblad, M. S.; Sjoeberg, J.; Albertsson, A. C.; Hartman, J. *ACS Symp. Ser.* 2007, 954, 153-167.
[51] Raid, G. J. S. in Biochemistry of Storage Carbohydrates,. Day, P. M.; Dixon, R. A. Eds., Academic Press: London, 1985, 265-288.
[52] McCleary, B. V.; Clark, A. H.; Dea, I. C. M.; Rees, D. A. *Carbohydrate Res.* 1981, 92, 269-285.
[53] Tahlawy, K. E.; Hudson, S. M. J. *Applied Polymer Science.* 2003, 89, 901–912.
[54] Shen, D.; Huang, Y. *Polymer.* 2004, 45 7091-7097.
[55] Coskun, M.; Temuz, M.M. *Polym. Int.* 2005, 54, 342-347.
[56] Shen, D.; Yu, H.; Huang, Y. Journal of Polymer Science: Part A: Polymer Chemistry. 2005, 43, 4099-4108.
[57] Vlcek, P.; Janata, M.; Latalova, P.; Kriz, J.; Adova, E. C.; Toman, L. *Polymer.* 2006, 47, 2587-2595.
[58] Bontempo, D.; Masci, G.; Leonardis, P. D.; Mannina, L.; Capitani, D.; Crescenzi, V. *Biomacromolecules.* 2006, 7, 2154-2161.
[59] Stmark, E. O. Harrisson, S.; Wooley, K. L.; Eva, E. M. *Biomacromolecules.* 2007, 8, 1138-1148.
[60] Vlcek, P.; Janata, M.; Latalova, P.; Dybal, J.; Spirkova, M.; Toman, L. *Journal of Polymer Science: Part A: Polymer Chemistry.* 2008, 46, 564-573.
[61] Yan, L.; Ishihara, K. Journal of Polymer Science: Part A: Polymer Chemistry. 2008, 46, 3306-3313.
[62] Chang, F.; Yamabuki, K.; Onimura, K.; Oishi, T. *Polymer Journal.* 2008, 40 (12), 1170-1179.
[63] Dupayage, L.; Save, M.; Dellacherie, E.; Nouvel, C.; Six, J. L. *Journal of Polymer Science: Part A: Polymer Chemistry*, 2008, 46, 7606-7620.
[64] Shinsuke I.; Kadla, J. F. *Biomacromolecules.* 2008, 9, 3308-3313.
[65] Sui, X.; Yuan, J.; Zhou, M.; Zhang, J.; Yang, H.; Yuan, W.; Wei, Y.; Caiyuan Pan. *Biomacromolecules.* 2008, 9, 2615-2620.
[66] Stmark, E. O.; Nystrom, D.; Malmstrom, E. Macromolecules. 2008, 41, 4405-4415.
[67] Li, Y.; Liu, R.; Liu, W.; Kang, H.; Wu, M.; Huang, Y. *Journal of Polymer Science: Part A: Polymer Chemistry.* 2008, 46, 6907-6915.
[68] Meng, T.; Gao, X.; Zhang, J.; Yuan, J.; Zhang, Y.; He, J. *Polymer.* 2009, 50, 447-454.
[69] Lin, C.; Zhan, H.; Liu, M.; Fu, S.; Zhang, J. *Carbohydrate Polymers.* 2009, 78 432-438.
[70] Munro, N. H.; Hanton, L. R.; Moratti, S. C.; Robinson, B. H. *Carbohydrate polymers.* 2009, 77, 496-505.
[71] Xu, F. J.; Zhang, Z. X.; Ping, Y.; Li, J. Kang, E. T.; Neoh, K. G. *Biomacromolecules.* 2009, 10, 285-293.
[72] Li, J.; Guo, Z.; Xin, J.; Zhao, G.; Xiao, H. *Carbohydrate Polymers.* 2010, 79 277-283.
[73] Ping, Y.; Liu, C. D.; Tang, G.P.; Li, J. S.; Li, J.; Yang, W.T.; Xu, F.J. *Adv. Funct. Mater.* 2010, 20, 3106-3116.
[74] Tan, J.; Li, Y.; Liu, R.; Kang, H.; Wang, D.; Ma, L.; Liu, W,; Wu, M.; Huang Y. *Carbohydrate Polymers.* 2010, 81, 213-218.
[75] Tian, W.; Fan, X.; Kong, J.; Liu, Y.; Huang, Y. *Polymer.* 2010, 51, 2556-2564.

[76] Xu, F. J.; Zhu, Y.; Liu, F. S.; Nie, J.; Ma, J.; Yang, W. T. *Bioconjugate Chem.* 2010, 21, 456-464.

[77] Wang, Z. H.; Li, W. B.; Ma, J.; Tang, G. P.; Yang, W. T.; Xu, F. J. *Macromolecules.* 2011, 44, 230-239.

[78] Mishra, V.; Kumar, R. *Carbohydrate Polymers*. 2011, 83, 1534-1540

[79] Yuan, W.; Li, X.; Gu, S.; Amin Cao.; Ren, J. *Polymer.* 2011, 52, 658-666.

[80] Jin-Feng Zhong, Xin-ShengChai, Xiao-LiQin, Shi-YuFu Carbohydrate Polymers xxx (2011) xxx–xxx doi:10.1016/j.carbpol.2011.04.050

[81] Mishra, V.; Rajesh Kumar. Carbohydrate Polymers. xxx (2011) xxx–xxx, doi:10.1016/j.carbpol.2011.04.052

[82] Lepoittevin, B.; Elhiri, A.; Bech, L.; Belleney, J.; Baltaze, J. P.; Capron, I.; Planchot, V.; Roger, P. *Carbohydrate Polymers.* 2011, 83, 1174-1179.

[83] Carlmark, A.; Malmstro E. *J. Am. Chem. Soc.* 2002, 124, 6-7.

[84] Plackett, D.; Jankova, K.; Egsgaard, H.; Hvilsted, S. *Biomacromolecules*. 2005, 6, 2474-2484.

[85] Liu, P.; Su, Z. *Carbohydrate Polymers*. 2005, 62, 159-163.

[86] Lindqvist, J.; Malmstrom, E. *Journal of Applied Polymer Science.* 2006, 100, 4155-4162.

[87] Liu, P.; Su, Z.; *Materials Letters.* 2006, 60, 1137-1139.

[88] Lei, Z.; Bi, S.; Yang, H. *Food Chemistry.* 2007, 104, 577-584.

[89] Lindqvist, J.; Nystrom, D.; Ostmark, E.; Antoni, P.; Carlmark, A.; Johansson, M.; Hult, A.; Malmstrom, E. *Biomacromolecules.* 2008, 9, 2139-2145.

[90] Liu, Y. L.; Chen, W. H.; Chang, Y. H. *Carbohydrate Polymers*. 2009, 76, 232-238.

[91] Xiao, M. M.; Li, S.; Chanklin, W.; Zheng, A.; Xiao, H. *Carbohydrate Polymers*. 2011, 83, 512-519.

[92] Ma, L.; Liu, R.; Tan, J.; Wang, D.; Jin, X.; Kang, H.; Wu, M.; Huang, Y. *Langmuir.* 2010, 26 (11), 8697-8703.

[93] Jiang, J.; Hua, D.; Jiang, J.; Tang, J.; Zhu, X. *Carbohydrate Polymers.* 2010, 81, 358-364.

[94] Bao, H.; Li, L.; Gan, L.H.; Ping, Y.; Li, J.; Ravi, P. *Macromolecules.* 2010, 43, 5679-5687.

[95] Yuan, W.; Zhao, Z.; Gu, S.; Ren, T.; Ren, *J. Materials Letters.* 2011, 65, 793-796.

[96] Wang, D.; Tan, J.; Kang, H.; Lin Ma.; Jin, X.; Liu, R.; Huang, Y. *Carbohydrate Polymers.* 2011, 84, 195-202.

[97] Chiefari, J.; Chong, Y. K.; Ercole, F.; Krstina, J.; Jeffery, J.; Le, T. P. T.; Mayadunne, R. T. A.; Meijs, G. F.; Moad, C. L.; Moad, G.; Rizzardo, E.; Thang, S. H. *Macromolecules.* 1998, 31, 5559-5562.

[98] Mayadunne, R. T. A.; Rizzardo, E.; Chiefari, J.; Chong, Y. K.; Moad, G.; Thang, S. H. *Macromolecules*. 1999, 32, 6977-6980.

[99] Rizzardo, E.; Chiefari, J.; Chong, B. Y. K.; Ercole, F.; Krstina, J.; Jeffery, J.; Le, T. P. T.; Mayadunne, R. T. A.; Meijs, G. F.; Moad, C. L.; Moad, G.; Thang, S. H. *Macromol. Symp.* 1999, 143, 291-307.

[100] Charmot, D.; Corpart, P.; Adam, H.; Zard, S. Z.; Biadatti, T.; Bouhadir, G. *Macromol. Symp.* 2000, 150, 23-32.

[101] Chong, Y. K.; Le, T. P. T.; Moad, G.; Rizzardo, E.; Thang, S. H. *Macromolecules.* 1999, 32, 2071-2074.

[102] Stenzel, M. H.; Davis, T. P.; Barner-Kowollik, C. *Chem. Commun.* 2004, 1546-1547.

[103] Mayadunne, R. T. A.; Jeffery, J.; Moad, G.; Rizzardo, E. *Macromolecules.* 2003, 36, 1505-1513.

[104] Quinn, J. F.; Chaplin, R. P.; Davis, T. P. J. *Polym. Sci., Part A: Polym. Chem.* 2002, 40, 2956-2966.

[105] Stenzel-Rosenbaum, M.; Davis, T. P.; Chen, V.; Fane, A. G. *J. Polym. Sci., Part A: Polym. Chem.* 2001, 39, 2777-2783.

[106] Fijten, M. W. M.; Meier, M. A. R.; Hoogenboom, R.; Schubert, *U. S. J. Polym. Sci., Part A: Polym. Chem.* 2004, 42, 5775-5783.

[107] Chong, Y. K.; Krstina, J.; Le, T. P. T.; Moad, G.; Postma, A.; Rizzardo, E.; Thang, S. H. *Macromolecules.* 2003, 36, 2256-2272.

[108] Adamy, M.; van Herk, A. M.; Destarac, M.; Monteiro, M. J. *Macromolecules.* 2003, 36, 2293-2301.

[109] Destarac, M.; Bzducha, W.; Taton, D.; Gauthier-Gillaizeau, I.; Zard, S. Z. *Macromol. Rapid Comm.* 2002, 23, 1049-1054.

[110] Hua, D.; Tang, J.; Cheng, J.; Deng, W.; Zhu, X. *Carbohydrate Polymers.* 2008, 73, 98-104.

[111] Jiang, J.; Hua, D.; Tang, J. International Journal of Biological Macromolecules. 2010, 46, 126-130.

[112] Tang, J.; Hua, D.; Cheng, J.; Jiang, J.; Zhu, X. *International Journal of Biological Macromolecules.* 2008, 43, 383-389.

[113] Roy, D.; Guthrie, J.T.; Perrier S. *Macromolecules.* 2005, *38,* 10363-10372.

[114] Martina, H.; S, Thomas, P. D.; Anthony, G. F. *J. Mater. Chem.,* 2003, 13, 2090-2097.

[115] Solomon, D. H.; Rizzardo, E.; Cacioli, P. EP0135280, 27-03-1985.

[116] Veregin, R. P. N.; Georges, M. K.; Hamer, G. K.; Kazmaier, P. M. *Macromolecules.* 1995, 28, 4391-4398.

[117] Georges, M. K.; Veregin, R. P. N.; Kazmaier, P. M.; Hamer, G. K. *Macromolecules.* 1993, 26, 2987-2988.

[118] Veregin, R. P. N.; Georges, M. K.; Kazmaier, P. M.; Hamer, G. K. *Macromolecules.* 1993, 26, 5316-5320.

[119] Burguiere, C.; Dourges, M. A.; Charleux, B.; Vairon, J. P. *Macromolecules.* 1999, 32, 3883-3890.

[120] Ananchenko, G. S.; Fischer, H. J. *Polym. Sci., Part A: Polym. Chem.* 2001, 39, 3604-3621.

[121] Ananchenko, G. S.; Souaille, M.; Fischer, H.; Le Mercier, C.; Tordo, P. *J. Polym. Sci., Part A: Polym. Chem.* 2002, 40, 3264-3283.

[122] Charleux, B.; Nicolas, J.; Guerret, O. *Macromolecules.* 2005, 38, 5485-5492.

In: Radical Polymerization: New Developments ISBN 978-1-62100-406-6
Editors: I. O. Paulauskas, L. A. Urbonas, pp. 199-221

Chapter 6

ADVANCEMENTS IN CONTROLLED RADICAL POLYMERIZATION FOR FUNCTIONAL POLYMERS

Harshad R. Patil, Hiren M. Bhajiwala and Virendrakumar Gupta *

Reliance Technology Group, Reliance Industries Ltd,
Hazira Complex,
Surat, Gujarat 394510, India.

ABSTRACT

Design and synthesis of structurally well defined polymer is one of the areas of current technology interest for niche product applications. Major research efforts are focused to move from coordination-addition to advanced radical polymerization methodology to produced new polymeric structures using polar and/or non-polar monomer for different end use applications. Evolution of controlled radical polymerization (CRP) lead to synthesis of defined architecture (star, comb, cyclic, dentritic etc) polymer with controlled molecular weights under mild reaction conditions. In the present chapter, various CRP s [Stable Free Radical Polymerization- SFRP, Atom Transfer Radical Polymerization -ATRP and Degenerative Transfer Process- DT] approaches are discussed for the synthesis of functional polymers. The emphasis is given for incorporation of olefin monomer into functional polymer matrix along with structure – property – performance relationship.

1. INTRODUCTION

Radical polymerization (RP) is one of the major process used for the synthesis of synthetic plastic. This technique is commercially successful due to large number of monomer

* Corresponding author. Tel.: +912616635381; Fax: +912616635879.
E-mail address: Virendrakumar_Gupta@ril.com.

facile to be polymerized by this technique which is less affected by impurities and water. It can be used with many different types of monomers like acidic, hydroxyl and amino etc. RP is suitable with bulk, suspension, solution, emulsion & dispersion etc. Special characteristic of RP is continuous formation and continuous disappearance of radicals leading to very fast reaction and formation of higher molecular weight polymer having broad polydispersity. One of the major drawbacks of the RP is poor control over the microstructure and broad range of Polydispersity index PDI. It is also practically impossible to design advance structure like block, star, etc [1-2].

To overcome the above problem, many scientists worked on polymerization technique based on reversible deactivation of polymer radicals or a degenerative transfer process, called 'living' or controlled radical polymerizations (CRP). Synthetic approach involved equilibrium between growing radical with dormant species, which minimized termination process called - control reversible deactivation radical polymerization. First living radical polymerization (LRP) was demonstrated by Szwarc M. through living anionic polymerization [3-5]. Otsu *et al.* also worked on the use of disulfides to polymerize styrene and methacrylates. These species initiate, transfer and reversibly terminate the chain during polymerization mechanism [6-7].

Above invention was the foundation of modern CRP techniques to synthesize well define precisely controlled polymers [5]. Drawback of anionic living polymerization involved its sensitivity towards moisture, carbon dioxide and acidic monomer. Improvement in the CRP techniques was done by many scientists to overcome above problem. Nitroxide mediated polymerization (NMP) [8-10], ATRP (atom transfer radical polymerization) [11,12] and reversible addition-fragmentation chain transfer (RAFT) [13,14] techniques were developed, which can be used polymerize wide range of monomer for various specialty applications such as thermoplastic elastomer [15], compounding, footwear, toys, medical, lubricants, gels, coating, surfactant [16], pigments, dispersant, health and beauty product [17-18], chromatographic device [19], controlled drug released in cardiovascular stents [20] etc.

Today CRP is one of the most attractive area of research to design structures for specialty applications and still many opportunities are available for designing of material with composition, topology, functionality etc. to expand polymer applications to human life. In this book chapter, we have discussed types of CRP techniques and new development which lead to expand its horizon.

2. Historical Evolution of CRP

Free radical polymerization is one of the most widely used techniques for preparing commercial polymers. Major drawback of this technique is uncontrolled polymerization can be achieved-for specialty applications. To overcome the above problem in late 1950s, Szwarc developed first anionic living polymerization technique [3] which is governed on the principle of reversible deactivation of active radical. Otsu and co-workers also reported reversible polymerization capable of synthesizing narrow PDI block polymer [7, 21].

In the mid 1980s, Rizzardo and co-workers developed for the first time reversible termination of radicals with a stable free radical at CSIRO in Australia. They found that nitroxide-stable free radicals were able to add to carbon-centered radicals to form alkoxy amines [8]. Initially 2,2,6,6,Tetramethyl piperidine (TEMPO) type compound was used for polymerization of styrene , but it required higher temperature to establish equilibrium. To overcome the problem of higher temperature of polymerization, in 1990s, Hawker developed new nitroxides which allowed polymerization to be carried out below 100°C. [9-10]. At the end of 1990s, ATRP technique was parallely developed by Sawamoto and Matyjaszweski [11-12]. They independently reported transition-metal– catalyzed polymerization of various vinyl monomers. In this technique, activated alkyl halide used as initiator, transition-metal complex as catalyst and nitrogen based ligand as stabilizing agent.

ATRP is used as versatile technique for the controlled polymerization of styrene derivatives, acrylates, methacrylates, etc. Use of activated alkyl halides as initiators allowed the easy introduction of functional end groups in the polymer [22-23]. Many different polymer architectures have been synthesized via ATRP. These architectures include block copolymers [11], graft copolymers [24-25] star branched polymers, etc. Another development was based on reversible addition fragmentation polymerization (RAFT) which was developed by Rizzardo and co-workers in 1990s using thiocarbonylthio compound. The activating substituent lead to polymerization of vinyl monomer having low PDI with high rate of chain transfer [13-14]. The historical evolution is depicted in Figure 1. The typical characteristics of RP and CPR are shown in Figure 2.

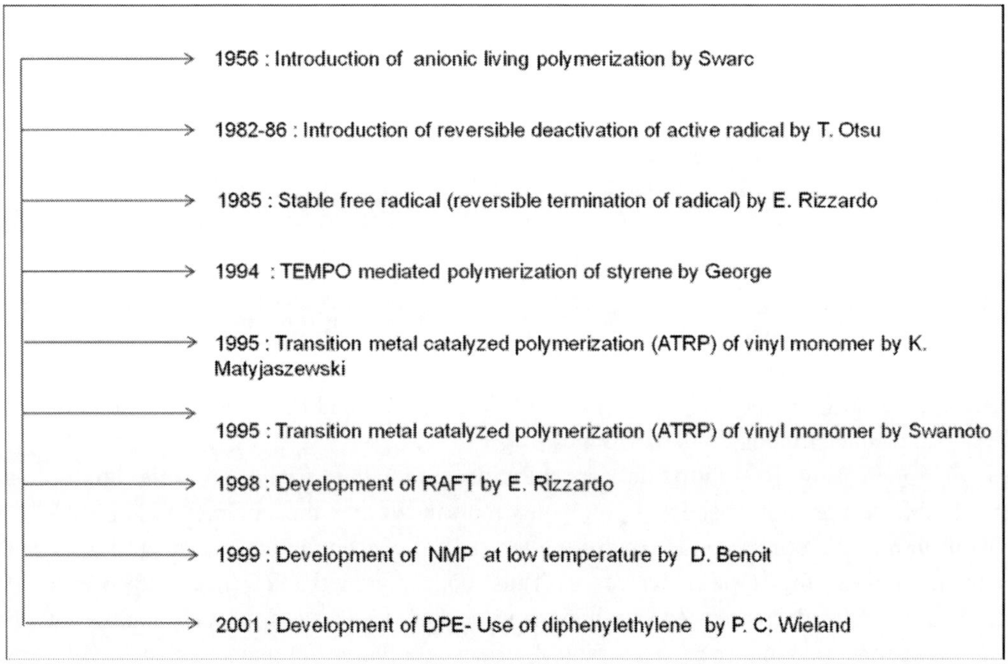

Figure 1. Historical evolution of control radical polymerization processes.

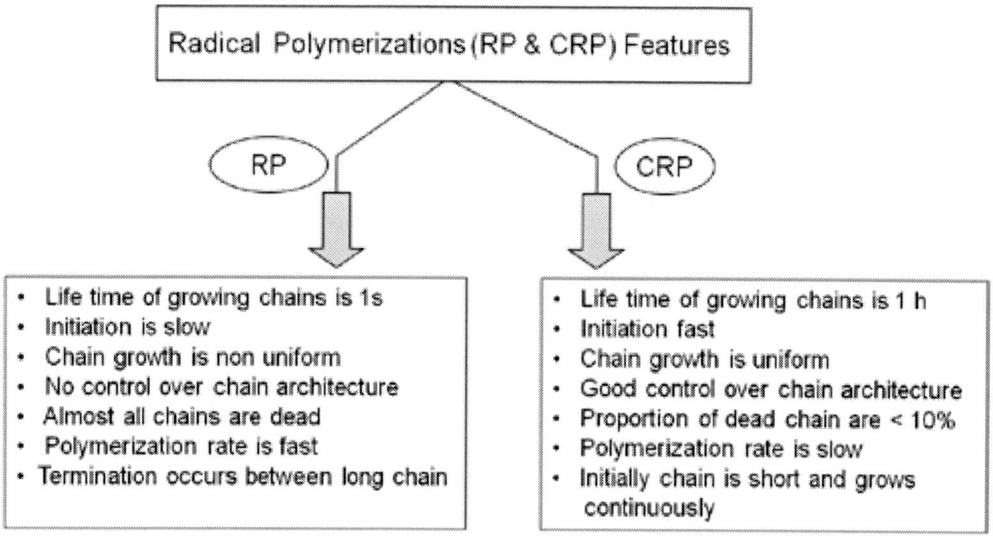

Figure 2. RP and CRP typical characteristics.

3. Types of Control Radical Polymerization

3.1. Nitroxide Mediated Polymerization (NMP)

Nitroxide-mediated polymerization (NMP) was initially reported in 1985 by Solomon and Rizzardo [8]. This technique was brought to attention in 1993 by Georges *et al* [26] and Hawker [27]. NMP is based on the reversible trapping of carbon-centered radicals by nitroxide species (Figure 3). The carbon-oxygen bond is thermally labile at higher temperature which undergoes homolytic cleavage forming a nitroxide radical. The radical can propagate or participate in other reactions. The rate of the polymerization is controlled by the radical concentration and which is dominated by activation and deactivation rate coefficients (k_a, k_d) and concentration of alkoxyamines.

The increasing free nitroxide concentration originating from inevitable bimolecular irreversible termination reaction of polymer radicals during the polymerization shifts the equilibrium to the dormant side controlling the radical concentration. As a result, the radical concentration in the system decreases. Thus both polymerization rate and irreversible termination rate decreases and polymerization kinetics results in self-control of the polymerization reaction (persistent radical effect). Polymers prepared using NMP show controlled architectures, a low polydispersity index (PDI) and can be used as macro-initiators to prepare more complex architectures e.g. block copolymers. The major disadvantage of NMP comes from usage of limited number of monomers whose polymerization can be controlled in a satisfactory manner: mainly styrene and its derivates. Another limitation is from high temperature requirement for polymerization (110°C). Lately nitroxides have been developed that can be employed for lower temperature polymerizations reaction also. These nitroxides may also be used in the polymerization of acrylates and 1,3-dienes [9,28,29]. The presence of a labile carbon-oxygen bond at the end of the polymer chains and higher cost inhibit commercial usage of NMP technique use for polymerization.

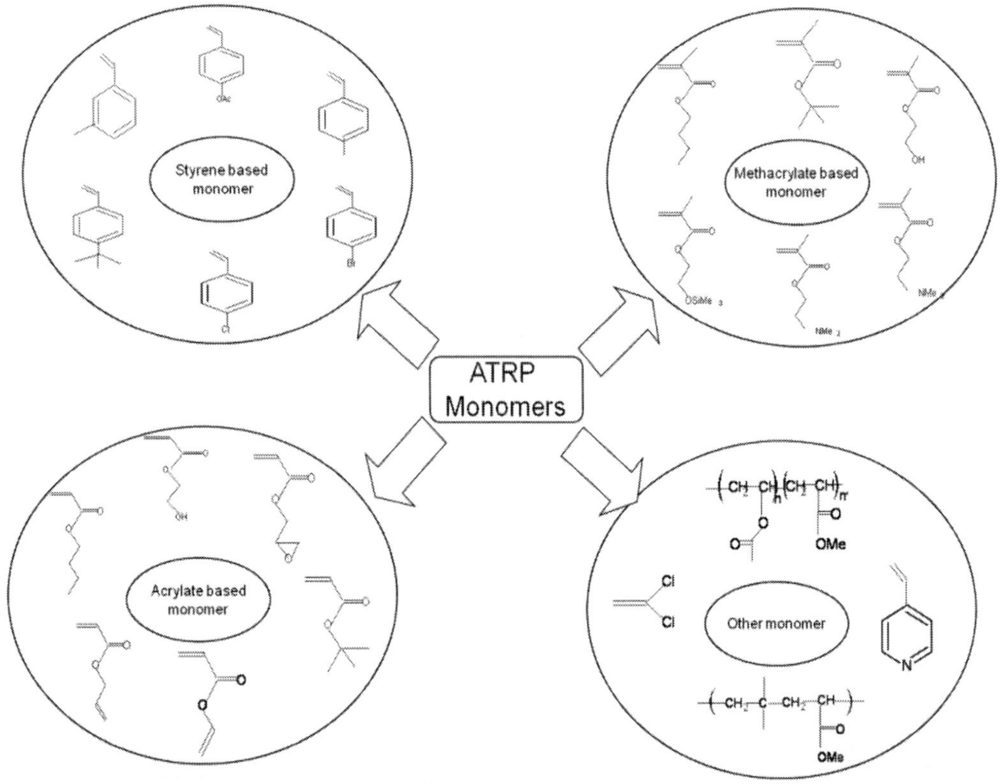

Figure 3. Functional monomers used during ATRP.

3.2. Reversible Addition-Fragmentation Chain Transfer Polymerization

In 1998, living free-radical polymerization by reversible addition-fragmentation chain transfer (RAFT) was reported[14]. RAFT polymerization is based on the reaction of a polymer radical with a chain transfer agent, e.g. dithioesters resulting in an intermediate radical. The intermediate radical will homolytically cleave in a polymeric radical (derived from the RAFT agent or the initial polymer radical) and a polymeric chain transfer agent (Figure 6). The polymerization has to be started by adding a small amount of radical source to the solution of monomer and RAFT agent. The main advantage of RAFT includes its capability to polymerize wide range of monomers including vinyl acetate and acid-functional monomers [30].

The disadvantage of polymers prepared by RAFT polymerization includes presence of the thermal and UV sensitive RAFT end-group present at the majority of the polymer chain ends. RAFT agents are expensive, often colored and are not readily available. Sulfur containing molecules of low molecular weight fragments also generate a strong odor.

3.3 . Atom Transfer Radical Polymerization

ATRP is one of the most popular technique for controlled radical polymerization for polymerization of various polar monomer like styrene, acrylate, methacrylate, acrylonitrile, (meta)acrylic acid etc. and nonpolar-polar monomer like dines & ά-olefins [31]. Basic component for ATRP are transition metal, ligand, solvent, alkyl halide initiator and various monomers.

ATRP works on the concept of reversible termination reaction. Basic mechanism involve generation of active alkyl radical ($R^{\bullet}$) due to homolytic cleavage of alkyl halide (R-X) bond by transition metal. During the reaction transition metal Mt^n/L move to higher oxidation stage Mt^{n+1}/L. Monomer reacts with alkyl radical and propagation reaction starts (K_p) which is followed by termination by coupling or disproportionate (K_t) leading to reversibly deactivate $X\text{-}Mt^{n+1}/L$ (K_{da}) to form Mt^n/L . Radical termination is moderated as a result of persistent radical [32-33] and ATRP equilibrium is shifted towards dormant species. Reversible deactivation reaction of $X\text{-}Mt^{n+1}/L$ with $P_n^{\bullet}$ forms halide capped polymer chain and metal with lower oxidation state. Generally overall activation rate constant ($K_{act)}$ is lower than deactivation constant (K_{da}). The control of ATRP reaction totally depends on forward and backward reaction rate constant. Concentration of radical, rate of polymerization, termination and polydispersity depends on equilibrium rate constant and their ratio. The value of K_{ATRP}, K_{act}, and K_{dact} depends on the catalyst, initiator, monomer structure, solvent and reaction conditions. K_{act} should be larger than K_{dact} to have good control over polydispersity [31].

Basic Component of ATRP

Catalyst

Catalyst is the main component of ATRP reaction. Metal with different oxidation state determine the position of atom transfer equilibrium and dynamics of dormant & active species. Requirements of catalyst for ATRP are : (1) It should have more than one oxidation state (2) It should have affinity towards halogen (3) It should strongly form complex with ligand. Among different transition metals, Cu can bepolymerized almost all types of monomers. It is easily available, cheap and forms dynamic equilibrium between dormant and persistent radical easily. It is compatible with different types of ligands [34]. It is most efficient ATRP catalyst due to easy conversion to next oxidation state with ligand. ATRP was successfully carried out using various metals of different group of periodic table from Groups 4 (Ti [35]), 6 (Mo [36-38]), 7 (Re [39]), 8 (Fe, [40-41], Ru[42-43] Os [44]), 9 (Rh, [45] Co [46]), 10 (Ni, [47-48] Pd [49]), and 11 (Cu [50]).

Monomer

ATRP allows to polymerize various polar monomer such as styrene, methacrylate, (meth)acrylamide vinyl pyridine, acrylonitrile, vinyl acetate, vinyl chloride, and others. Non polar monomer such as diene and alpha olefins were also successfully polymerized using ATRP. The method is tolerant to many functional groups like ester, hydroxyl, amino, epoxy and others except carboxylic acid group because it protonate amine based ligand necessitating protection before polymerization. Polymerization reaction depends on magnitude of K_{ATRP}. If K_{ATRP} is low or very large, in both cases the reaction rate will be slow [51] . The monomers widely used for ATRP are depicted in Figure 3.

Initiators

The role of the initiator is to determine the number of growing polymer chains. In ATRP, generally alkyl halides (RX) are used as the initiators. Hemolytic cleavage of alkyl halide radical takes place due to transition metal catalyst and polymerization undergoes at active alkyl species.The halide group (X) rapidly and selectively migrate between the growing chain and the transition-metal complex. Generally -Cl, -Br & -I are used as halide. Formation of narrow molecular weight distribution and well define microstructure of polymer are only due to equilibrium between persistent and dormant species in the reaction. Different halogen compound are selected based on compatibility with monomer and other components of reaction system. Iodine works well for acrylate polymerizations in copper-mediated ATRP [52]. Fluorine is generally not used because the C-F bond is strong enough & prevent to undergo hemolytic cleavage. Some pseudo halogens, specifically thiocyanates and thiocarbamates, have been used successfully in the polymerization of acrylates and styrenes [52-54]. The initiators widely used for ATRP are shown in Figure 4.

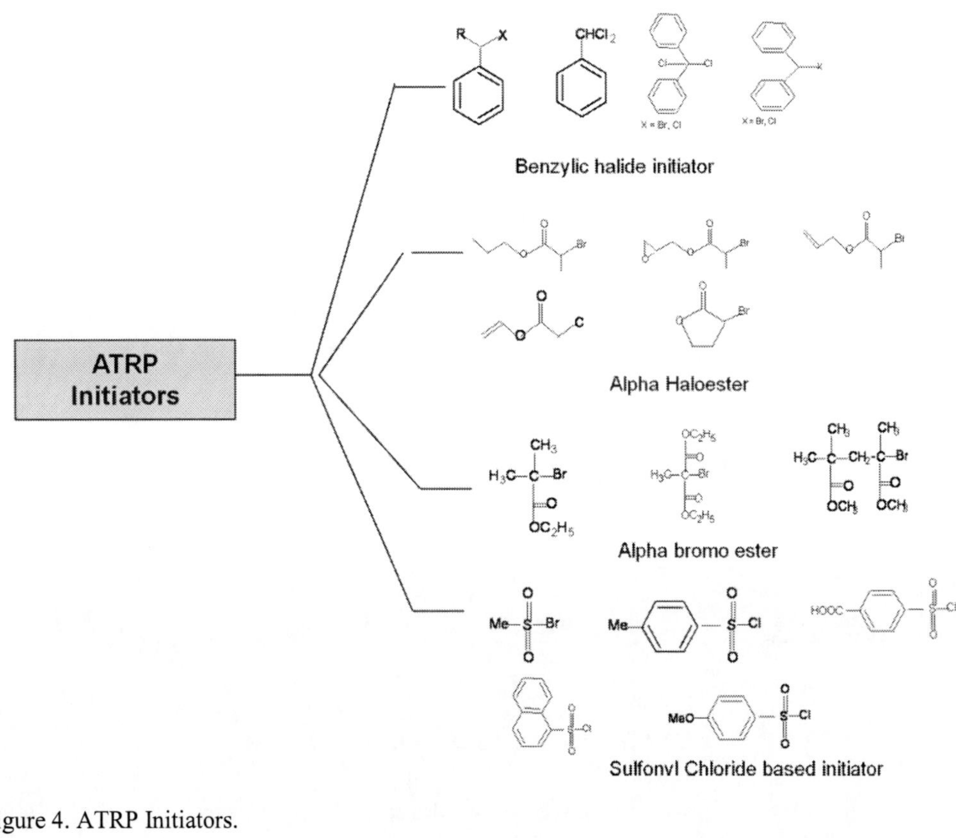

Figure 4. ATRP Initiators.

Ligand

The role of the ligand in ATRP is to solubilize the transition-metal salt in the organic media and to adjust the redox potential of the metal center for appropriate reactivity and dynamics for the atom transfer. Generally nitrogen based organic compounds are used as ligands because it makes weak bond with metal & form organometallic compound. Reactivity of the organometallic compound depends on various factors such as structure, number of nitrogen and carbon atom etc. Reactivity of different ligand affects reaction rate and microstructure of the polymer. Polydented ligand with Cu are well known for ATRP [55]. General rules for activity of ligands are : (1) ligand topology (cyclic ~ liner < branch) (2) nature of N based ligand (aryl amine < aryl imine < alkyl imine < alkyl amine < pyridine) (3) Steric effect e.g Me_6TREN is 100000 times more active than Et_6TREN., (4) side reactions such as proton abstraction or reduction/ oxidation of the radicals. The ligands used for ATRP are shown in Figure 5.

The mechanism of all three CPR technique are depicted in Figure 6. The advantage and disadvantage of all three methods are also shown in Figure 7.

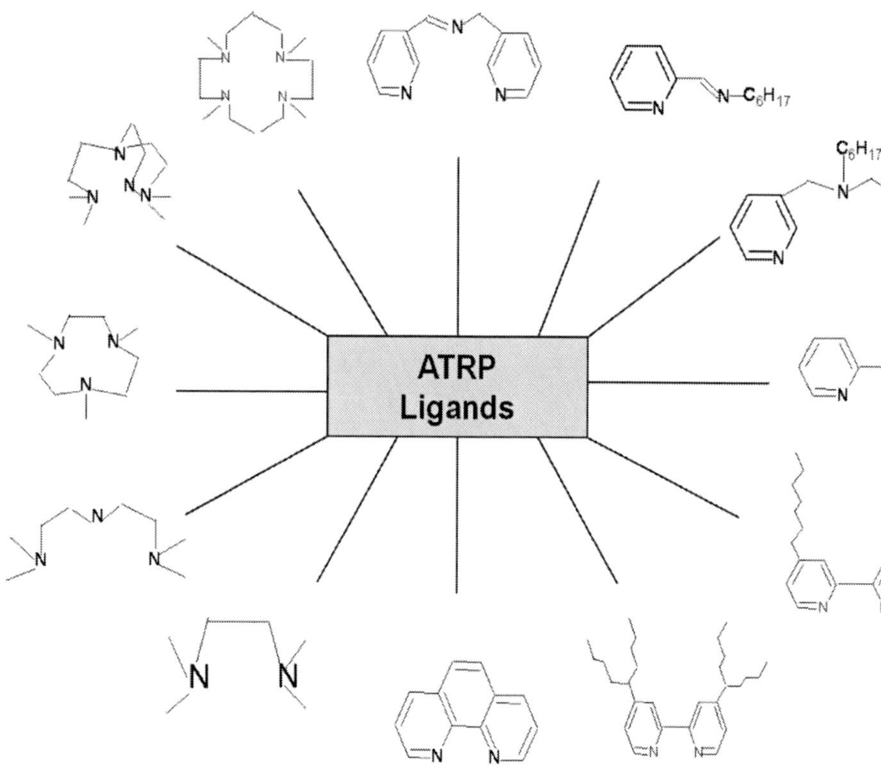

Figure 5. ATRP Ligands.

4. FUNCTIONALIZED POLYOLEFINS

Variety of olefinic polymers have been developed in modern life. Polyolefins (polyethylene, polypropylene etc) are widely used commercial polymers due to their well balanced physical and chemical properties. But they possess poor adhesion, stiffness, heat resistance, printability and dyabiliy. The summary of properties For polar and nonpolar polymer are described in Table 1. Functionalization of polyolefins is one of the approaches to overcome this problem through improving properties of polyolefin in that direction. It enables functional polyolefins application in wider area which is currently not achievable by commodity plastics. These polymers also possess very high potential for value addition.

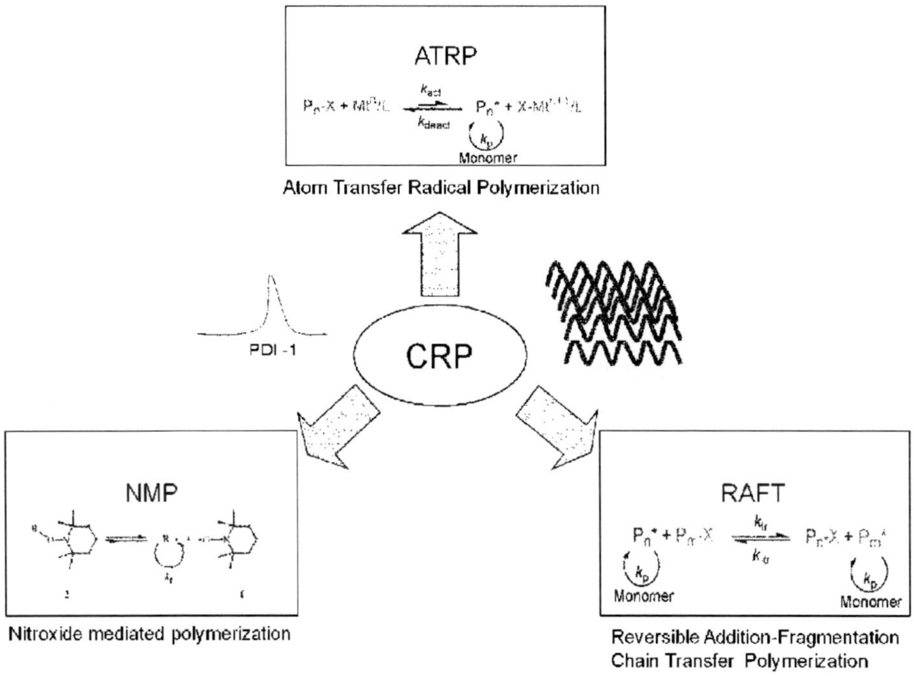

Figure 6. Types of control radical polymerization processes.

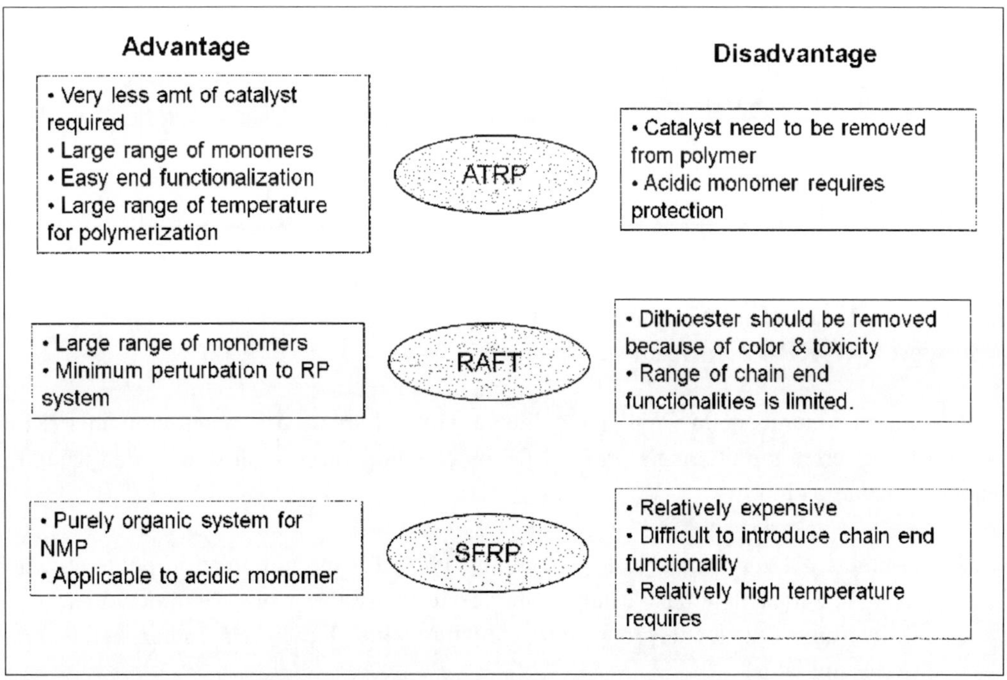

Figure 7. Advantage and disadvantage of CRP techniques.

Functionality of polyolefin can be explained as introduction of polar functional group in polyolefin backbone. Maleic anhydride (MAH) modified polypropylene is widely known as commercially available functional polyolefin with wide range of MAH, olefin contents and molecular weight. The application of polyolefins after functionalization can be extended to areas such as active species supporting in catalysis, medication, photoelectronic materials, biomaterials, photo materials and environmental protection which have never been explored with polyolefin before Addition of functional polar group lead to catalyst deactivation of Ziegler Natta (titanium supported magnesium dichloride) catalyst. Hence, functional polyolefin cannot be produced using early transition metal ZN catalyst.

Currently synthesis of functionalized polyolefin have been studied widely as a basic science by many scientists with various polar monomers like styrene, methyl methacrylates, glycidyl methacrylates etc. Several approaches have been studied in prior art which includes (a) Macromonomer / macroinitiator approach followed by CRP technique (b) Grafting of polar monomer from polyolefin by surface activation (c) Direct polymerization of olefin with polar monomer though CRP technique (d) Direct radical polymerization of olefin with polar monomer (e) direct polymerization of olefin with polar monomer through late transition metal or single site catalyst (e) Cracking of polyolefin to form active site for polar monomer incorporation. We have briefly discussed the approach (a) to (d).

Table 1. Properties of typical Polyolefin and Polar monomer containing polymer

Polyolefin Properties	Polar monomer containing polymer
• Durability • Weather Resistance • Chemical Resistance • Light weight • Hardness/softness • Hydrophobic	• Reactivity • Adhesion • Impact resistance • Compatibility • Heat Resistance • Ionic • Biocompatible

4.1. Macro Monomer/ Macro Initiator Approach Followed by CRP Technique

Polyolefin which contain vinyl or vinylidene terminal are used as macromonomer (MM) which are converted to initiating sites for CRP by chemical reaction. It was carried using the general concepts as follows

1) Synthesis of functional group containing polyolfin by coordination polymerization using late transition metal catalyst/ single site catalyst system (Macromonomer)
2) Converting functional group into initiating sites like – Br (other halides) etc (Macroinitiator)
3) Introducing comonomer into polyethylene backbone by ATRP techniques

However, this approach requires additional steps leading to cumbersome process with higher cost. It allows wide range to polar monomer incorporation and stringent control over polymerization to produce block, graft and star copolymer. This sections describes some of the works carried out in that direction. Several reviews are written in prior art on this approach [56]. PE-g-PS was first produced by Matyjaszewski and coworkers in 1999 followed by synthesis of PBA-g-PE using Pd based catalyst compound and subsequent ATRP technique for copolymerization [57].

Pd based catalyst was used to produced PE macroinitiator having relatively low M_w (~1000) compared to conventional PE. PE—MI was polymerized wih nBA (n-butyl acrylates) using CuBr/Ligand system to produced PE-g-PBA. Metallocene catalyst are generally used to produce macroinitiator in addition to use of late transition metal catalyst system. One of limitation faced by scientist to commercialize this approach is to have low solubility of PE macroinitiator in solvents which are used in ATRP technique. Kallitsis and coworkers reported reaction of PE-g-PGMA (poly glycidyl methacrylates) and end functionalized polystyrene to produce PS-g-[PE-g-PGMA]. Such blends are widely used as compatabilizer for PE and PS blends. [58]

Hwu *et al.* synthesized graft polymer of MMA and BA with PE-g-PMS (p-methyl styrene) via ATRP. PE-g-PMS was prepared through constrain geometry zirconium catalyst followed by bromination process using N-bromosuccinimide [59]. PA (polyacrylates)-g-PE (polyethylene) using DT and ATRP was synthesized by Matyjaszewski and coworkers using macro initiator of PE. PE-OH was first prepared by hydrolysis of PE-Zn produced by bis (imino) pyridine iron/diethyl zinc/MAO system with ethylene as monomer. PE-MI was further reacted with n-butyl acrylates and t-butyl acryalate to synthesize PE-b-PnBA and PE-b-PtBA. [60]. Vinyl (double bond) containing PE (PE Macromonomer) was reacted by Inoue *et al.* with α-halo carboxylic acid to produced PE-macroinitiator. PE-MI was then reacted with n-BA, MMA and styrene by ATRP technique to produce block copolymer [61]. Keneyosh et al. synthesized PE-MI (PE macroinitiator) converting vinyl of vinyl terminated PE to –OH followed by –Br moieties. Further it was copolymerized with styrene using CuBr/Me6TREN catalyst system to produce PE-g-PS [62].

Park *et al* synthesized copolymer of PE (vinyl monomer containing PE macromonomer) with styrene, butyl acrylate and glycidyl methacrylate (GMA) using CuBr/bPy as catalyst. The PE macromonomer was converted to PE initiator (PE-Br) by bromination process using HBr. The synthesized polymer showed PDI of 3. PE-MI was reacted with styrene, butyl acrylates and glycidyl methacrylate by ATRP technique to produce graft polymer. [63]. Guan and coworkers synthesized dendritic nano particle of ethylene with branched –Br containing comonomer using palladium catalyst. The M_w of dendrimer can be controlled simply by controlling polymerization time [64]. Zhang *et al.* synthesized hyperbranched PMMA with PE as core with PE-MI concept. PE-MI were produced using Pd-diimine catalyst for polymerization of ethylene with 2-(2-bromooisobutyryloxy) ethyl acryalates (BIFA). [65].

Matyjaszewaski and coworkers synthesized different polyethylene microinitiators like PE-Cl, PE-Br through reaction of oxyrane ring of P(E-co-GMA) with either chloroacetic acid and 2-bromoisobutyric acid. These microinitiator were later used for ATRP using CuCl/dNbpy wih styrene and MMA. Mw of such polymer increased linearly with time [66].

4.2. Grafting of Polar Monomer from Polyolefin by Surface Activation

Grafting method of polar segment by ionizing radiation in presence of air, ozone or UV as accelators or radical initiators is well known method for producing polyolefin hybrid.

However, it is difficult to obtain well defined polymer structure because of side chain reaction by radical species such as recombination, disproportions, cross linking and chain scission resulting in loss of properties. Other approach involve surface activation of polyolefins.

HDPE film was brominated or functionalized using 2,2,2-tribromoethne or benzophenon by Yamamoto *et al.* followed by ATRP reaction which yielded the grafted PMMA chain from already produced HDPE film [67]. Studies have also been conducted for usage of initiator used in radical polymerization like AIBN (α, α'- azobizisobutoxy nitrile), benzoyl peroxide (BPO) or 1, 1, 2, 2- tetraphenyl-1,2-ethandiol (TPED) for CRP. [68-70]. Grafting on PE surface were also performed by Shimada and coworkers using irradiation process on PE film followed by MMA grafting using $FeCl_3/PPh_3$/AIBN system [71]. PE-g-AA (acrylic acid) were further functionalized by Luo *et al* with nitrogen containing initiators. Later, ATRP of such functionalized/initiating species containing polymer was performed using $CuCl/Me_6TREN$ to graft polyacrylamide on surface to increase surface hydrophilicity of PE-g-AA[72].

4.3. Direct Polymerization of Olefin with Polar Monomer though CRP Technique

Matyjaszewaski and coworkder synthesized n-BA with 1-Octene copolymer using ARGET ATRP directly using $CuCl_2$/TMPA [Tris (2-pyridyl) methyl] amine system which resulted in 16-18 % incorporation of 1-Octene in polymer back bone. The polydispersity was relatively high (1.52-1.62) due to the formation of 1-Octene dormant species with Cu(II) deactivator which could not be reactivated. Use of ARGET technique helped to reduce Cu concentration to ppm level, thereby reducing concentration/population of dormant species which could not be regenerated. Tin (II) 2-ethylhexanoated [$Sn(EH)_2$] was used as reducing agent of Cu (II) to Cu(I). [73].

Klumerman and coworders synthesized copolymer of 1-Octene and MMA with 20% alkene incorporation using ATRP: CuCl/PMDETA system and 2,2,2-trichloroethane or phenyl sulfonyl chloride as initiator. Narrow MWD of polymer indicated that normal ATRP took place. Increase in alkene % in feed lead to increase of alkene % in polymer backbone. [74]. Synthesis of functionalized polyolefin which are generally produced via metallocene or late transition metal catalyst system is widely studied. Oxygen, nitrogen and halogen containing polyolefin, silane, dienes, cyclic olefins etc functional group containing monomer were studied by Hussaina and workers, with ethylene or propylene to produce functional polyolefin by metallocene catalyst [75]. Controlled random and alternating copolymer of MA and 1-octene was prepared through RAFT (reversible addition-fragmentation transfer) by Sen and coworkers. Benzyl 1-pyrrocarbodithioate was used as transfer agent, $AlCl_3$ as complexing agent along with AIBN as initiator for such polymerization. Maximum 18-20 % alkene (1-hexene / 1-octene) was incorporated in copolymer [76].

Sen and coworkers further reported synthesis of MA with ethylene and propylene (separately) by copper mediated controlled copolymerization. Earlier approach were to use higher α-olefin with mild condition. Free and control radical polymerization was performed using AIBN and CuBr/PMDETA/Ethyl 2-bromopropionate (ATRP) at 60°C and 90°C respectively [77]. Appx 21-22 wt% of propylene and 6-8 wt% of ethylene could be incorporated by such technique at high pressure conditions. Terpolymer of methyl acrylate, ethylene and propylene was synthesized with 23% ethylene and 7-11 % propylene. Late transition metal catalysts are used for olefin-polar copolymer due to their oxyphilicity and tolerance to other functionalities. Palladium containing catalyst are widely used for such application along with catalyst from other transition metal like Ni, Co, Fe and Mn etc. [78]. Sen and coworker also reported synthesis of fluorinated olefin containing acrylates polymer by ATRP technique : CuBr/PMDETA and ethyl 2-bromopropionate with 9-12 % alkene incorporation [79].

Sen and coworkers also reported synthesis of MA with 1-alkenes and norbornenes using nitorxide mediated copolymerization. These technique overcome the disadvantage of ATRP: removal of transition metal from polymer after synthesis. They have used 2,2,5-trimethyl-3-(1-phenyl ethoxy)-4-phenyl-3-azahexene as an initiator and 2,2,5-trimethyl-4-phenyl-3-azahexene-3-nitroxide as control agent. Ethylene, propylene, 1-hexene and 1-octene was copolymerized (separately) with MA with 13.6, 15.5, 9.7 (max) and and 8 (max) wt% was incorporation resepctively – which followed living characteristics of polymerization indicated by linear increase in M_w with time. Sen and coworkers also studied norbornene incorporation upto 28 wt% (max) in MA backbone with controlled nature of polymerization as confirmed by narrow MWD. Weight average molecular weight of synthesized copolymer was in the range of 5000 to 9000 [80]. One of the application of PE-g-PMMA is to use for compatibilizer for mixing of two immiscible phase of PE and PMMA. In most of CRP cases, random incorporation of monomers was observed in copolymer unlike the lewis mediated radical polymerization approach used for copolymerization of polar monomer with higher α-olefins.

4.4. Direct Radical Polymerization of Olefin with Polar Monomer

Methyl acrylates and 1-octene was copolymerized using AIBN with acidic alumina (lewis acid) by Gupta and coworkers [81]. Max 61 wt% of 1-octene can be incorporated in MA backbone as confirmed through ^{1}H NMR with PDI of 1.75-2.29. Sen and coworkers studied 1-alkene (ethylene, propylene, 1-hexene and 1-decene) and methyl acrylates/methyl methacrylates copolymerization using $Sc(OTf)_3$ (lewis acid) and AIBN (initiator). 30-47 wt% 1-alkene were incorporated with 4 mole % addition of lewis acid [82]. The summary of direct polymerization of MA with 1-alkene by different approach is represented in Table 2. The compilation shows that 1-alkene incorporation is more for Lewis mediated RP compared RP and CRP techniques. It is due to complexation of strong lewis acid to ester group of MA resulting in highly electron deficient modified MA. The modified MA allows more 1-octene incorporation in polymer backbone [83].

Table 2. Summary on polymerization of MA with 1-alkene by different CRP and RP techniques

Sr No	1-alkene	1-alkene incorporation (%)	Mw/Mn	polymerization technique
1	Ethylene	8.6-14.6	1.5	Cu mediated ATRP
		5.7	9.0	Radical polymerization
		13.6	1.2	Nitroxide mediated
		50.3	2.4	Lewis mediated RP
2	Propylene	21.7	1.4	Cu mediated ATRP
		21.5	2.0	Radical polymerization
		15.5	1.2	Nitroxide mediated
3	1-Hexene	11.8-21.3	1.3	Cu mediated ATRP
		11.6-17.9	1.5-1.7	Radical polymerization
		6.8-9.7	0.7-2.3	Nitroxide mediated
		50-73.1	1.2-2.2	Lewis mediated RP
		10.8-20.2	1.4-20.0	RAFT
4	1-octene	23.6	1.2	Cu mediated ATRP
		12.9	1.6	Radical polymerization
		4.8-8.9	1.1-1.2	Nitroxide mediated
		13.5-17.1	1.2-2.0	RAFT

5. Macromolecular Architectures of Polymer Produced By CRP

Polymers with targeted molecular weights, polydispersity and architectures can be synthesized by ATRP. Polymers with different compositions (block copolymers, homo polymers, random/statistical copolymers, graft copolymers), functionalities (macromonomer, multifunctional stars, multifunctional/hyperbranched, side functional group) and topology (linear, star, graft, branched, network, cyclic) can be produced by various CRP techniques.

Topology

Star type polymer can be synthesized by different approaches which involved use of multifunctional initiators from which several arms were grown [84]. It is also called as *core first approach*. In *arm first* approach, chain attaches to functional core. RAFT technique generally provide possibility of synthesizing such structure by attaching Z group of RAFT agent to functional core [85]. Comb like polymers were synthesized by various approaches such as *grafting from*, *grafting onto* and *grafting through*. In *graft from* approach, chains are grown from polymer backbone like *core first* approach [86]. *Grafting into* approach involved attachment of chain to polymer backbone like *arm first* approach [87]. Grafting through approach involved use of vinyl terminated macromonomers as comonomer together with low molecular weight monomers. Hyper branched monomers are produced by monomers which also works as initiator for initiation of new branches. The degree of branching is controlled by degree of polymerization [88].

Compositions

Block, graft, alternating etc can be produced due to control on propagation that can be achieved by CRP techniques. This is possible by manipulation of chain end group synthesized by cationic, anionic, coordination and step growth polymerization. Random copolymers are formed when comonomer have a similar reactivity. Alternate patten of polymer is obtained when monomer with strong electron donor group are polymerized with monomer with strong electron acceptor groups. When more reactive monomer (MAH, acrylates) and less reactive monomer (olefins) are copolymerized with lewis acid to produce alternating nature of polymer is formed. Hybrid material are formed by attaching synthetic polymer to natural or inorganic material [89]. Functional group (-OH) in natural produced are converted to bromine containing initiating sites for ATRP. Subsequently, ATPR was conducted.

Functionality

Propagating chain allows incorporation of many functionalities into polymer chain i.e side functional group can be directly introduced to a polymer backbone. One of way end functional initiators can be introduced ~~is~~ by using functional initiator or by converting chain end to another functional group i.e alkoxy amine in NMP to alkyl halide, dithioeser to thiols and alkyl halide to hydroxyl or amine etc.

6. Applications

Evolution of CRP technique in last two decades lead to opening up of some of the area which was unexplored due to limiting characteristics of polar and nonpolar polymers. Recent advances in PO chemistry have lead to creation of polyolefin bases "Polymer Hybrids" consisting of polyolefin segment and other polyolefin or polar segment. Surfactant type polymers which has hydrophilic and hydrophobic segments finds its application in pigment dispersion, various additives and health/cosmetics product. Some of copolymers finds its application as microelectronic devices [90]. Graft polymers are widely used as compatibilizer for mixing two immiscible polymer phases. Polymers which has some ascending/descending pattern of comonomers (gradient polymer) are finding its application in vibration dampening also [91]. Comb, star, branched polymes are widely used as viscosity modifiers or lubricants.

Chain end functional polyacrylates can also be used for reactive blending. Hybrid material (synthetic/natural or organic/inorganic) can lead to numerous possible applications wherein some of disadvantage of natural products can be addressed by synthetic polymer by hybridization. One of the area where such material with desired topography and functionalities can be used effectively is in separation science by membranes. Membrane can be designed with target functionalities and topography to achieve desire performance. The commercial polyolefin grades which have polar monomer are described in Table 3. Most of these grades are produced by radical polymerization techniques. The commercial grades produced by CRP techniques are now gaining momentum for niche specialty applications. Degussa, IBM, Ciba, RohMax Oil Additives, Arkema are exploring CRP for application in area of memory chip, pigment, dispersion, additives and coating etc. Kaneka, Japan announced large pilot scale plant for producing polymers by ATRP technique [92].

Table 3. Commercial polymers having olefinics/polar moieties

Sr No	Company	Trade Nam	Polymer detail
1	Dupont	Vamac™	Ethylene-acrylic elstomer Ethylene-acrylate rubber Ethylene-Methacrylates terpolymer
2	Dow	Amplify™ EA Amplify™ GR Amplify™ VA	Ethylene-ethyl acryales copolymer Maleic anhydride grafted copolymer Ethylene-Vinyl acetate (EVA)
3	Arkema	EVATANE™ LOTRYL™ LOTADER™ OREVAC™	Ethylene-Vinyl acetate (EVA) Ethylene-acrylate copolymer Ethylene-acrylate terpolymer Olefin- Maleic anhydride
4	BASF	LUCALEN™	Ethylene-acrylate copolymer
5	Westlake	EBSC™	Ethylene-butyl acrylates copolymer
6	Mitsui	ADMER™	Functionally modified Polyolefin
7	Borealis	Super Smooth™ Visco™	Semi conductive polyethylene Silane containing polyethylene for crosslinking

7. Future Perspective

CRP technique allowed manipulation of chain growth mechanism with various functionalities and topography. It provided the tool to target polymer performance by design of various morphologies and functionalities to achieve macromolecular properties alteration. CRP is gaining momentum since it requires mild reaction condition, simple reaction step up and it covers wide range of monomers. Some of the area where focus is required for commercialization point of view are as follows:

- Focused research will be useful for simplifying copolymerization of functional/polar monomer with olefinic (less reactive) monomer. This area has very high untapped potential from application point of view. ATRP/SFRP need to be clubbed in one step with coordination polymerization to simplify the process. Macromonomer and macroinitiator approach requires more steps and adds to cost.
- Structures of initiators, transition metals catalyst and ligands can be studied by computational techniques to design such structures which can provide thermodynamically best process for maximum conversions. Application of fundamentals of organic chemistry and polymer chemistry can lead to change to the level of molecular mechanism.

- Some of areas can be looked involves membrane separation science, characterization tools and techniques etc which can guide synthetic work at lab scale. Deep understanding of structure-properties-application will lead to widen the scope of applications. Understanding natural material and natural phenomena can give indication to start new studies.
- The major road block for commercialization of CPR is availability of initiators, ligands, catalyst, controlling agent etc at economical and affordable prices. Heterophasic and solution free processes (current one is homogenous/solution based process) need to be explores to make process more economical. Cross functional team approach – organic chemistry, polymer science, polymer engineering, processing science, chemical engineering – can be used to solve the problem.

REFERENCES

[1] Encyclopedia of polymer science and technology. John Wiley & Sons, 6, 696-697.

[2] Muller, H.E.; Matyjaszewski, K. Controlled and Living polymerization. Willey-VCH Verlag GmbH & Co. KGaA, *Weinheim*. 2009, 103-104.

[3] Szwarc, M. Living polymers. *Nature. (London*). 1956, 178, 1168.

[4] Szwarc, M.; Levy, M.; Milkovich, R. Polymerization initiated by electron transfer to monomer. A new method of formation of block polymers. *J. Am. Chem. Soc*. 1956, 78, 2657.

[5] Braunecker, W.A.; Matyjaszewski K.; Controlled/living radical polymerization: Features, developments, and perspectives, *Prog. Polym. Sci*. 2007, 32, 93–146.

[6] Otsu, T.; Yoshida, M. Role of initiator-transfer agent-terminator (iniferter) in radical polymerizations: Polymer design by organic disulfides as iniferters. *Macromol. Rapid Commun*. 1982, 3, 127-132.

[7] Otsu, T.; Yoshida, M.; Tazaki, T. A model for living radical polymerization. *Macromol. Rapid Commun*. 1982, 133-140.

[8] Solomon, D. H.; Rizzardo, E.; and Cacioli, P. New polymerization process and polymers produced there by. *Eur. Pat*. EP 135280 A2, 1985.

[9] Benoit, D.; Chaplinski, V.; Braslau , R.; Hawker C. J. Development of a Universal Alkoxyamine for “Living” Free Radical Polymerizations. *J. Am. Chem. Soc*. 1999, 121, 3904– 3920.

[10] Benoit, D.; Grimaldi, S.; Robin S, J.; Finet , and Gnanou. Y. Kinetics and Mechanism of Controlled Free-Radical Polymerization of Styrene and n-Butyl Acrylate in the Presence of an Acyclic β-Phosphonylated Nitroxide. *J. Am. Chem. Soc*. 2000, 122,5929–5939.

[11] Wang J.-S.; Matyjaszewski, K. Controlled/"living" radical polymerization. atom transfer radical polymerization in the presence of transition-metal complexes, *J. Am. Chem. Soc*. 1995, 117, 5614–5615.

[12] Kato, M.; Kamigaito, M.; Sawamoto, M.; Higashimura, T. Polymerization of Methyl Methacrylate with the Carbon Tetrachloride/Dichlorotris- (triphenylphosphine) ruthenium(II)/Methylaluminum Bis(2,6-di-tert-butylphenoxide) Initiating System: Possibility of Living Radical Polymerization. *Macromolecules. 1995*, 28, 1721–1723.

[13] Le, T. P.; Moad, G.; Rizzardo, E.; Thang, S. H. Polymerization with living characteristics with controlled dispersity, polymers prepared thereby, and chain-transfer agents used in the same. *PCT Int. Appl.*WO9801478 A1 (1998), (Commonwealth Scienti.c and Industrial Research Organization).
[14] Chiefari, J.; Chong, Y. K.; Ercole, F.; Krstina, J.; Jeffery, J.; Le, T. P. T.; Mayadunne, R. T. A.; Meijs, G. F.; Moad, C. L.; Moad, G.; Rizzardo, E.; Thang, S. H. Living Free-Radical Polymerization by Reversible Addition−Fragmentation Chain Transfer: The RAFT Process. *Macromolecules*. 1998, 31, 5559–5562.
[15] Matyjaszewski K.; Shipp D.A.; McMurtry, G. P.; Gaynor, S. G.; Pakula, T. Simple and Effective One-Pot Synthesis of (Meth)Acrylic Block Copolymers Through Atom Transfer Radical Polymerization. *J. Polym. Sci. Part A: Polym. Chem.* 2000, 38, 2023–31.
[16] Burguiere, C,; Pascual, S.; Coutin, B.; Polton, A.; Tardi, M.; Charleux, B. Matyjaszewski, K.; Vairon, J P. Amphiphilic block copolymer prepared via controlled radical polymerization as surfactants for emulsion polymerization. *Macromol. Symp.* 2000, 150, 39–44.
[17] Auschra, C.; Eckstein, E.; Knischka, R.; Pirrung, F.; Harbers, P. Controlled polymers for pigment dispersants. *Eur. Coatings J.* 2004, 26, 31.
[18] Auschra, C.; Eckstein, E.; Knischka, R.; Pirrung, F.; Harbers, P. Tailor-made additives controlled polymer structures enhance performance. *Eur. Coatings J.* 2005, 156, 8, 60,62–63.
[19] McCarthy, P.; Tsarevsky, N.V.; Bombalski, L.; Matyjaszewski, K.; Pohl, C.; Grafting chromatographic stationary phase substrates by atom transfer radical polymerization. *ACS Symp. Ser*, 2006, 944, 252–68.
[20] Richard, R.E.; Schwarz, M.; Ranade, S.; Chan, A. K.; Matyjaszewski, K.; Sumerlin, B. Evaluation of Acrylate-Based Block Copolymers Prepared by Atom Transfer Radical Polymerization as Matrices for Paclitaxel Delivery from Coronary Stents. *Biomacromolecules*. 2005, 6, 3410–3418.
[21] Otsu, T.; Tazaki, T. Living radical polymerization in homogeneous system with phenylazotriphenylmethane as a thermal iniferter. *Polym. Bull.* 198, 16, 277–84.
[22] Coessens, V.; Pintauer, T.; Matyjaszewski, K. Functional polymer by atom transfer radical polymerization. *Prog. Polym. Sci.* 2001, 26, 337.
[23] Bon, S. A. F.; Steward, A. G.; Haddleton, D. M. Modification of the ω-bromo end group of poly(methacrylate)s prepared by copper(I)-mediated living radical polymerization. *J. Polym. Sci., Part A: Polym. Chem.* 2000, 38, 2678–2686.
[24] Paik, H. J.; Gaynor, S. G.; Matyjaszewski, K. Synthesis and characterization of graft copolymers of poly(vinyl chloride) with styrene and (meth)acrylates by atom transfer radical polymerization. *Macromol. Rapid Commun.* 1998, 19, 47–52.
[25] Beers, K. L.; Gaynor, S. G.; Matyjaszewski, K.; Sheiko, S. S.; Moeller, M. The Synthesis of Densely Grafted Copolymers by Atom Transfer Radical Polymerization. *Macromolecules.* 1998, 31, 9413–9415.
[26] Georges, M. K.; Veregin, R. P. N.; Kazmaier, P. M.; Hamer, G. K. Narrow molecular weight resins by a free-radical polymerization process *Macromolecules.* 1993, 26, 2987-2988.
[27] Hawker, C. J. Molecular Weight Control by a "Living" Free-Radical Polymerization Process. *J. Am. Chem. Soc.* 1994, vol. 116, 11185-11186.

[28] Grimaldi, S.; Finet, J. P.; Le Moigne, F.; Zeghdaoui, A.; Tordo, P.; Benoit, D.; Fontanille, M.; Gnanou, Y. Acyclic β-Phosphonylated Nitroxides: A New Series of Counter-Radicals for "Living"/Controlled Free Radical Polymerization *Macromolecules.* 2000, 33, 1141-1147.

[29] Benoit, D.; Harth, E.; Fox, P.; Waymouth, R. M.; Hawker, C. J. Accurate Structural Control and Block Formation in the Living Polymerization of 1,3-Dienes by Nitroxide-Mediated Procedures *Macromolecules.* 2000, 33, 363-370.

[30] Hong, C. Y.; You, Y. Z.; Bai, R. K.; Pan, C. Y.; Borjihan, G. Controlled polymerization of acrylic acid under 60Co irradiation in the presence of dibenzyl trithiocarbonate. *J. Pol. Sc., Part A: Pol. Chem.* 2001, 39, 3934-3939.

[31] Lena, Fabio.; Matyjaszewski. Transition metal catalyst for controlled radical polymerization *Prog. Polym. Sci.* 2010, 35, 959-1021.

[32] Fischer, H. The persistent radical effect: a principle for selective radical reactions and living radical polymerizations. *Chem. Rev.* 2001, 101, 3581–610.

[33] Tang, W,; Tsarevsky, N V.; Matyjaszewski, K. Determination of equilibrium constants for atom transfer radical polymerization. *J. Am. Chem. Soc.* 2006, 128, 1598–604.

[34] Matyjaszewski, K.; XiaJ. Atom transfer radical polymerization. *Chem. Rev.* 2001, 101, 2921-90

[35] Kabachii, Y.A.; Kochev, S.Y.; Bronstein, L.M.; Blagodatskikh, I.B.; Valetsky, P.M. Atom transfer radical polymerization with Ti(III) halides and alkoxides. *Polym. Bull.* 2003, 50, 271–8.

[36] Le Grognec, E.; Claverie, J.; Poli, R.; Radical polymerization of styrene controlled by half-sandwich Mo(III)/Mo(IV) couples: all basic mechanisms are possible. *J. Am. Chem. Soc.* 2001, 123, 9513–24.

[37] Brandts, J.A.M.; van de Geijn, P.; van Faassen, E.E.; Boersma, J.; Van Koten, G. Controlled radical polymerization of styrene in the presence of lithium molybdate(v) complexes and benzylic halides. *J. Organomet. Chem.* 1999, 584, 246–53.

[38] Maria, S.; Stoffelbach, F.; Mata, J.; Daran, J.C.; Richard, P.; Poli, R. The radical trap in atom transfer radical polymerization need not be thermodynamically stable. A study of the MoX3(PMe3)3. catalysts. *J. Am. Chem. Soc.* 2005, 127, 5946–56.

[39] Kotani, Y.; Kamigaito, M.; Sawamoto, M. Re(V)-mediated living radical polymerization of styrene:1 ReO2I(PPh3)2/ R-I initiating systems. *Macromolecules.* 1999, 32, 2420–4.

[40] Matyjaszewski, K.; Wei, M.; Xia, J.; McDermott, N.E.. Controlled/''living'' radical polymerization of styrene and methyl methacrylate catalyzed by iron complexes. *Macromolecules.* 1997, 30, 8161–4.

[41] Teodorescu, M.; Gaynor, S.G.; Matyjaszewski, K. Halide anions as ligands in iron-mediated atom transfer radical polymerization. *Macromolecules.* 2000, 33, 2335–9.

[42] Kato, M.; Kamigaito, M.; Sawamoto, M.; Higashimura, T. Polymerization of methyl methacrylate with the carbon tetrachloride/dichlorotris- (triphenylphosphine) ruthenium(II)/methylaluminum bis(2,6-di-tert-butylphenoxide) initiating system: possibility of living radical polymerization. *Macromolecules.* 1995, 28, 1721–3.

[43] Simal, F.; Demonceau, A.; Noels, A.F. Highly efficient ruthenium-based catalytic systems for the controlled freeradical polymerization of vinyl monomers. *Angew Chem. Int. Ed.* 1999, 38, 538–40.

[44] Braunecker, W.A.; Itami, Y.; Matyjaszewski, K. Osmiummediated radical polymerization. *Macromolecules.* 2005, 9402–4.

[45] Percec, V.; Barboiu, B.; Neumann, A.; Ronda, J.C.; Zhao, M. Metal-catalyzed ''living'' radical polymerization of styrene initiated with arenesulfonyl chlorides. From heterogeneous to homogeneous catalysis. *Macromolecules.* 1996, 29, 3665–8.

[46] Wang, B.; Zhuang, Y.; Luo, X.; Xu, S.; Zhou, X. Controlled/ ''living'' radical polymerization of MMA catalyzed by cobaltocene. *Macromolecules.* 2003, 36, 9684–6.

[47] Granel, C.; Dubois, P.; Jerome, R.; Teyssie, P. Controlled radical polymerization of methacrylic monomers in the presence of a bis(ortho-chelated) arylnickel(II) complex and different activated alkyl halides. *Macromolecules.* 1996, 29, 8576–82.

[48] Uegaki, H.; Kotani, Y.; Kamigaito, M.; Sawamoto, M. Nickelmediated living radical polymerization of methyl methacrylate. *Macromolecules.* 1997, 30, 2249–53.

[49] Lecomte, P.; Drapier, I.; Dubois, P.; Teyssie, P.; Jerome, R. Controlled radical polymerization of methyl methacrylate in the presence of palladium acetate, triphenylphosphine, and carbon tetrachloride. *Macromolecules.* 1997, 30, 7631–3.

[50] Wang, J-S.; Matyjaszewski, K. Controlled/''living'' radical polymerization. atom transfer radical polymerization in the presence of transition-metal complexes. *J. Am. Chem. Soc.* 1995, 117, 5614–5.

[51] Queffelec, J.; Gaynor, S. G.; Matyjaszewski, K. Optimization of Atom Transfer Radical Polymerization Using Cu(I)/Tris(2-(dimethylamino)ethyl)amine as a Catalyst *Macromolecule.* 2000, 33, 8629.

[52] Natanya, M. L.; Hansen, Katja, Jankova.; Søren, Hvilsted. Fluoropolymer materials and architectures prepared by controlled radical polymerizations. *European Polymer Journal*, 2007,43,(2),*255-293.*

[53] Kwak, Y.; Matyjaszewski, K. Effect of Initiator and Ligand Structures on ATRP of Styrene and Methyl Methacrylate Initiated by Alkyl Dithiocarbamate. *Macromolecules.* 2008, 41 (18), 6627–6635.

[54] Tsarevsky, N. V.; Matyjaszewski , K. Combining Atom Transfer Radical Polymerization and Disulfide/Thiol Redox Chemistry: A Route to Well-Defined (Bio)degradable Polymeric Materials. *Macromolecules.* 2005, *38,* 3087-3092.

[55] Weitzer ,M.; Schindler, S.; Brehm, G.; Schneider, S.; Hörmann, E.; Jung, B.; Kaderli, S., Zuberbühler, A. D. Reversible Binding of Dioxygen by the Copper(I) Complex with Tris(2-dimethylaminoethyl)amine (Me_6tren) Ligand. *Inorg. Chem.* 2003, *42* (6),1800–1806

[56] Yulai, Z.; Xiao,A.; Yu, H. The synthesis of modified polyethylene via coordination polymerization followed by ATRP ,RAFT ,NMRP or ROP; *Prog. Polym. Sci.* 2010, 35(10) , 1195-1216

[57] Miller, P.J.; Teodorescu, M., Peterson, M. L. Matyjaszewski, K.; Graft-ing styrene and acrylates fromfunctionalized polyethylene by atom transfer radical polymerization. *Abstr. Papers. Am. Chem. Soc.* 1999, 218, 541- XXX.

[58] Karavia, V.; Deimede, V.; Kallitsis, J.K.Synthesis of end-functionalized polystyrenes using ATRP and their grafting onto polyethylene copolymers. *J. Macromol. Sci. Pure Appl. Chem.* 2004, A41, 115–31.

[59] Hwu, J.M.; Chang, M.J.; Lin, J.C.; Cheng, H.Y.; Jiang, G.J. Synthesis and application of functional polyethylene graft copolymers by atom transfer radical polymerization. *J. Organomet. Chem.* 2005, 690:6300.

[60] Kaneyoshi, H.; Inoue, Y; Matyjaszewski, K. Synthesis of block and graft copolymers with linear polyethylene segments by combination of degenerative transfer coordination polymerization and atom trans- fer radical polymerization. *Macromolecules.* 2005, 38, 5425–35.

[61] Inoue, Y.; Matyjaszewski, K. Preparation of polyethylene block copolymers by a combination of post metallocene catalysis of ethylene polymerization and atom transfer radical polymerization. *J. Polym. Sci. Part A Polym. Chem.* 2004, 42, 496–504.

[62] Kaneyoshi, H.; Matyjaszewski, K. Synthesis of a linear polyethylene macromonomer and preparation of polystyrene-graft- polyethylene copolymers via grafting-through atom transfer radical polymerization. *J. Appl. Polym. Sci.* 2007, 105, 3–13.

[63] Park, E.S. Preparation and applicability of functionalized polyethy- lene with an ethylene /1,7-octadiene copolymer. *J. Appl. Polym. Sci.* 2008, 109, 3631–8.

[64] Chen, G.; Huynh, D.; Felgner, P.L.; Guan, Z. Tandem chain walking polymerization and atom transfer radical polymerization for efficient synthesis of dendritic nanoparticles for bioconjugation. *J. Am. Chem. Soc.* 2006, 128, 4298–302.

[65] Zhang, K., Wang, J., Subramanian, R., Ye, Z., Lu, H., Yu, Q. Chain walking ethylene copolymerization with an ATRP inimer for one-pot synthesis of hyperbranched polyethylenes tethered with ATRP initiating sites. *Macromol. Rapid Commun.* 2007, 28, 2185–91.

[66] Matyjaszewski, K.; Teodorescu, M.; Miller, P.J.; Peterson, M.L. Graft copolymers of polyethylene by atom transfer radical polymerization. *J. Polym. Sci. Part A Polym. Chem.* 2000, 38, 2440–8.

[67] Yamamoto, K.; Miwa, Y.; Tanaka, H.; Sakaguchi, M.; Shimada, S. Living radical graft polymerization of methyl methacrylate to polyethylene film with typical and reverse atom transfer radical polymerization. *J. Polym. Sci. Part A Polym. Chem.* 2002, 40, 3350–9.

[68] Matyjaszewski, K.; Patten, T.E.; Xia, J.; Controlled/living radical polymerization : kinetics of the homogeneous atom transfer radical polymerization of styrene. *J. Am. Chem. Soc.* 1997, 119, 674–80.

[69] Xia, J.; Matyjaszewski, K. Controlled/"living" radical polymerization. Homogeneous reverse atom transfer radical polymerization using AIBN as the initiator. *Macromolecules.* 1997, 30, 7692–6.

[70] Xia, J.; Matyjaszewski, K.; Homogeneous reverse atom transfer radical polymerization of styrene initiated by peroxides. *Macromolecules.* 1999; 32, 5199–202.

[71] Yamamoto, K.; Tanaka, H.; Sakaguchi, M.; Shimada, S.; Well-defined poly(methyl methacrylate) grafted to polyethylene with reverse atom transfer radical polymerization initiated by peroxides. *Polymer.* 2003, 44, 7661–9.

[72] Luo, N.; Husson, S.M.; Hirt, D.E.; Schwark, D.W. Surface grafting of poly- acrylamide from polyethylene-based copolymer film. *J. Appl. Polym. Sci.* 2004, 92, 1589–95.

[73] Tanaka, K.; Matyjaszewaski, K.; Controlled copolymerization of n-butyl acrylate with nonpolar 1-alkene using activators regenerated by electron transfer radical polymerization, *Macromolecule.* 2007, 40,5255-5260

[74] Venkatrsh R.; Harrision,S.; Haddleton, D.M.; Klumperman B. Olefin Copolymerization via Controlled Radical Polymerization: Copolymerization of Acrylate and 1-Octene, *Macromolecule.* 2004, 37, 4406-4416

[75] Atiqullaha, M.; Tinklb,M.; Pfaendnerc,R.; Akhtara,M.N.; Hussaina, I.; Synthesis of Functional Polyolefins using Metallocenes: A Comprehensive Review. *Polymer Review.* 2010, 50, 178-230

[76] Liu,S.; Gu, B.; Rowlands, H. A.; Sen, A. Controlled random and alternating copolymerization of methyl acrylates with 1-alkenes, *Macromolecules.* 2004, 37, 7924-7929.

[77] Liu, S.; Elysashiv, S.; Sen, A.; Copper mediated controlled copolymerization of methyl acrylates with 1-alkene under mild condition, *J. Am. Chem. Soc.* 2001, 123, 12738-12739.

[78] Nakamura, A.; Ito, S.; Nozaki, K. Coordination-Insertion copolymerization of functional polar monomers, *Chem. Rev.* 2009, 109, 5215-5244.

[79] Poli, D.; Nagel, M.; Sen, A. Lewis Acid mediated copolymerization of methyl acrylates with 1-alkene, *Macromolecules.* 2005, 38, 7262-7265.

[80] Gu, B.; Shengsheng, L.; Leber, J.D.; Sen, A. Nitroxide mediated copolymeization of methyl acrylates and nonbornenes, *Macromolecules.* 2004, 37, 5142-5144.

[81] Kaur, S.; Singh, G.; Kothari, A.V.; Methyl acrylate/1-Octene: Lewis mediated polymerization, *J. Appl. Polym. Sci.* 2009, 111, 87-93.

[82] Nagel M.; Poli D.; Sen, A.; Lewis mediated copolymerization of methyl acrylates and methyl methacrylates wih 1-alkene, *Macromolecule.* 2005, 38, 7262-7265

[83] Liu, S.; Sen, A. Living/controlled copolymerization of acrylates with non activated alkenes, *J. Polym. Sci A: Polym. Chem.* 2004, 43, 6175-6192.

[84] Angot, S.; Murthy, S.; Taton, D.; Gnanou, Y. Atom transfer radical polymerization of styrene using a novel octafunctional initiator: synthesis of well-defined polystyrene stars, *Macromolecules* 1998, 31, 7218.

[85] Mayadunne, R.T.A.; Jeffery, J.; Moad, G.; Rizzardo, E. Living free radical polymerization with reversible addition–fragmentation chain transfer (RAFT polymerization): approaches to star polymers. *Macromolecules.* 2003, 36, 1505–13.

[86] Paik, H.J.; Gaynor, S.G.; Matyjaszewski, K. Synthesis and characterization of graft copolymers of poly(vinyl chloride) with styrene and (meth)acrylates by atom transfer radical polymerization. *Macromol. Rapid Commun.* 1998, 19, 47–52.

[87] Sumerlin BS, Tsarevsky NV, Louche G, Lee RY, Matyjaszewski, K. Highly efficient ‘‘Click’’ functionalization of poly(3-azidopropyl methacrylate) prepared by ATRP. *Macromolecules.* 2005, 38, 7540–5.

[88] Baudry, R.; Sherrington, D.C. Synthesis of highly branched poly(methyl methacrylate)s using the ‘‘Strathclyde Methodology’’ in aqueous emulsion. *Macromolecules.* 2006, 39, 1455–60.

[89] Pyun, J., Matyjaszewski, K. Synthesis of nanocomposite organic/inorganic hybrid materials using controlled/‘‘living’’ radical polymerization. *Chem. Mater.* 2001,13, 3436–48.

[90] Lindner, S.M.; Thelakkat, M. Nanostructures of n-type organic semiconductor in a p-type matrix via self-assembly of block copolymers. *Macromolecules.* 2004, 37, 8832–5.

[91] Matyjaszewski, K.; Ziegler, M.J.; Arehart, S.V.; Greszta, D.; Pakula, T.; Gradient copolymers by atom transfer radical copolymerization. *J. Phys. Org. Chem.* 2000, 13, 775–86.

[92] Spanswick, J.; Matyjazewski, K. Controlled/living radical polymerization, *Material today*. March 2005, 26-32.

In: Radical Polymerization: New Developments ISBN 978-1-62100-406-6
Editors: I. O. Paulauskas, L. A. Urbonas, pp. 223-238

Chapter 7

THERMAL REDOX AND PHOTOINDUCED RING OPENING POLYMERIZATION REACTIONS: INITIATING SYSTEMS BASED ON ORGANOSILANES BEARING SI-SI BOND

***M-A. Tehfe*[*a], *J. Lalevée*[* a&], *M. El-Roz*[a], *N. Blanchard*[b], *F. Morlet-Savary*[a&], *B. Graff,*[a&] and *J.P. Fouassier*[a]**

[a]Department of Photochemistry, University of Haute Alsace, ENSCMu, 3 rue Alfred Werner, 68093 Mulhouse Cedex, France
[&] Present address: Institut de Science des Matériaux de Mulhouse IS2M, LRC 7228 CNRS, UHA, 15 rue Jean Starcky, 68057 Mulhouse, France
[b]Laboratory of Organic and Bioorganic Chemistry, University of Haute Alsace, ENSCMu, 3 rue Alfred Werner, 68093 Mulhouse Cedex, France

ABSTRACT

Highly efficient Type I photoinitiators (PI) based on the silyl radical chemistry are reviewed here for Free Radical Photopolymerization (FRP) or Free Radical Promoted Cationic Photopolymerization (FRPCP). The ability of these structures to initiate the polymerization of acrylates (FRP) or epoxy (FRPCP) monomers is presented here in detail. These compounds (10,10'-bis 10-phenyl-10H-phenoxasilin *Si-1* and 9,9'-dimethyl-9,9'-bis 9H-9-silafluorene *Si-2*) and poly(methyl phenyl silane) *Si-3* directly generate silyl radicals under light irradiation. A low oxygen inhibition in the photopolymerization is outlined. In any case, the silylium cations are the initiating species for the ring opening polymerization. A thermal redox process at RT between these organosilanes and oxidizing agents (silver salts) also generates initiating species for the cationic polymerization of epoxy monomers. The present results evidence the high potential of this new class of (photo)initiators based on organosilanes for the thermal redox or photoinitiated epoxy monomer polymerization.

[*] Authors E-mail address: mohamad-ali.tehfe@uha.fr.
[*] Authors E-mail address: j.lalevee@uha.fr.

INTRODUCTION

The search for new photoinitiating systems PI for photopolymerization reactions in the Radiation Curing and Laser Imaging areas (coatings, varnishes, paints, adhesives, graphic arts, stereolithography, microelectronics, laser direct imaging, dentistry and medicine, optics…) has continuously received a considerable attention. Recent examples can be found in [1-15] and references therein. In Free Radical Photopolymerization (FRP), the initiating radicals are mostly generated from organic compounds in Type I cleavable one-component PI (1), Type II two-component PI (based on a hydrogen transfer reaction between a photoinitiator and a co-initiator) and even multi-component (3 to 5 compounds) PI.

$$PI \rightarrow {}^{1,3}PI \rightarrow R^{\bullet} \rightarrow\rightarrow\rightarrow R\text{-}M_n^{\bullet} \quad (1)$$

The production efficiency of these species as well as their reactivity towards the monomer will obviously govern the efficiency of the initiation step and as a consequence the efficiency of the polymerization reaction (rate of polymerization and final conversion). Both aspects are related to the kinetics of the excited state processes and the thermodynamics of the involved reactions. Various time resolved laser spectroscopy techniques (e.g. laser flash photolysis), analytical techniques (e.g. ESR) and quantum mechanical calculations allow to get spectral (transient absorption spectra), kinetic (interaction rate constants), thermodynamic (addition reaction enthalpy, formation enthalpy) information, quantum yields of the different involved processes, potential energy surfaces… The huge progress made in the experimental and theoretical approach really improves the interpretation of the photochemical reactivity/practical efficiency relationships.

In FRP, a drawback concerns the well known oxygen inhibition [1c,1m,16]. Indeed, according to their lifetimes, excited states are more or less quenched by O_2 (2a). Both the initiating $R^{\bullet}$ (2b) and propagating radicals $R\text{-}M_n^{\bullet}$ (2c) are scavenged by O_2. They yield i) highly stable peroxyl radicals (2b, 2c) which cannot participate in any further polymerization initiation reactions and ii) hydroperoxides (PH being a polymer chain or a hydrogen donor) that can accumulate in the film (2d). In very low viscosity or thin samples where the light intensity is reduced, the oxygen inhibition has a larger dramatic effect on the polymerization profile.[1c,1m,15-18]

$${}^{1,3}PI + O_2 \rightarrow \text{quenching} \quad (2a)$$

$$R^{\bullet} + O_2 \rightarrow ROO^{\bullet} \quad (2b)$$

$$R\text{-}M_n^{\bullet} + O_2 \rightarrow R\text{-}M_nOO^{\bullet} \quad (2c)$$

$$ROO^{\bullet}(+R\text{-}M_nOO^{\bullet}) + PH \rightarrow ROOH\ (+R\text{-}M_nOOH) + P^{\bullet} \quad (2d)$$

Up to now, in FRP, several methods have been proposed to limit the oxygen inhibition (e.g. see a review in [1c,1m,16-18 and references herein]). To overcome the oxygen inhibition in particular FRP reactions where severe experimental conditions are imposed (low intensity, low viscosity, and relatively low thickness), the basic idea proposed here consists in the destruction of $ROO^\bullet$ and the concomitant formation of new initiating radicals. This has been already achieved using silanes and the specific input of the silyl radical chemistry to various Type I or Type II PI photoinitiating systems: several mechanistic aspects as well as some acrylate polymerization profiles in particular systems have been reported and discussed in previous papers [14-15,17-18]. The key radical is a silyl either formed through the reaction of a silane with the radicals generated in a Type I PI (3) or created in a Type II PI (4).

$$PI \rightarrow R^\bullet \quad (h\nu) \tag{3a}$$

$$R'^\bullet + \text{silane} \rightarrow \text{silyl} \tag{3b}$$

$$^*PI + \text{silane} \rightarrow \text{silyl} \tag{4}$$

Ring opening polymerization ROP is very important for the synthesis of different polymer derived from epoxides... The search for new initiating systems for reactions under soft conditions such as i) working at low temperature (ideally RT) or ii) using convenient light sources ($\lambda > 300$ nm) remains important. Usual cationic photoinitiating systems are usually based on iodonium salts which only absorb wavelengths in the UV range e.g. < 300 nm. An elegant process referred as free radical promoted cationic photopolymerization FRPCP has been proposed [19]. This latter approach consists in producing a radical $R^\bullet$ (from usual Type I or Type II radical photoinitiatior PI) which in turn is oxidized by the iodonium salt, this latter species (R^+) being able to initiate the ring opening polymerization of epoxy monomers. One drawback in FRPCP is the strong sensitivity of the system to the presence of oxygen (or more generally of any free radical inhibitor) as the radicals generated react with O_2.

Our recent interest for Si centered radicals [20-21] prompted us to search for cleavable Si-Si bond based photoinitiators. Up to now, the available addition rate constants of silyl radicals to double bonds are rather scant, the main values being gathered in ref [22]. Twenty years ago, the photodegradation of polysilane polymers or oligomers occurring from a Si-Si cleavage has been reported to be efficient. On the opposite, the photoinitiation ability of these silyl radicals is low as exemplified by the determined low initiation quantum yields Φ_i ($\Phi_i \sim 0.001$ compared to 0.25 for a benzoin ether) [19,23,24]. Recently new polysilane structures exhibiting higher photoinitiation efficiencies were proposed [24d]. Organosilanes bearing (at least) a Si-Si bond are usable as initiating systems for the ROP of epoxides. For thermal (redox) or light induced processes, silylium cations can be generated by oxidation of the organosilanes by silver salts and oxidation of the silyl radicals, respectively.

According to our works using the silyl radical chemistry to design various photoinitiating systems less sensitive to the oxygen inhibition [15,18,25], we select here the Si-1, Si-2 and Si-3 compounds (10,10'-bis 10-phenyl-10H-phenoxasilin; 9,9'-dimethyl-9,9'-bis 9H-9-silafluorene and poly(methyl phenyl silane) *Scheme 1*) that should meet the desired properties. These molecules based on Si-Si bonds were already proposed for FRP [14]. The cleavage of Si-Si bond was also already well known for poly(organosilane) (Si-3) [24b,26-32]. These structures that generate silyl radicals will be used here as i) photoinitiating systems for the FRPCP of epoxides and ii) thermal initiators in redox processes to generate cationic structures (for the first time) when combined with oxidizing agents.

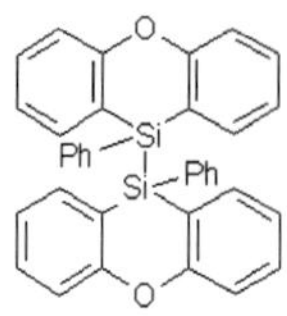

Si-1

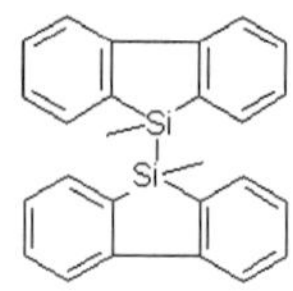

Si-2

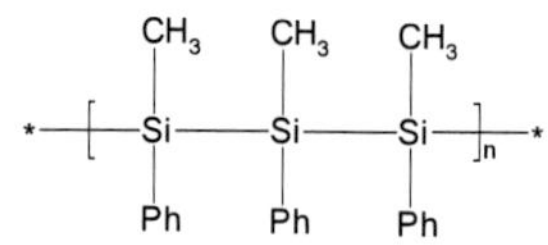

Si-3

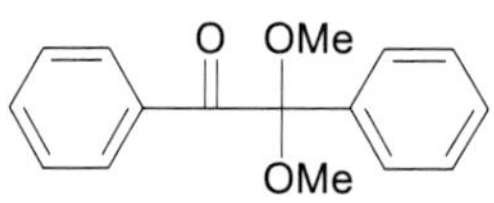

DMPA

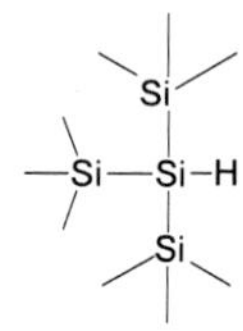

TTMSS

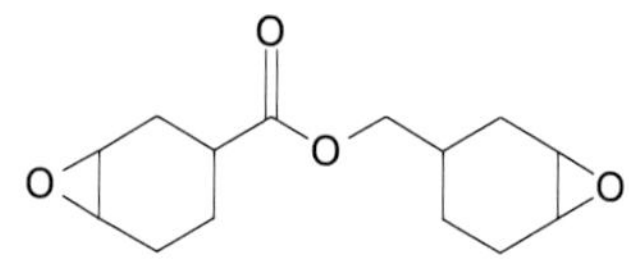

EPOX

Scheme 1.

The Chemical Systems and Polymerization Procedures

The selected compounds are depicted in *Scheme 1*. The initiators 10,10'-bis 10-phenyl-10H-phenoxasilin Si-1; 9,9'-dimethyl-9,9'-bis 9H-9-silafluorene Si-2 were obtained from Aldrich and poly(methyl phenyl silane) Si-3 was obtained from Gelest. 2,2'-dimethoxyphenyl acetophenone (DMPA) used as a reference photoinitiator, diphenyl iodonium hexafluorophosphate (Ph_2I^+), tris(trimethylsilyl)silane (TTMSS) and silver hexafluoroantimonate ($AgSbF_6$) were obtained from Aldrich. (3,4-epoxycyclohexane)methyl 3,4-epoxycyclohexylcarboxylate (EPOX) was a gift from Cytec (Uvacure 1500), limonene dioxide (LDO) from Millennium Speciality Chemicals and Ebecryl 605 (Eb 605) from Cytec.

I) Free Radical Photopolymerization (FRP)

For film polymerization experiments, a given photoinitiator was dissolved in the polymerizable medium (Ebecryl 605 (Cytec)/1,6-hexanediol diacrylate 70%/30% w/w). For comparison, DMPA was used in the same conditions. The laminated or aerated films (50 µm thick) deposited on a BaF_2 pellet were irradiated with the glass plate filtered polychromatic light delivered by a Xe-Hg lamp (Hamamatsu, L8252, 150 W). The evolution of the double bond content was continuously followed by real time FTIR spectroscopy (Nexus 870, Nicolet) as reported in [33-34]. The rates of polymerization Rp were calculated from the linear part of the conversion vs. time curves.

II) Free Radical Promoted Cationic Photopolymerization (FRPCP)

The two-component photoinitiating systems are based (except otherwise stated) on **PI**/Ph_2I^+ (0.5%/1% w/w). These conditions were selected for good polymerization profiles. The EPOX films (25 µm thick) deposited on a BaF_2 pellet were irradiated under air inside the

IR spectrometer cavity. The evolution of the epoxy group content at about 790 cm^{-1} is continuously followed by real time FTIR spectroscopy (Nexus 870, Nicolet) as reported in [35]. A polychromatic light from a Xe-Hg lamp (filtered light at $\lambda > 300$ nm; Hamamatsu, L8252, 150 W) was used for the irradiation. The stability of the prepared formulations was excellent (see below).

III) Redox Initiating Systems at RT

The silver salt (1% w/w) is dissolved in the selected monomer (EPOX or LDO) and the organosilane (Si-1, Si-2 and Si-3) is introduced into the formulation. The polymerization is carried out in pill-box for a sample of 1g. The progress of the exothermic polymerization is followed by monitoring the sample temperature using a thermocouple connected to a DaqPro-5300 (resolution 0.1°C) [20b].

1. Backgrounds: The Organosilanes in FRP

1. A. Excited States and Radical Properties

The absorption properties of Si-1 and Si-2 are gathered in Table 1. The absorption coefficients of the UV transition mainly observed below 330 nm are representative of □□* transitions. Under laser excitation at 266 nm of Si-1 and Si-2, new transients were observed [14]. The spectra being similar to that found for Ph3Si. (maximum at about 330 nm) [22], these transients can be ascribed to the silyls generated through the homolytic cleavage of the Si-Si bond. The rise time of the silyl radicals is less than the resolution time of the laser set-up (<10 ns): for both compounds, the dissociation rate constants are thus > 108 s-1. For Si-3, a similar behavior was found in [36].

The very high reactivity of the silyl radicals towards the addition to MA was recently outlined [14]: it is noticeably higher than that reported for the well known phosphinoyl, hydroxy isopropyl or benzoyl initiating radicals (for example, the reaction is about 4 000 and 20 times more efficient for Si-1 and Si-2 than the similar addition of the benzoyl radical [1c,1f,1k,1m,37]). The Si-1 derived silyl radical should be considered as one of the most efficient powerful initiating agent reported so far [37].

Table 1. Absorption and polymerization ability of Si-1 and Si-2

	Absorption	Photoinitiating Properties[b]
	λ_{max} (nm)	Rp_{rel} (laminate)
Si-1	305 (17 700[a])	0.96 (3.37[c])
Si-2	276 (40 050[a]) 287 (37 920[a])	0.35 (1.11[c])

[a] The extinction coefficients are given in bracket (in $mol^{-1}\,l\,cm^{-1}$) in acetonitrile.

[b] The Rps are relative to DMPA.

[c] The Rps in bracket are corrected from the amount of light absorbed (see text).

1.B. The Photoinitiating Ability

The photoinitiating abilities of Si-1 and Si-2 were investigated and compared to that of the well known Type I photoinitiator DMPA (*Figure 1*). The relative polymerization rates with respect to DMPA are given in *Table 1* as well as the relative rates corrected from the absorbed light intensity. Under a polychromatic light exposure, Si-1 exhibits a practical efficiency close to that of DMPA. For this compound, however, a significant absorption is only noted for the 313 nm band. Therefore, the overall absorption is obviously lower than for the reference. Considering the same amount of absorbed light, the intrinsic photochemical reactivity appears as about 3 times higher (in laminated conditions or under air) than that of DMPA. Si-1 is a remarkable photoinitiator. For Si-2, a lower reactivity is found (about 1 in laminated condition and 0.35 under air). For both compounds Si-1 and Si-2, the obtained final conversions were higher than 75 % for polymerization time of 120s. For Si-3, excellent polymerization profiles were also obtained (see part 4).

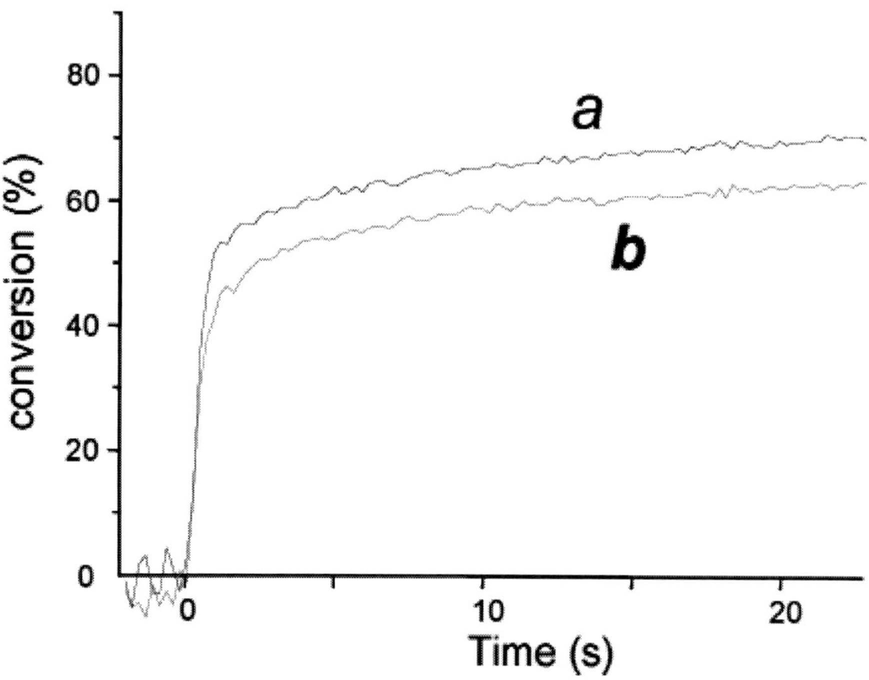

Figure 1. Photopolymerization profiles of Ebecryl 605/HDDA (70/30% w/w) in the presence of the different photoinitiating systems: (a) DMPA and (b) Si-1 systems (0.1 % w).

1. C. Initiating Mechanism in FRP

The reactivity difference between Si-1 - Si-3 can be ascribed to the lower reactivity of the silyl radicals generated and/or the different cleavage quantum yields [14,33]. Some side reactions of the silyl radicals (recombination with the growing polymer chains, transfer reaction through hydrogen abstraction…) can also be invoked.

2. The Photoinitiating Ability of the PI/Ph_2I^+ Combinations for FRPCP

The photoinitiating abilities of Si-1/Ph_2I^+ and Si-2/Ph_2I^+ in for FRPCP were investigated and compared to that of the well known reference system: DMPA/Ph_2I^+ (*Figure 2*). Remarkably, Si-1, Si-2 and DMPA exhibit a similar performance to initiate the ring opening polymerization of the epoxy monomer. With Si-3, a remarkable efficiency under air is noted in combination with the iodonium salt in FRPCP [36] (*Figure 3*).

Using Ph_2I^+ alone, the polymerization is very slow at $\lambda > 300$ nm showing that a sensitization process is requested to ensure a good efficiency in our conditions. Silyl radicals were generated in Si-1-Si-3 under light irradiation from the Si-Si bond cleavage process (5). It has been already found that the silyl radicals can be easily oxidized by iodonium salts into silylium cations which are highly efficient structures to initiate the ring opening polymerization of epoxy monomers (6-7). The high dissociation quantum yields found previously for Si-1 and Si-2, – 1.0 and 0.8, respectively – ensure a very efficient production of $R_3Si^\bullet$ [14].

$$R_3Si\text{-}SiR_3 \rightarrow 2\ R_3Si^\bullet \quad (h\nu) \tag{5}$$

$$R_3Si^\bullet + Ph_2I^+ \rightarrow R_3Si^+ + Ph^\bullet + Ph\text{-}I \tag{6}$$

$$R_3Si^+ + M \rightarrow R_3Si\text{-}M^+ \tag{7}$$

Interestingly, the addition of a silane (TTMSS) to the PI/Ph_2I^+ systems leads in any case to a dramatic increase of both the polymerization rate and the final conversion under air (*Figure 4*). A decrease of the inhibition time for the polymerization under air is also observed in the presence of TTMSS. The high conversion of the Si-H functions (~ 55%) is in line with the overall efficiency. Such a behavior has been previously observed for the FRPCP of usual epoxides [15d]. In the presence of TTMSS, the phenyl radical ($Ph^\bullet$) produced in (6) is converted into a new silyl radical $(TMS)_3Si^\bullet$. This process corresponds to a hydrogen transfer reaction (8). The high conversion of the Si-H functions noted above demonstrates that TTMSS does not only react with the phenyl radical ($Ph^\bullet$) but also with the peroxyl radicals generated under air (8b). The increase of the silyl radical yield is due to the generation of phenyl radicals in the medium which enhances the silyl radical generation by the TTMSS/$Ph^\bullet$ hydrogen abstraction (8a).

$$Ph^\bullet + (TMS)_3Si\text{-}H \rightarrow PhH + (TMS)_3Si^\bullet \tag{8a}$$

$$ROO^\bullet + (TMS)_3Si\text{-}H \rightarrow ROOH + (TMS)_3Si^\bullet \tag{8b}$$

Therefore, the main interest of the TTMSS addition is the formation of radicals which can be easily oxidized (6), the resulting silylium cations being excellent polymerization initiating structures as already noted in other systems e.g. such as camphorquinone/iodonium salt/TTMSS [15d].

The storage stability of the different formulations was checked for most of the different systems. A good stability is generally found for some weeks at RT without using an inert atmosphere.

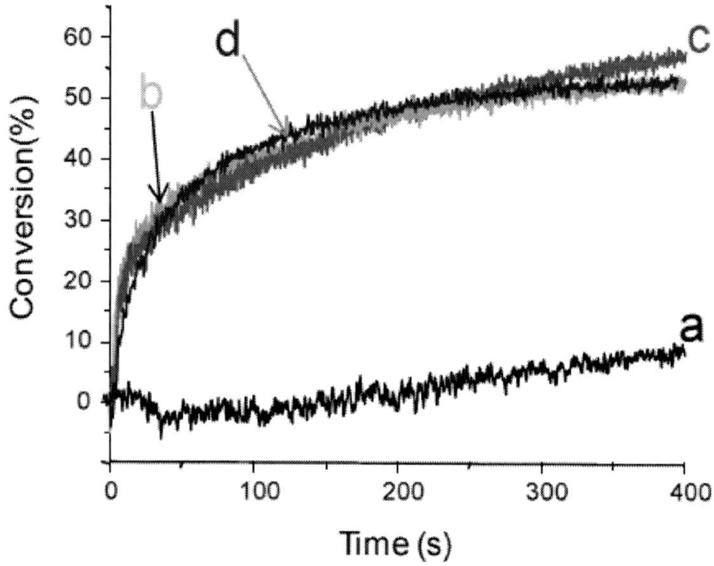

Figure 2. Polymerization profiles of EPOX under air. Upon a Xe-Hg lamp irradiation ($\lambda > 300$ nm; 220 mW/cm²) in the presence of (a) Ph_2I^+ (1% w/w); (b) DMPA/Ph_2I^+ (0.5 %/1% w/w); (c) Si-1/Ph_2I^+ (0.5%/1% w/w); (d) Si-2/Ph_2I^+ (0.5%/1% w/w).

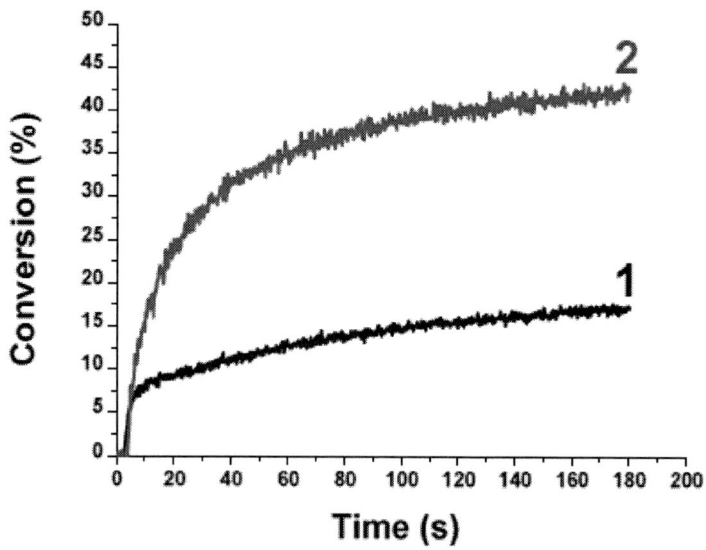

Figure 3. Polymerization profiles of EPOX upon a Xe-Hg lamp irradiation ($\lambda > 300$ nm; 220 mW/cm²) in the presence of Si-2/Ph_2I^+ (0.5%/1% w/w). Under air (2) and in laminate (1).

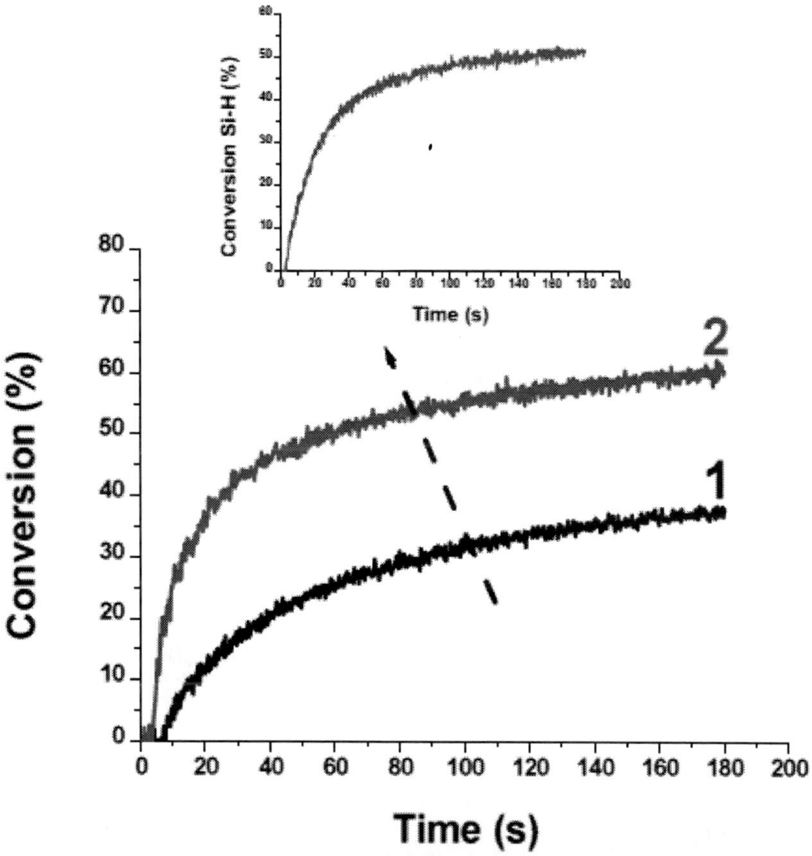

Figure 4. Polymerization profiles of EPOX under air. Upon a Xe-Hg lamp irradiation ($\lambda > 300$ nm; 66 mW/cm²) in the presence of (1) Si-2/Ph_2I^+ (0.5%/1% w/w) and (2) Si-2/TTMSS/Ph_2I^+ (0.5 %/3%/1% w/w). Insert: conversion of the Si-H content in the case of (2).

3. The Redox Initiating Systems at RT

Using Si-2/$AgSbF_6$ and Si-3/$AgSbF_6$, the thermal polymerization of EPOX occurs at RT (*Figure 5*). The polymerization does not occur using these organosilanes or $AgSbF_6$ independently, thereby demonstrating that this combination is necessary to initiate the process. A complete polymerization is obtained after about 300 s.

Interestingly, the proposed new systems are quite versatile as they are also able to initiate the polymerization of other epoxy monomers such as limonene dioxide LDO (*Figure 6*). This behavior indicates that the formation of the cationic species is also efficient in LDO.

The formation of Ag(0) in the film during the polymerization process is well shown through the formation of a new absorption band centered at about 400 nm (*Figure 7*). This can be safely ascribed to the surface plasmon resonance (SPR) absorption band of the silver nanoparticles as it has been reported in [38]. A redox process highlighting the reducing agent character of organosilanes derived from R_3Si-SiR_3 (9) is thus evidenced.

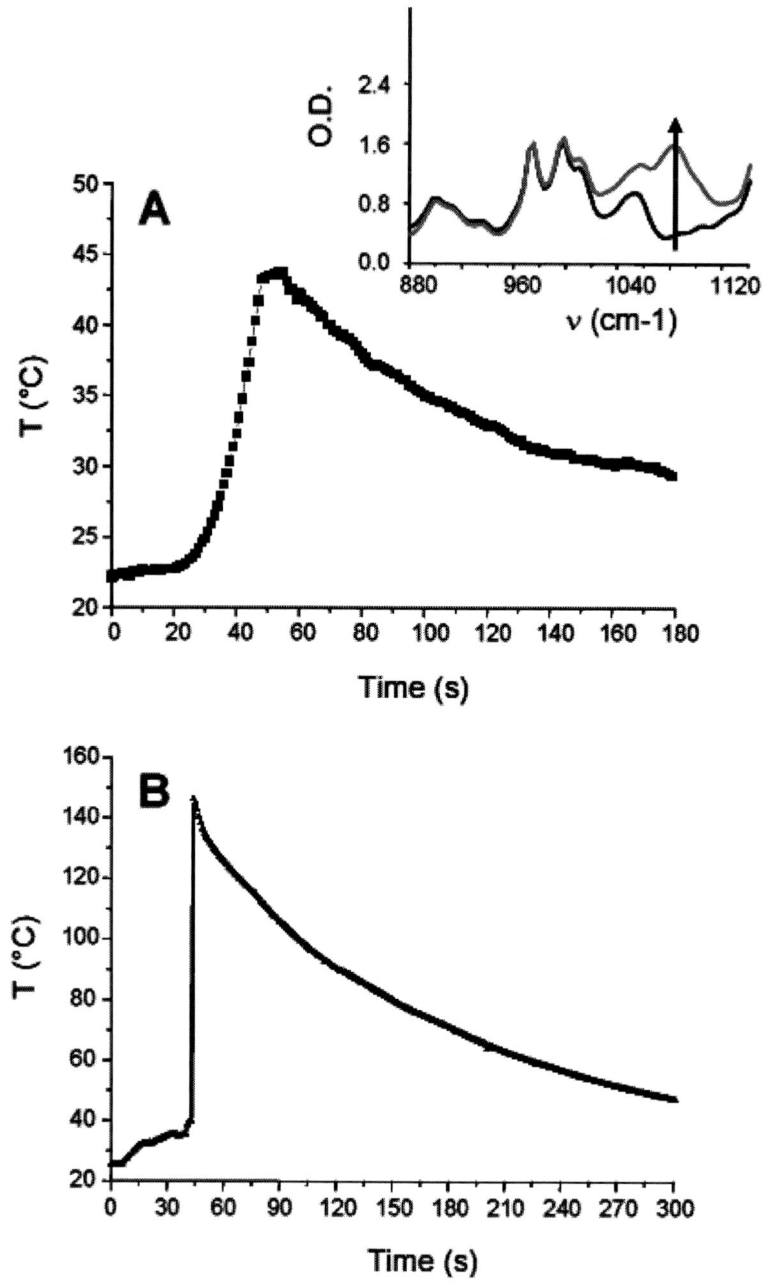

Figure 5. Role of the organosilane in the thermal cationic polymerization of EPOX under air. (A) Si-2/$AgSbF_6$ (0.5%/1% w/w) and (B) Si-3/$AgSbF_6$ (0.5%/1% w/w). Insert: IR spectra recorded during the photopolymerization of EPOX under air.

Recently, it has been found that polysilanes act as reducing agents for Ag(I) in organic solvent [39]. In the present work, it is found that compounds being Si-Si bond in combination with $AgSbF_6$ acts as a very powerful initiating system. This is a highly worthwhile alternative as polysilanes have two main drawbacks: i) a poor solubility and ii) a rather hard synthetic access. In this reaction, silylium cations are probably generated to initiate the ring opening polymerization of the epoxy monomer (10).

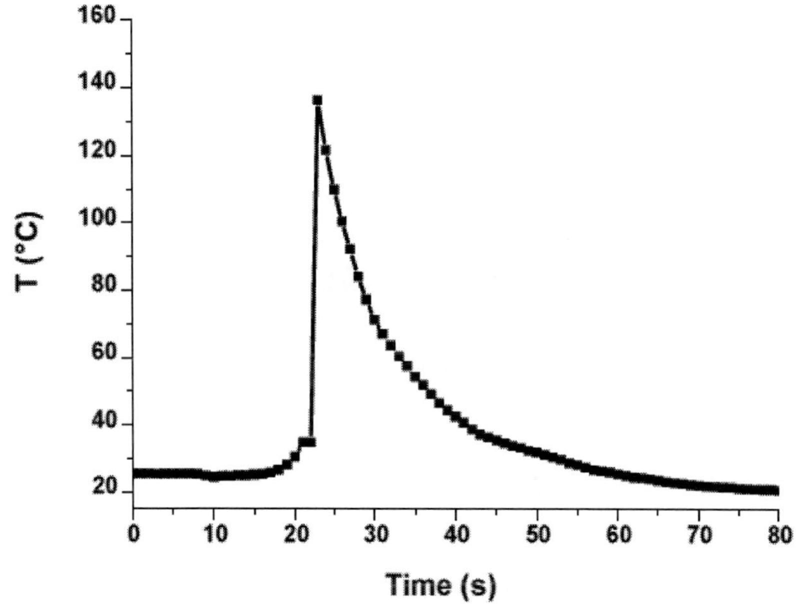

Figure 6. Role of the monomer in the thermal cationic polymerization of LDO under air in the presence of Si-3/$AgSbF_6$ (0.5%/1% w/w).

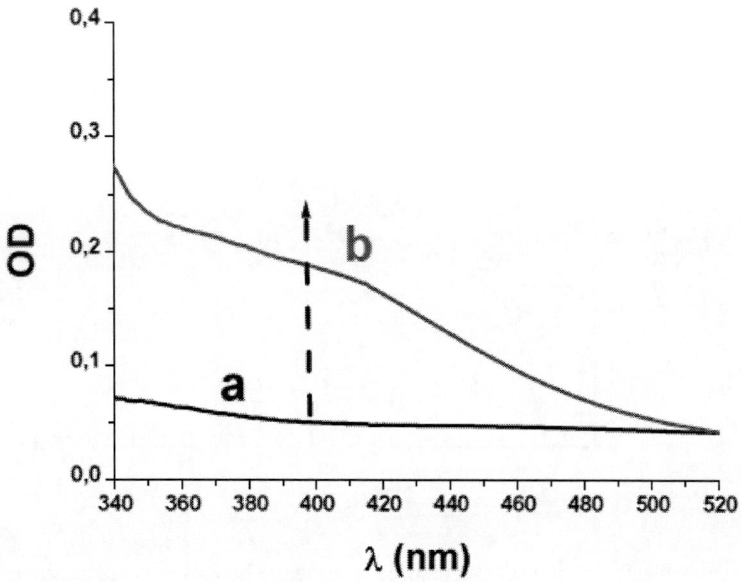

Figure 7. UV-visible absorption spectra of a polymerized EPOX film using $AgSbF_6$ (1% w/w) in the absence (a) and in the presence (b) of Si-2 (0.5 % w/w). Upon a Xe-Hg lamp irradiation ($\lambda > 300$ nm; 220 mW/cm²). Polymerization time ~ 180 s.

$$R_3Si\text{-}SiR_3 + Ag^+ \rightarrow [R_3Si\text{-}SiR_3]^{\bullet+} + Ag(0) \quad (9)$$

$$[R_3Si\text{-}SiR_3]^{\bullet+} + M \rightarrow\rightarrow\rightarrow \text{polymer} \quad (10)$$

4. Formation of Nanoparticles in an Acrylate Matrix

The silyl radicals can also be oxidized by silver salt leading to a Ag(0) nanoparticle NP synthesis (*Figure 8A*). The formation of the polymer network is associated with an in-situ generation of Ag(0) NPs. The thermal initiation of the FRP by the silyl radicals (the addition to the monomer) is in competition with their oxidation by the silver salt i.e. the initiating ability of *Si-3*/$AgBF_4$ is slightly lower than that of *Si-3* (*Figure 8*).

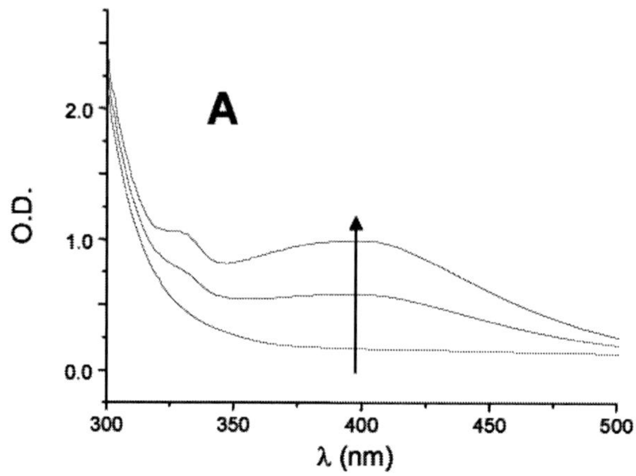

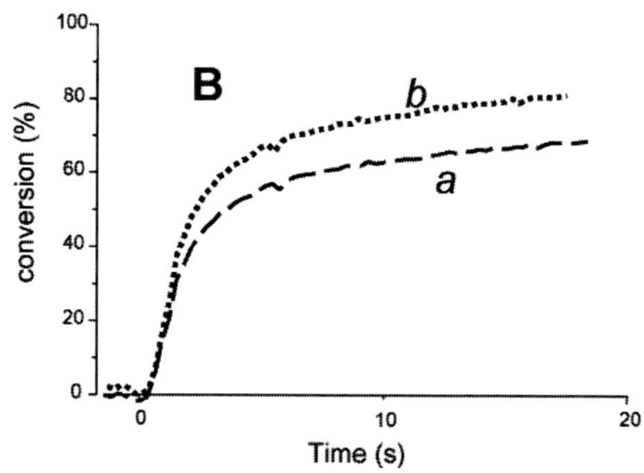

Figure 8. (A) UV-visible absorption spectra recorded during the FRP of trimethylolpropane triacrylate (TMPTA) using Si-3/$AgBF_4$ as an initiating system. (B) Polymerization profiles of TMPTA in laminated conditions. Upon a Xe-Hg lamp irradiation ($\lambda > 300$ nm; 66 mW/cm²) in the presence of (a) Si-3/$AgBF_4$ (0.5%/0.2% w/w) and (b) Si-3 (0.5 % w/w).

CONCLUSION

The examples shown in this chapter outline the interest of different organosilanes (Si-1 - Si-3) as radical sources in FRP and FRPCP. The key point is the presence of the silyl radicals. New redox systems were also proposed for the initiation of the thermal cationic polymerization of epoxides at RT. The development of other systems based on organosilanes for thermal as well as photochemical free radical or cationic polymerization processes will be presented in forthcoming works.

REFERENCES

[1] a) Radiation Curing of Polymeric Materials, Hoyle, C.E.; Kinstle, J.F. Eds., ACS Symposium series, 417, 1990; b) Lasers in Polymer Science and Technology, Fouassier, J.P.; Rabek, J.F. Eds., CRC Press, Boca Raton, 1990; c) Radiation Curing in Polymer Science and Technology, Fouassier, J.P.; Rabek, J.F., Eds., Elsevier Science Publishers LTD, London, 1993; d) Chemistry and technology of UV and EB formulation for coatings, inks and paints, Oldring, PKT Ed., Sita Technology, London, 1994; e) Current Trends in Polymer Photochemistry, Allen, N.S.; Edge, M.; Bellobono, I.R.; Selli, E. Eds., Eds., Ellis Horwood, New-York, 1995; f) Fouassier, J.P., Photoinitiation, Photopolymerization and Photocuring: Fundamental and Applications, Hanser Publishers, New-York, 1995; g) Photopolymerization: Fundamentals and Applications, ACS Symposium series 673, Scranton, A.B.; Bowman, A.; Peiffer, R.W. Eds., Washington DC, 1997; h) Crivello, J.V., Photoinitiators for Free Radical, Cationic and Anionic Photopolymerization, 2nd Ed.; Bradley, G., Ed.; New York, 1998; i) Davidson, S.; Exploring the Science, Technology and Application of UV and EB Curing, Sita Technology Ltd, London, 1999; j) Neckers, D.C.; Jager, W., UV and EB at the Millenium, Sita Technology, London, 1999; k) Dietliker K., A Compilation of Photoinitiators commercially available for UV today, Sita Technology Ltd, Edinburgh, London, 2002; l) Photoinitiated Polymerization, Belfied, K.D.; Crivello, J.V. Eds., ACS Symposium series 847, Washington DC, 2003; m) Photochemistry and UV Curing, Fouassier, J.P. Ed., Research Signpost, Trivandrum, India, 2006; n) Schwalm, R., UV coatings: basics, recent developments and new applications, Elsevier, Oxford UK, 2007; o) Schnabel, W., Polymers and Light: Fundamentals and Technical Applications, Wiley-VCH, Weinheim, 2007; p) Macromolecular engineering: from precise macromolecular synthesis to macroscopic materials properties and applications, Matyjaszewski, K.; Gnanou, Y.; Leibler, L. Eds, Wiley-VCH, Weinheim, 2007; q) Lasers in Chemistry, Lackner, M. Ed., Wiley-VCH., Weinheim, 2008; r) Handbook on photochemistry and photophysics of polymer materials, Allen, N.S. Eds, Wiley, USA, in press; s) Basics of photopolymerization reactions, Fouassier J.P.; Allonas X., Eds., Researchsignpost, Trivandrum India, 2010, in press.

[2] a) Fouassier J.P., Allonas X., Lalevée J., in "Macromolecular engineering: precise synthesis, materials properties, applications", Matyjaszewski K., Gnanou Y., Leibler L., Eds, vol 1, 643-672, Wiley-VCH, Weinheim, 2007; b) Allonas X., Crouxte-Barghorn C., Fouassier J.P., Lalevée J., Malval J.P., Morlet-Savary F., in "Lasers in Chemistry",

M. Lackner Ed., Wiley-VCH., under press; c) Fouassier J.P., Allonas X., Lalevée J., Dietlin C., in "Handbook on photochemistry and photophysics of polymer materials", Allen N.S., Eds, Wiley, 2010, New York USA.

[3] a) Liska R., Seidl B., *J. Polym. Sci., Part A : Polym. Chem.,* 2005, 43, 101; b) Liska R., Herzog D., *J. Polym. Sci., Part A : Polym. Chem.,* 2004, 42, 752; c) Seidl B., Kalinyaprak-Icten K., Fuss N., Hoefer M., Liska R., *J. of Polym. Sci., Part A: Polym. Chem.,* 2007, 46, 289; d) Ganster B., Fischer U.K., Moszner N., Liska R., *Macromolecular. Rapid Comm.,* 2008, 29, 57.

[4] Fuchs A., Bolle T., Ilg S., Husler R., *Proc. Rad. Tech. Europe.* 2003, 1, 507.

[5] a) Visconti M., Cattaneo M., *Proc. Rad. Tech. Europe.* 2003, 1, 205; b) Visconti M., Belloti E., Cattaneo M., *Proc. Rad. Tech. Europe.* 2005, 2, 313; c) Dietlin C., Morlet-Savary F., Allonas X., Fouassier J.P., Visconti M., Norcini G., Romagnano S., *J. Appl. Polym. Sci.,* 2008, 109, 825; d) Dietlin C., Lalevée J., Allonas X., Fouassier J.P., Visconti M., Li Bassi G., Norcini G., *J. App. Polym. Sci.,* 2008, 107, 246; e) Casiraghi A., Norcini G., Visconti M., Li Bassi G., Dietlin C., Allonas X., Fouassier J.P., IT Patent VA2007A000023, Lamberti, March 2007.

[6] a) Fedorov A.V., Ermoshkin A.A., Mejiritski A., Neckers D.C., Macromolecules, 2007, 40, 3554; b) Fedorov A.V., Ermoshkin A.A., Neckers D.C., *J. Applied. Polym. Sci.,* 2008, 107, 147.

[7] a) Aydin M., Arsu N., Yagci Y., Jockusch S., Turro N.J., Macromolecules, 2005, 38, 4133; b) Balta D.K., Arsu N., Yagci Y., Jockusch S., Turro N.J., *Macromolecules,* 2007, 40, 4138; c) Gacal B., Akat H., Balta D.K., Arsu N., Yagci Y., *Macromolecules,* 2008, 41, 2401; d) Durmaz Y.Y., Moszner N., Yagci Y., *Macromolecules,* 2008, 41, 6714; e) Keshin S., Jockusch S., Turro N.J., Arsu N., *Macromolecules,* 2008, 41, 4631.

[8] Bartoszewicz J., Hug G.L., Pietrzak M., Kozubek H., Paczkowski J., Marciniak B., *Macromolecules*, 2007, 40, 8642.

[9] Corales T., Catalina F., Allen N.S., Peinado C., *J. Photochem. Photobiol., A: Chem.,* 2005, 169, 95.

[10] Wu G., Nie J., J. Photochem. Photobio., A: Chem., 2006, 183, 154.

[11] Ye G., Yang J., Zeng Z., Chen Y., J. *of Appl. Polym. Sci.,* 2006, 99, 3417.

[12] a) Hoyle C.E., Lee T.Y., Joensson S., *Eur. Polym. J.,* 2007, 43, 172; b) Lee T.Y., Guymon C.A., Jonsson E.S., Hait S., Hoyle C.E., *Macromolecules,* 2005, 38, 7529.

[13] Kim K., Scranton A., *J. Polym. Sci., Part A: Polym. Chem.,* 2004, 42, 5863.

[14] Lalevée J., El Roz M., Morlet-Savary F., Graff G., Allonas X., Fouassier J.P., *Macromolecules,* 2007, 40, 8527.

[15] a) Lalevée J., Blanchard N., El Roz M., Allonas X., Fouassier J.P., *Macromolecules,* 2008, 41, 2347; b) Lalevée J., Zadoina L., Allonas X., Fouassier J.P., *J. Polym. Sci., Polym. Chem.,* 2007, 45, 2494; c) Lalevée J., Dirani A., El Roz M., Graff B., Allonas X., Fouassier J.P., *Macromolecules,* 2008, 41, 2003; d) Lalevée J., El Roz M., Allonas X., Fouassier J.P., *J. Polym. Sci. Polym. Chem.,* 2008, 46, 2008; e) Lalevée J., Dirani A., El Roz M., Allonas X., Fouassier J.P., *J. Polym. Sci., Polym. Chem.,* 2008, 46, 3042; f) Lalevée J., Blanchard N., Allonas X., Fouassier J.P., *J. Organomet. Chem.,* 2008, 693, 3643; g) Lalevée J., Blanchard N., Chany A.C., El-Roz M., Souane R., Graff B., Allonas X., Fouassier J.P., *Macromolecules,* 2009, 42, 6031; h) Tehfe M.A., Lalevée J., Gigmes D., Fouassier J.P., *Macromolecules,* 2010, 43, 1364.

[16] a) Hoyle C.E., Kim K.J., J. *Appl. Polym. Sci.,* 2003, 33, 2985; b) Decker C. Makromol. Chem., 1979, 180, 2027; b) Bolon D.A., Webb K.K., *J. Appl. Polym. Sci.,* 1978, 9, 2543; c) Wight F.R., *J. Polym. Sci.: Polym Lett.* Ed., 1978, 16, 121; d) Studer K., Decker C., Beck E., Schwalm R., *Prog. Org. Coat.,* 2003, 48, 101; e) "Radiation Curing of Polymeric Materials" ACS Symposium Ser. 417, Ed. Hoyle C.E.; Kinstle J.F. American Chemical Society, Washington, 1989.
[17] El-Roz M., Tehfe M.A., Lalevée J., Graff B., Allonas X., Fouassier J.P., *Macromolecules,* 2010, 43, 2219.
[18] Tehfe M.A., Lalevée J., Gigmes D., Fouassier J.P., *J. Polym. Sci. Part A: Chem,* 2010, 48, 1830.
[19] a) Arsu N., Hizai G., Yagci Y., *Macromol. Rep.* 1995, 1257; b) Yagci Y., Schnabel W., *Makromol. Chem. Rapid Commun.,* 1987, 8, 209 ; c) Yagci Y., Kminek I., Schnabel W., *Polymer,* 1993, 34, 426 ; d) Yagci Y., Ledwith A., *J. Polym. Sci. Polym. Chem.,* 1988, 26, 1911 ; e) Crivello J.V., Sangermano M., *J. Polym. Sci. Part A : Chem,* 2001, 39, 343 ; f) Dursun C., Degirmenci M., Yagci Y., Jockusch S., Turro N.J., *Polymer,* 2003, 44, 7389.
[20] a) Tehfe M.A., Lalevée J., Allonas X., Fouassier J.P., *Macromolecules,* 2009, 42, 8669; b) Souane R., Tehfe M.A., Lalevée J., Gigmes D., Fouassier J.P., *Macromol. Chem. Phys.* 2010, 211, 1441.
[21] Lalevée, J.; Allonas, X.; Fouassier, J.P. *J. Org. Chem.,* 2007, 72, 6434.
[22] Chatgilialoglu, C. in "Organosilanes in Radical Chemistry". John Wiley & Sons, 2004.
[23] Kminek, I.; Yagci, Y.; Schnabel, W.; *Polym. Bull.,* 1992, 29, 277.
[24] a) Peinado, C.; Alonso, A.; Catalina, F. ; Schnabel, W.; *J. Photochem. Photobio. A: Chem.* 2001, 141, 85; b) West, R.; Wolff, A.R.; Peterson, D.J.; *J. Radiation.* Curing, 1986, 13, 35; c) Michl, J.; Downing, J.W. ; Karatsu, T. ; *Pure & Appl. Chem.,* 1988, 60, 959; d) Peinado, C.; Alonso, A.; Catalina, F.; Schnabel, W.; *Macromol. Chem. Phys.,* 2000, 201, 1156.
[25] Lalevée J., Tehfe M.A., Gigmes D., Fouassier J.P., *Macromolecules,* 2010, 43, 6608.
[26] Hiraoka H., Hofer D.C., Miller R.D., Pederson L.A., Willson C.G., U.S. 4464460, *IBM Corp., Invs., C. A.* 1984, 101, 201564g.
[27] Kepler R.G., Zeigler J.M., Harrah L.A., Kurtz S.R., *Bull. Am. Phys. Soc.* 1983, 28 362.
[28] Wolff A.R., West R., *Appl. Organomet. Chem.,* 1987, 1, 7.
[29] West R., *J. Organomet. Chem.,* 1986, 300, 327.
[30] Trefonas P., Damewood J.R., West R., Miller R.D., *Organometallics,* 1985, 4, 1318.
[31] Trefonas P., Miller R.D., West R.J., *J. Am. Chem. Soc.,* 1985, 107, 2737.
[32] Ishikawa M., Kumada M., *Adv. Organometal. Chem.,* 1981, 19, 51.
[33] Lalevée, J.; Allonas, X.; Jradi, S.; Fouassier, J.P. *Macromolecules,* 2006, 39, 1872.
[34] Pappas, S.P. "UV Curing: Science and Technologie", II, Technologie Marketing *Corporation,* 1985, p 3.
[35] Allonas, X.; Grotzinger, C.; Lalevee, J.; Fouassier, J.P.; Visconti, M. *Eur. Polym. J.* 2001, 37, 897.
[36] El-Roz M., Lalevée J., Morlet-Savary F., Allonas X., Fouassier J.P., *Macromolecules,* 2009, 42, 4464.
[37] Lalevée, J.; Allonas, X.; Fouassier, J.P.; *J. Phys. Chem. A.* 2004, 108, 4326.
[38] a) Taleb A., Petit C., M.P., *J. Phys. Chem. B,* 1998, 102, 2214; b) Link S., El-Sayed M.A., *Annu. Rev. Phys. Chem.* 2003, 54, 331; c) Pastoriza-Santos I., Liz-Marzán L.M.,

Pure Appl. Chem., 2000, 72, 83; d) Van Hyning D.L., Klemperer W.G., Zukoski C.F., *Langmuir,* 2001, 17, 3120; e) He S., Yao J., Jiang P., Shi D., Zhang H., Xie S., Pang S., Gao H., *Langmuir,* 2001, 17, 1571.

[39] Shankar R., Sahoo U., Shahi V., *Macromolecules,* 2011, 44, 3240.

INDEX

D

E

F

G

H

N

Q

R

S

T

U

V

W

X

Y

Z